NANOFLUIDS

The Wiley Bicentennial–Knowledge for Generations

Each generation has its unique needs and aspirations. When Charles Wiley first opened his small printing shop in lower Manhattan in 1807, it was a generation of boundless potential searching for an identity. And we were there, helping to define a new American literary tradition. Over half a century later, in the midst of the Second Industrial Revolution, it was a generation focused on building the future. Once again, we were there, supplying the critical scientific, technical, and engineering knowledge that helped frame the world. Throughout the 20th Century, and into the new millennium, nations began to reach out beyond their own borders and a new international community was born. Wiley was there, expanding its operations around the world to enable a global exchange of ideas, opinions, and know-how.

For 200 years, Wiley has been an integral part of each generation's journey, enabling the flow of information and understanding necessary to meet their needs and fulfill their aspirations. Today, bold new technologies are changing the way we live and learn. Wiley will be there, providing you the must-have knowledge you need to imagine new worlds, new possibilities, and new opportunities.

Generations come and go, but you can always count on Wiley to provide you the knowledge you need, when and where you need it!

William J. Pesce
President and Chief Executive Officer

Peter Booth Wiley
Chairman of the Board

NANOFLUIDS

Science and Technology

Sarit K. Das
Indian Institute of Technology Madras, Chennai, India

Stephen U. S. Choi
University of Illinois at Chicago, Chicago, Illinois
Korea Institute of Energy Research, Daejeon, Korea

Wenhua Yu
Argonne National Laboratory, Argonne, Illinois

T. Pradeep
Indian Institute of Technology Madras, Chennai, India

BICENTENNIAL 1807 WILEY 2007 BICENTENNIAL

WILEY-INTERSCIENCE

A JOHN WILEY & SONS, INC., PUBLICATION

Published by John Wiley & Sons, Inc., Hoboken, New Jersey.
Published simultaneously in Canada.

Wiley Bicentennial Logo: Richard J. Pacifico

Library of Congress Cataloging-in-Publication Data:

Nanofluids : Science and Technology / Sarit K. Das ... [et al.].
p. cm.
Includes index.
ISBN 978-0-470-07473-2 (cloth)
1. Microfluidics. 2. Nanofluids. I. Das, Sarit K.
TJ853.N36 2007
620′.5—dc22

2007012094

10 9 8 7 6 5 4 3 2

To all our faithful families, who value excellence in education and respect in relationship, and our treasured teachers, who inspired our quest for new horizons in science and technology

CONTENTS

PREFACE

In 1959, the celebrated physicist Richard Feynman presented the idea of micromachines at the annual meeting of the American Physical Society. Today, it is worth looking back at those predictions to find that reality has overtaken imagination. However, this journey to the present ultrathin devices is not likely to continue unabated. Already, designers of electronic and computing devices are feeling the bottleneck that they have reached. Surprisingly, the bottleneck is not electronic but thermal. The movement toward smaller devices that operate with increasing speed brings about ever-increasing heat flux. Interestingly, the challenge of dissipating the heat lies not only at the micro but also at the mega level. Large transport vehicles, high- and medium-temperature fuel cells, and controlled bioreactors pose a similar challenge to heat transfer technology. Thus, today, with heat transfer technology standing at a critical juncture, the cooling needs of cutting- edge technologies are demanding a paradigm shift in approach.

All past efforts to improve cooling technology were in a sense "penny wise and pound foolish," due to the fact that although every effort has been made to improve transport processes, very little attention has been paid to the fact that cooling fluids themselves are very poor conductors of heat. This inherent inadequacy of cooling fluids provides an expectation that the present level of heat removal can be enhanced significantly by designing fluids that are more conducting. *Nanofluids,* in which nano-sized particles (typically less than 100 nanometers) are suspended in liquids, has emerged as a potential candidate for the design of heat transfer fluids. A study by a group at Argonne National Laboratory showed that these fluids enhance thermal conductivity of the base liquid enormously, which is beyond the explanation of theories on suspensions. More than a century ago, Maxwell presented a theory for effective conductivity of slurries. However, major problems such as sedimentation, erosion, and high pressure drop prevented the usual microparticle slurries to be used as heat transfer fluids. Nanofluids, on the other hand, were found to be very stable, devoid of such problems, due to the small size of the particles and the small volume fraction of the particles needed for heat transfer enhancement.

This discovery brought about a wave of studies in this area, predominantly experimental confirmation of the huge potential of nanofluids as well as efforts to theorize the phenomenon. The enthusiasm of the research community in this area was evident not only from the number of papers published during the first few years of the twenty-first century, but also from the number of queries the present authors received from researchers all over the globe. Thus, the need for an introductory text in this nascent field of research was felt very strongly. However, the feeling remained abstract until an offer to publish came from John Wiley & Sons. This offer gave us an opportunity to come together to fulfill the need for a text, particularly in view of the difficulty faced by young interdisciplinary researchers. Wiley must be complimented for taking this bold step.

The decision to write a book on nanofluids was courageous but also had its problems. First, with the variety of aspects of nanofluid research pouring in every day, it was difficult to set a direction and evolve a unified approach. Also, there was difficulty in determining the prerequisites for the book, due to the highly interdisciplinary nature of nanofluids. Finally, there was the need to provide a lucid journey into the science and technology of nanofluids rather than a glossary of published articles. After considerable deliberation among authors located around the globe, it was decided that the book should be written for researchers in all areas of science and technology, without prerequisites. For this reason, some elementary information and analyses have been incorporated in Chapters 3, 4, and 6 describing conduction, convection, and boiling of nanofluids, keeping in mind that many readers might not have adequate background in these areas. The other important issue was the incorporation of basic chemical and physical aspects of the synthesis and characterization of nanofluids; in Chapter 2 the focus is on various techniques available for the synthesis of nanoparticles as well as the tools required to characterize them. The large number of methods and references related to this chapter have been presented as an appendix which can serve as a glossary for the research community.

With the continuously increasing archive of research articles on nanofluids, it is difficult to present a treatise that includes all the important research work. Although every efforts has been made to include the available literature, we had to limit ourselves to journal publications as authentic research works, and only pulications preceding the third quarter of 2006 have been included.. If there are omissions, it is simply ignorance of the work on the part of the authors, which we will be happy to correct in the future.

It goes without saying that such an effort needs support from all corners. The first is obviously the editorial and production departments of John Wiley & Sons, in particular Darla P. Henderson, Rebekah Amos, Andrew Prince, and Angioline Loredo, who had been extremely cooperative in our endeavor. The institutions we belong to –the Indian Institute of Technology, Argonne National Laboratory, University of Illinois at Chicago, and Korea Institute of Energy Research –have been extremely supportive, providing a sound infrastructure for research and for writing the book. The families of all the authors have always been supportive and merit special mention for their patience and understanding.

The best judge of any book is the reader. If the present text can elicit a few new ideas toward a better cooling technology with nanofluids, the authors will consider their efforts to be well rewarded.

SARIT K. DAS
STEPHEN U. S. CHOI
WENHUA YU
T. PRADEEP

May 17, 2007

1 Introduction

Ultrahigh-performance cooling is one of the most vital needs of many industrial technologies. However, inherently low thermal conductivity is a primary limitation in developing energy-efficient heat transfer fluids that are required for ultrahigh-performance cooling. Modern nanotechnology can produce metallic or nonmetallic particles of nanometer dimensions. Nanomaterials have unique mechanical, optical, electrical, magnetic, and thermal properties. Nanofluids are engineered by suspending nanoparticles with average sizes below 100 nm in traditional heat transfer fluids such as water, oil, and ethylene glycol. A very small amount of guest nanoparticles, when dispersed uniformly and suspended stably in host fluids, can provide dramatic improvements in the thermal properties of host fluids. *Nanofluids* (*nano*particle *fluid* suspensions) is the term coined by Choi (1995) to describe this new class of nanotechnology-based heat transfer fluids that exhibit thermal properties superior to those of their host fluids or conventional particle fluid suspensions. *Nanofluid technology*, a new interdisciplinary field of great importance where nanoscience, nanotechnology, and thermal engineering meet, has developed largely over the past decade. The goal of nanofluids is to achieve the highest possible thermal properties at the smallest possible concentrations (preferably $< 1\%$ by volume) by uniform dispersion and stable suspension of nanoparticles (preferably < 10 nm) in host fluids. To achieve this goal it is vital to understand how nanoparticles enhance energy transport in liquids.

Since Choi conceived the novel concept of nanofluids in the spring of 1993, talented and studious thermal scientists and engineers in the rapidly growing nanofluids community have made scientific breakthrough not only in discovering unexpected thermal properties of nanofluids, but also in proposing new mechanisms behind enhanced thermal properties of nanofluids, developing unconventional models of nanofluids, and identifying unusual opportunities to develop next-generation coolants such as smart coolants for computers and safe coolants for nuclear reactors. As a result, the research topic of nanofluids has been receiving increased attention worldwide. The recent growth of work in this rapidly emerging area of nanofluids is most evident from the exponentially increasing number of publications. Figure 1.1 shows clear evidence of the significance of nanofluids research.

Since 1999 the nanofluids community has published more than 150 nanofluid-related research articles. In 2005 alone, 71 research articles were published in

Nanofluids: Science and Technology, By Sarit K. Das, Stephen U. S. Choi, Wenhua Yu, and T. Pradeep

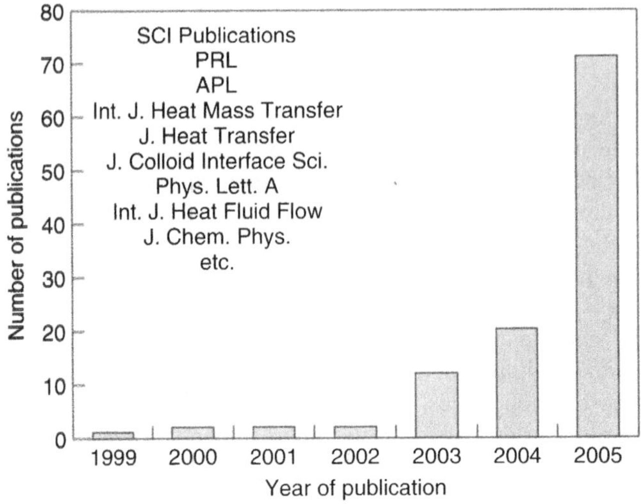

Fig. 1.1 Annual SCI publications on nanofluids.

Science Citation Index (SCI) journals such as *Nature Materials Physical Review Letters*, and *Applied Physics Letters*. In addition to the increasing number of articles published per year, there are two more indicators that give weight to the argument that nanofluid research is getting more and more active and important. First, prestigious institutions worldwide, including the Massachusetts Institute of Technology (MIT), the University of Leeds, and the Royal Institute of Technology, Sweden have established nanofluid research groups or interdisciplinary centers that focus on nanofluids. Several universities have graduated Ph.D.s in this new area of nanofluids. Second, small businesses and large multinational companies in different industries and markets are working on these promising coolants for their specific applications. Escalating interest in nanofluids is based on the realization that it is possible to develop ultrahigh-performance coolants whose thermal properties are drastically different from those of conventional heat transfer fluids, because in the nanoscale range, fundamental properties of nanomaterials such as nanofluids depend strongly on particle size, shape, and the surface/interface area.

The main objective of this introductory chapter is to sketch out a big picture of the small world of nanofluids through a brief review of some historically major milestones such as the concept of nanofluids, the production and performance of nanofluids, the mechanisms and models of nanofluids, and potential applications and benefits of nanofluids. Finally, future research on the fundamentals and applications of nanofluids is addressed. The future research directions described in this chapter are not inclusive but illustrate how to undertake the challenges inherent in developing theory of nanofluids and in scaling up production of nanofluids. Nanofluids are being developed to achieve ultrahigh-performance cooling and have the potential to be next-generation coolants, thus representing a very significant and far-reaching cooling technology for cross-cutting applications.

1.1. FUNDAMENTALS OF COOLING

1.1.1. Cooling Challenge

Cooling is indispensable for maintaining the desired performance and reliability of a wide variety of products, such as computers, power electronics, car engines, and high-powered lasers or x-rays. With the unprecedented increase in heat loads (in some cases exceeding 25 kW) and heat fluxes (in some cases exceeding $2000\,W/cm^2$) caused by more power and/or smaller feature sizes for these products, cooling is one of the top technical challenges facing high-tech industries such as microelectronics, transportation, manufacturing, metrology, and defense. For example, the electronics industry has provided computers with faster speeds, smaller sizes, and expanded features, leading to ever-increasing heat loads, heat fluxes, and localized hot spots at the chip and package levels. These thermal problems are also found in power electronics or optoelectronic devices. Air cooling is the most basic method for cooling electronic systems. However, heat fluxes over $100\,W/cm^2$ in electronic devices and systems will necessitate the use of liquid cooling. Recently, single-phase liquid cooling technologies such as the microchannel heat sink, and two-phase liquid-cooling technologies such as heat pipes, thermosyphons, direct immersion cooling, and spray cooling for chip- or package-level cooling have emerged. Nanofluid technology offers a great potential for further development of high-performance, compact, cost-effective liquid cooling systems.

In the transportation industry, cooling is a crucial issue because the trend toward higher engine power and exhaust-gas regulation or hybrid vehicles inevitably leads to larger radiators and increased frontal areas, resulting in additional aerodynamic drag and increased fuel consumption. A pressing need for cooling also exists in ultrahigh–heat-flux optical devices with brighter beams, such as high-powered x-rays.

1.1.2. Conventional Methods to Enhance Heat Transfer

The conventional way to enhance heat transfer in thermal systems is to increase the heat transfer surface area of cooling devices and the flow velocity or to disperse solid particles in heat transfer fluids. However a new approach to enhancing heat transfer to meet the cooling challenge is necessary because of the increasing need for more efficient heat transfer fluids in many industries, such as the electronics, photonics, transportation, and energy supply industries.

Conventional Soild–Liquid Suspensions and Their Limitations The century-old technique used to increase cooling rates is to disperse millimeter- or micrometer-sized particles in heat transfer fluids. The major problem with suspensions containing millimeter- or micrometer-sized particles is the rapid settling of these particles. If the fluid is kept circulating to prevent particle settling, millimeter- or micrometer-sized particles would wear out pipes, pumps, and bearings. Furthermore, such particles are not applicable to microsystems because they can clog microchannels. These conventional solid fluid suspensions are not practical because they require the addition of a large number of particles (usually, >10 vol%), resulting in significantly greater pressure drop and pumping power.

Microchannel Cooling and Its Limitations Another way to increase heat rejection rates is to use extended surfaces, such as fins and microchannels, for air or liquid cooling. The present-day manufacture of microchannel structures with characteristic dimensions of less than $100\,\mu m$ and the application of these microchannel structures to heat exchangers (Tuckerman and Peace, 1981) represents an engineering breakthrough in heat transfer technology because microscale heat-exchangers have the potential to reduce the size and effectiveness of various heat-exchange devices.

Microscale heat exchangers have numerous attributes, including high thermal effectiveness, high heat transfer surface/volume ratio, small size, low weight, low fluid inventory, and design flexibility. Because their microchannel systems are extremely compact and lightweight compared to conventional systems, materials and manufacturing costs could be lowered, an attractive advantage that would draw the interest of many manufacturing firms. For example, the electronics industry has applications in cooling advanced electronic packages; for the automotive industry, the weight difference between conventional and microchannel systems (such as in air conditioners) could lead to significant gains in fuel economy; in the heating, ventilation, and air-conditioning (HVAC) industry, refrigeration and air-conditioning equipment volumes could be reduced, and this would save space in buildings; and in chemical and petroleum plants, plant size could be reduced through process intensification. Minimizing the size and weight of cooling systems based on microchannel cooling technology is also crucial in the military–avionics industry. Unfortunately, current designs of thermal management systems have already adopted this extended surface technology to its limits. Therefore, with continued miniaturization and increasing heat dissipation in new generations of products, the cooling issue will intensify in many industries: from electronics and photonics to transportation, energy supply, defense, and medical. Nanofluids are being developed in response to these pressing needs for more efficient heat transfer fluids in many industries.

1.2. FUNDAMENTALS OF NANOFLUIDS

Heat transfer is one of the most important processes in many industrial and consumer products. The inherently poor thermal conductivity of conventional fluids puts a fundamental limit on heat transfer. Therefore, for more than a century since Maxwell (1873), scientists and engineers have made great efforts to break this fundamental limit by dispersing millimeter- or micrometer-sized particles in liquids. However, the major problem with the use of such large particles is the rapid settling of these particles in fluids. Because extended surface technology has already been adapted to its limits in the designs of thermal management systems, technologies with the potential to improve a fluid's thermal properties are of great interest once again. The concept and emergence of nanofluids is related directly to trends in miniaturization and nanotechnology. Maxwell's concept is old, but what is new and innovative in the concept of nanofluids is the idea that particle size is of primary importance in developing stable and highly conductive nanofluids.

1.2.1. Miniaturization and Nanotechnology

Since Nobel prize winner Richard P. Feynman presented the concept of micromachines in his seminal talk, "There's Plenty of Room at the Bottom—An Invitation to Enter a New Field of Physics," in December 1959 at the annual meeting of the American Physical Society at the California Institute of Technology (available on the Web at http://nano.xerox.com/nanotech/feynman.html), miniaturization has been a major trend in modern science and technology. Almost 40 years later, another Nobel prize winner, H. Rohrer, presented the chances and challenges of the nano-age and declared that nanoscience and nanotechnology had entered the limelight in the 1990s from virtual obscurity in the 1980s (Rohrer, 1996). *Nano* is a prefix meaning one-billionth, so a *nanometer* is one-billionth of a meter. *Nanotechnology* is the creation of functional materials, devices, and systems by controlling matter at the nanoscale level, and the exploitation of their novel properties and phenomena that emerge at that scale.

Early reviews of research programs on nanotechnology in the United States, China, Europe, and Japan show that nanotechnology will be an emerging and exciting technology of the twenty-first century and that universities, national laboratories, small businesses, and large multinational companies have established nanotechnology research groups or interdisciplinary centers that focus on nanotechnology (Fissan and Schoonman, 1998; Hayashi and Oda, 1998; Li, 1998; Roco, 1998).

Just as downsizing is a fashion in the world of business, downscaling such as microelectromechanical system (MEMS) technology and nanotechnology is a clear fashion in the world of science and technology. One feature of these rapidly emerging technologies is that they are strongly interdisciplinary. In the coming nano-age, nanotechnology with unforeseen applications is expected to revolutionize many industries. Nanotechnology is expected to affect society in the twenty-first century as much as the silicon transistor, plastics, and antibiotics did in the twentieth century. It is estimated that nanotechnology is at a level of development similar to that of computer/information technology in the 1950s (Roco, 1998).

Engineers now fabricate microscale devices such as microchannel heat exchangers and micropumps that are the size of dust specks. Further major advances would be obtained if the coolant flowing in the microchannels were to contain nanoscale particles to enhance heat transfer. Nanofluid technology will thus be an emerging and exciting technology of the twenty-first century. With the continued miniaturization of technologies in many fields, nanofluids with a capability of cooling high heat fluxes exceeding $1000\,W/cm^2$ would be paramount in the advancement of all high technology.

1.2.2. Emergence of Nanofluids

The emergence of nanofluids as a new field of nanoscale heat transfer in liquids is related directly to miniaturization trends and nanotechnology. Here a brief history of the Advanced Fluids Program at Argonne National Laboratory (ANL) is described to show that the program has encompassed a wide range (meters to nanometers) of size regimes and how a wide research road has become narrow, starting with large scale and descending through microscale to nanoscale in this program, culminating in the invention of nanofluids.

Large-Scale Heat Transfer Experiments In 1985, ANL started a long-term research program to develop advanced energy transmission fluids. Sufficient funding for this program was provided through the Buildings and Community Systems staff of the U.S. Department of Energy (DOE). Early efforts focused on the development of advanced energy transmission fluids for use in district heating and cooling (DHC) systems. These systems are characterized by long distribution pipes of large diameter that convey pumped energy transmission fluids between the source and sink heat exchangers. These systems operate with small temperature differences, and therefore large volumes of fluids must be pumped to satisfy load demands. The Advanced Fluids Program for DHC applications included friction-reducing additives and phase-change materials. Friction-reducing additives have been tested in a large-scale DHC system simulator with a pipe diameter of 0.15 m and a length of 21.34 m.

Realizing that large-scale experiments are very costly, the advanced fluids team had to find an exit from large-scale tests. Choi learned that mirror cooling was an important issue at ANL's new advanced photon source (APS). His proposal was funded by the APS Laboratory Directed Research and Development (LDRD) Program. This project represented a dramatic downscaling, from 0.15-m pipe to 50-μm channels. However, he did not stop in this microworld but continued his downscaling journey until his research culminated in the invention of nanofluids.

Microscale Heat Transfer Project The APS is a user facility for synchrotron radiation research. The first optical elements of the APS beamlines absorb a tremendous amount of energy that is rapidly transformed to heat as the elements reflect the beam. Cooling these high-heat-load x-ray optical elements proved to be a formidable task that could not be handled by conventional cooling technologies, and thus a new and innovative cooling method was needed. In 1991, Choi developed a new project to design and analyze a microchannel heat exchanger that uses liquid-nitrogen as the cooling fluid. The work by Choi et al. (1992) on microchannel liquid-nitrogen cooling of high-heat-load silicon mirrors represents a milestone in the area of microscale forced-convection heat transfer (Duncan and Peterson, 1994). For Choi, this project had another significance: It was crucial in positioning him for bridging microtechnology with nanotechnology, as described in the next section.

Nanoscale Heat Transfer as a New Heat Transfer Enhancement Approach When Choi worked on microchannel liquid-nitrogen cooling, he noted its limit: that the pressure drop in the microchannel heat exchanger increases significantly as the diameter of the flow passage decreases and that a cryogenic system is needed for liquid-nitrogen cooling. In a microchannel liquid-nitrogen heat exchanger, the heat transfer would be excellent, but at the cost of high pumping power and an expensive cryogenic system. Furthermore, continuing cooling demands from future x-ray source intensities at the APS have driven him to think of a new heat transfer enhancement approach. He wanted to develop a new heat transfer fluid concept that enables heat transfer enhancement without a large pumping power increase and without cryogenic coolants. So he focused on the thermal conductivity of the fluid itself rather than on channel size.

Although Maxwell's idea of using metallic particles to enhance the electrical or thermal conductivity of matrix materials is well known (Maxwell, 1873), Choi realized through his research project experience with suspensions of micrometer-sized particles and fibers in the 1980s that such conventional particles cannot be used in microchannel flow passages. However, modern nanotechnology provides great opportunities to process and produce materials with average crystallite sizes below 50 nm. Recognizing an opportunity to apply this emerging nanotechnology to established thermal engineering, Choi focused on a smaller world and while reading several articles on nanophase materials, wondered, what would happen if nanoparticles could be dispersed into a heat transfer fluid and visualized the concept of nanofluids: stable suspensions of dancing nanoparticles in liquids. Choi first thought of validating the idea when he read an article in the ANL publication *Logos* on nanocrystalline materials (Siegel and Eastman, 1993) and realized that ANL's Materials Science Division (MSD) has a unique capability to produce nanophase materials. DOE's Basic Energy Sciences office has funded MSD to work on the synthesis, microstructural characterization, and properties of nanophase materials, although all of that work was focused on producing nanoparticles and consolidating them to make solids and then characterizing the novel properties of these solid bulk nanophase materials.

When Choi received an ANL director's call for proposals in May 1993, he wrote a proposal in which he proposed that nanometer-sized metallic particles could be stably suspended in industrial heat transfer fluids to produce a new class of engineered fluids with high thermal conductivity. He submitted his first nanofluids proposal to an annual competition within the lab for startup funding. This proposal was not funded, however, nor was a second proposal developed with MSD's J. A. Eastman. A third proposal, in 1994, was successful. This first nanofluids project was funded for three years and ended in 1997. Since then,

Argonne's nanofluids research has received external funding from DOE to work on issues related to both fundamentals and applications of nanofluids.

In addition to the work at Argonne, investigators in Japan and Germany have published articles that describe fluids resembling those developed at ANL. However, it should be noted that ANL developed the concept of nanofluids independent of the work in Japan and Germany. Masuda et al. worked on the thermal conductivity and viscosity of suspensions of Al_2O_3, SiO_2, and TiO_2 ultrafine particles and published a paper written in Japanese (Masuda et al., 1993). Although there are similarities between the Japanese work and our own, there are also several important distinctions. For example, the Japanese investigators added an acid (HCl) or base (NaOH) to produce suspensions of oxide particles because their oxide particles did not form stable suspensions in fluids. However, we were able to make stable nanofluids with no dispersants at all. We discovered that our oxide nanoparticles have excellent dispersion properties and form suspensions that are stable for weeks or months. Furthermore, the unique thermal features of ANL's nanofluids are the principal distinction between the Japanese and ANL work.

In 1993, Arnold Grimm, an employee of R.-S. Automatis in Mannheim, Germany obtained a patent related to improved thermal conductivity of a fluid containing dispersed solid particles (Grimm, 1993). He dispersed Al particles measuring 80 nm to 1 μm into a fluid. He claimed a 100% increase in the thermal conductivity of the fluid for loadings of 0.5 to 10 vol%. The serious problem with these suspensions was rapid settling of the Al particles, presumably because in his study the particle size was much larger than in Argonne's nanofluids work.

1.2.3. Development of the Concept of Nanofluids

In the development of energy-efficient heat transfer fluids, the thermal conductivity of the heat transfer fluids plays a vital role. Despite considerable previous research and development efforts on heat transfer enhancement, major improvements in cooling capabilities have been constrained because traditional heat transfer fluids used in today's thermal management systems, such as water, oils, and ethylene glycol, have inherently poor thermal conductivities, orders-of-magnitude smaller than those of most solids. Due to increasing global competition, a number of industries have a strong need to develop advanced heat transfer fluids with significantly higher thermal conductivities than are presently available.

It is well known that at room temperature, metals in solid form have orders-of-magnitude higher thermal conductivities than those of fluids (Touloukian et al., 1970). For example, the thermal conductivity of copper at room temperature is about 700 times greater than that of water and about 3000 times greater than that of engine oil, as shown in Table 1.1. The thermal conductivity of metallic liquids is much greater than that of nonmetallic liquids. Therefore, the thermal conductivities of fluids that contain suspended solid metallic particles could be expected to be significantly higher than those of conventional heat transfer fluids.

Table 1.1 Thermal Conductivity of Various Materials

	Material	Thermal Conductivity (W/m · K)[a]
Metallic solids	Silver	429
	Copper	401
	Aluminum	237
Nonmetallic solids	Diamond	3300
	Carbon nanotubes	3000
	Silicon	148
	Alumina (Al_2O_3)	40
Metallic liquids	Sodium at 644 K	72.3
Nonmetallic liquids	Water	0.613
	Ethylene glycol	0.253
	Engine oil	0.145

[a] At 300 K unless otherwise noted.

For more than 100 years, scientists and engineers have made great efforts to enhance the inherently poor thermal conductivity of liquids by adding solid particles in liquids. Numerous theoretical and experimental studies of the effective thermal conductivity of suspensions that contain solid particles have been conducted since Maxwell presented a theoretical basis for predicting the effective conductivity of suspensions more than 100 years ago (Maxwell, 1873). However, all of the studies on the thermal conductivity of suspensions have been confined to millimeter- or micrometer-sized particles. This conventional approach has two major technical problems: (1) conventional millimeter- or micrometer-sized particles settle rapidly in fluids, and (2) the conductivities of these suspensions are low at low particle concentrations. Furthermore, these conventional suspensions do not work with the emerging "miniaturized" devices because they can clog the tiny channels of such devices.

Modern nanotechnology has enabled the production of metallic or nonmetallic nanoparticles with average crystallite sizes below 100 nm. The mechanical, optical, electrical, magnetic, and thermal properties of nanoparticles are superior to those of conventional bulk materials with coarse grain structures. Recognizing an excellent opportunity to apply nanotechnology to thermal engineering, Choi conceived the novel concept of nanofluids by hypothesizing that it is possible to break down these century-old technical barriers by exploiting the unique properties of nanoparticles. Nanofluids are a new class of nanotechnology-based heat transfer fluids engineered by dispersing nanometer-sized particles with typical length scales on the order of 1 to 100 nm (preferably, smaller than 10 nm in diameter) in traditional heat transfer fluids. At the 1995 annual winter meeting of the American Society of Mechanical Engineers (Choi, 1995) Choi presented the remarkable possibility of doubling the convection heat transfer coefficients using ultrahigh-conductivity nanofluids instead of increasing pumping power by a factor of 10.

1.2.4. Importance of Nanosize

As noted above the basic concept of dispersing solids in fluids to enhance thermal conductivity is not new; it can be traced back to Maxwell. Solid particles are added because they conduct heat much better than do liquids. The major problem with the use of large particles is the rapid settling of these particles in fluids. Other problems are abrasion and clogging. These problems are highly undesirable for many practical cooling applications. Nanofluids have pioneered in overcoming these problems by stably suspending in fluids nanometer-sized particles instead of millimeter- or micrometer-sized particles. Compared with microparticles, nanoparticles stay suspended much longer and possess a much higher surface area. The surface/volume ratio of nanoparticles is 1000 times larger than that of microparticles. The high surface area of nanoparticles enhances the heat conduction of nanofluids since heat transfer occurs on the surface of the particle. The number of atoms present on the surface of nanoparticles, as opposed to the interior, is very large. Therefore, these unique properties of nanoparticles can be exploited to develop nanofluids with an unprecedented combination of the two features most highly desired for heat transfer systems: extreme stability and ultrahigh thermal conductivity. Furthermore, because nanoparticles are so small, they may reduce erosion and clogging dramatically. Other benefits envisioned for nanofluids include decreased demand for pumping power, reduced inventory of heat transfer fluid, and significant energy savings.

Because the key building block of nanofluids is nanoparticles (1000 times smaller than microparticles), the development of nanofluids became possible simply because of the advent of nanotechnology in general and the availability of nanoparticles in particular. Researchers in nanofluids exploit the unique properties of these tiny nanoparticles to develop stable and high-thermal-conductivity heat transfer fluids. Stable suspension of small quantities of tiny particles makes conventional heat transfer fluids cool faster and thermal management systems smaller and lighter.

It should be noted that in today's science and technology, size matters. Size is also an important physical variable in nanofluids because it can be used to tailor nanofluid thermal properties as well as the suspension stability of nanoparticles. Maxwell's concept is old, but what is new and innovative with the concept of nanofluids is the idea of using nanometer-sized particles (which have become available to investigators only recently) to create stable and highly conductive suspensions, primarily for suspension stability (gravity is negligible) and for dynamic thermal interactions. Nanotechnogy offers excellent prospects for producing a new type of heat transfer fluid that has excellent thermal properties and cooling capacity, due primarily to novel nanoscale phenomena—phenomena that overturn our sense of familiarity. Therefore, the pioneers of nanofluids have taken the solid–fluid suspension concept to an entirely new level. Table 1.2 contrasts suspensions of microparticles and nanoparticles and shows the benefits of nanofluids containing nanoparticles.

Table 1.2 Comparison of the Old and the New

	Microparticles	Nanoparticles
Stability	Settle	Stable (remain in suspension almost indefinitely)
Surface/volume ratio	1	1,000 times larger than that of microparticles
Conductivity[a]	Low	High
Clog in microchannel?	Yes	No
Erosion?	Yes	No
Pumping power	Large	Small
Nanoscale phenomena?	No	Yes

[a] At the same volume fraction.

1.3. MAKING NANOFLUIDS

Materials for base fluids and nanoparticles are diverse. Stable and highly conductive nanofluids are produced by one- and two-step production methods. Both approaches to creating nanoparticle suspensions suffer from agglomeration of nanoparticles, which is a key issue in all technology involving nanopowders. Therefore, synthesis and suspension of nearly nonagglomerated or monodispersed nanoparticles in liquids is the key to significant enhancement in the thermal properties of nanofluids.

1.3.1. Materials for Nanoparticles and Fluids

Modern fabrication technology provides great opportunities to process materials actively at nanometer scales. Nanostructured or nanophase materials are made of nanometer-sized substances engineered on the atomic or molecular scale to produce either new or enhanced physical properties not exhibited by conventional bulk solids. All physical mechanisms have a critical length scale below which the physical properties of materials are changed. Therefore, particles smaller than 100 nm exhibit properties different from those of conventional solids. The noble properties of nanophase materials come from the relatively high surface area/volume ratio, which is due to the high proportion of constituent atoms residing at the grain boundaries. The thermal, mechanical, optical, magnetic, and electrical properties of nanophase materials are superior to those of conventional materials with coarse grain structures. Consequently, research and development investigation of nanophase materials has drawn considerable attention from both material scientists and engineers (Duncan and Rouvray, 1989).

1. *Nanoparticle material types.* Nanoparticles used in nanofluids have been made of various materials, such as oxide ceramics (Al_2O_3, CuO), nitride ceramics (AlN, SiN), carbide ceramics (SiC, TiC), metals (Cu, Ag, Au), semiconductors

(TiO_2, SiC), carbon nanotubes, and composite materials such as alloyed nanoparticles $Al_{70}Cu_{30}$ or nanoparticle core–polymer shell composites. In addition to nonmetallic, metallic, and other materials for nanoparticles, completely new materials and structures, such as materials "doped" with molecules in their solid–liquid interface structure, may also have desirable characteristics.

2. *Host liquid types.* Many types of liquids, such as water, ethylene glycol, and oil, have been used as host liquids in nanofluids.

1.3.2. Methods of Nanoparticle Manufacture

Fabrication of nanoparticles can be classified into two broad categories: physical processes and chemical processes (Kimoto et al., 1963; Granqvist and Buhrman, 1976; Gleiter, 1989). Currently, a number of methods exist for the manufacture of nanoparticles. Typical physical methods include inert-gas condensation (IGC), developed by Granqvist and Buhrman (1976), and mechanical grinding. Chemical methods include chemical vapor deposition (CVD), chemical precipitation, micro emulsions, thermal spray, and spray pyrolysis. A sonochemical method has been developed to make suspensions of iron nanoparticles stabilized by oleic acid (Suslick et al., 1996).

The current processes for making metal nanoparticles include IGC, mechanical milling, chemical precipitation, thermal spray, and spray pyrolysis. Most recently, Chopkar et al. (2006) produced alloyed nanoparticles $Al_{70}Cu_{30}$ using ball milling. In ball milling, balls impart a lot of energy to a slurry of powder, and in most cases some chemicals are used to cause physical and chemical changes. These nanosized materials are most commonly produced in the form of powders. In powder form, nanoparticles are dispersed in aqueous or organic host liquids for specific applications.

1.3.3. Dispersion of Nanoparticles in Liquids

Stable suspensions of nanoparticles in conventional heat transfer fluids are produced by two methods: the two-step technique and the single-step technique. The *two-step method* first makes nanoparticles using one of the above-described nanoparticle processing techniques and then disperses them into base fluids. The *single-step method* simultaneously makes and disperses nanoparticles directly into base fluids. In either case, a well-mixed and uniformly dispersed nanofluid is needed for successful production or reproduction of enhanced properties and interpretation of experimental data. For nanofluids prepared by the two-step method, dispersion techniques such as high shear and ultrasound can be used to create various particle–fluid combinations.

Most nanofluids containing oxide nanoparticles and carbon nanotubes reported in the open literature are produced by the two-step process. If nanoparticles are produced in dry powder form, some agglomeration of individual nanoparticles may occur due to strong attractive van der Waals forces between nanoparticles. This undesirable agglomeration is a key issue in all technology involving nanopowders. Making nanofluids using the two-step processes has remained a challenge because individual particles quickly agglomerate before dispersion, and nanoparticle agglomerates settle out in the liquids. Well-dispersed stable nanoparticle suspensions are produced by fully separating nanoparticle agglomerates into individual nanoparticles in a host liquid. In most nanofluids prepared by the two-step process, the agglomerates are not fully separated, so nanoparticles are dispersed only partially. Although nanoparticles are dispersed ultrasonically in liquid using a bath or tip sonicator with intermittent sonication time to control overheating of nanofluids, this two-step preparation process produces significantly poor dispersion quality. Because the dispersion quality is poor, the conductivity of the nanofluids is low. Therefore, the key to success in achieving significant enhancement in the thermal properties of nanofluids is to produce and suspend nearly monodispersed or nonagglomerated nanoparticles in liquids.

A promising technique for producing nonagglomerating nanoparticles involves condensing nanophase powders from the vapor phase directly into a flowing low-vapor-pressure fluid. This approach, developed in Japan 20 years ago by Akoh et al. (1978), is called the VEROS (vacuum evaporation onto a running oil substrate) technique. VEROS has been essentially ignored by the nanocrystalline-materials community because of subsequent difficulties in separating the particles from the fluids to make dry powders or bulk materials. Based on a modification of the VEROS process developed in Germany (Wagener et al., 1997), Eastman et al. (1997) developed a direct evaporation system that overcomes the difficulties of making stable and well-dispersed nanofluids. The direct evaporation–condensation process yielded a uniform distribution of nanoparticles in a host liquid. In this much-longed-for way to making nonagglomerating nanoparticles, they obtained copper nanofluids with excellent dispersion characteristics and intriguing properties. The thermal conductivity of ethylene glycol, the base liquid, increases by 40% at a Cu nanoparticle concentration of only 0.3 vol%. This is the highest enhancement observed for nanofluids except for those containing carbon nanotubes. However, the technology used by Eastman et al. has two main disadvantages. First, it has not been scaled up for large-scale industrial applications. Second, it is applicable only to low-vapor-pressure base liquids. Clearly, the next step is to see whether they can compete with the chemical one-step method described below.

Zhu et al. (2004) developed a one-step chemical method for producing stable Cu-in-ethylene glycol nanofluids by reducing copper sulfate pentahydrate ($CuSO_4{\cdot}5H_2O$) with sodium hypophosphite ($NaH_2PO_2{\cdot}H_2O$) in ethylene glycol under microwave irradiation. They claim that this one-step chemical method is faster and cheaper than the one-step physical method. The thermal conductivity enhancement approaches that of Cu nanofluids prepared by a one-step physical method developed by Eastman et al. (2001). Although the two-step method works well for oxide nanoparticles, it is not as effective for metal nanoparticles such as copper. For nanofluids containing high-conductivity metals, it is clear that the single-step technique is preferable to the two-step method.

The first-ever nanofluids with carbon nanotubes, nanotubes-in-synthetic oil (PAOs), were produced by a two-step method (Choi et al., 2001). Multiwalled carbon nanotubes (MWNTs) were produced in a CVD reactor, with xylene as the primary carbon source and ferrocene to provide the iron catalyst. MWNTs having a mean diameter of ~ 25 nm and a length of $\sim 50\mu m$ contained an average of 30 annular layers. Chopkar et al. (2006) used ball milling to produce $Al_{70}Cu_{30}$ nanoparticles and dispersed their alloyed nanoparticles in ethylene glycol.

1.4. EXPERIMENTAL DISCOVERIES

Experimental work in a growing number of nanofluids research groups worldwide has discovered that nanofluids exhibit thermal properties superior to those of base fluids or conventional solid–liquid suspensions. For example, thermal conductivity measurements have shown that copper and carbon nanotube (CNT) nanofluids possess extremely high thermal conductivities compared to those of their base liquids without dispersed nanoparticles (Choi et al., 2001; Eastman et al., 2001) and that CNT nanofluids have a nonlinear relationship between thermal conductivity and concentration at low volume fractions of CNTs (Choi et al., 2001). Soon, other distinctive features, such as strong temperature-dependent thermal conductivity (Das et al., 2003b) and strong size-dependent thermal conductivity (Chon et al., 2005) were discovered during the thermal conductivity measurement of nanofluids.

Although experimental work on convection and boiling heat transfer in nanofluids is very limited compared to experimental studies on conduction in nanofluids, revolutionary discoveries such as a twofold increase in the laminar convection heat transfer coefficient (Faulkner et al., 2004) and a threefold increase in the critical heat flux in pool boiling (You et al., 2003) are as unexpected as the discoveries related to conduction. The potential impact of these discoveries on heat transfer applications is large. Therefore, nanofluids promise to bring about a revolution in cooling technologies. As a consequence of these discoveries, research and development on nanofluids has drawn considerable attention from industry and academia over the past several years.

1.4.1. Milestones in Thermal Conductivity Measurements

Initial experimental work has focused on thermal conductivity measurements as a function of concentration, temperature, and size. Later experimental work on boiling and convection heat transfer of nanofluids has added another dimension to the superb heat transfer properties of nanofluids. The effective thermal conductivities of nanofluids were typically measured using a transient hot-wire (THW) method, as this is one of the most accurate ways to determine the thermal conductivities of materials (Lee et al., 1999). Other methods are the oscillating temperature method and the steady-state method.

Metallic Nanofluids with High Thermal Conductivity at Low Concentrations Although measurements of the thermal conductivity of nanofluids started with oxide nanoparticles (Masuda et al., 1993; Lee et al., 1999), nanofluids did not attract much attention until Eastman et al. (2001) showed for the first time that copper nanofluids, produced using the single-step direct evaporation method, have more dramatic conductivity increases than those of oxide nanofluids produced by the two-step method. For some nanofluids, a smalll amount of thioglycolic acid (< 1 vol%) was added to further improve the dispersion. Interestingly, Cu nanoparticles coated with thioglycolic acid gave a 40% increase in the thermal conductivity of ethylene glycol at a particle loading of only 0.3 vol%. This work has demonstrated that metallic nanoparticles whose surface is modified with surfactant molecules produce stable and highly conductive nanofluids at concentrations one order of magnitude lower than those of oxides. Furthermore, this work has shown that the measured thermal conductivities of the copper nanofluids greatly exceed the values predicted by currently available macroscopic theories. Thus, it can be concluded that studies on metallic nanofluids have opened a new horizon with highly enhanced thermal conductivity with low-particle-volume fractions.

Nonlinear Relationship between Thermal Conductivity and Concentration The high thermal conductivity multiwalled of carbon nanotubes (see Table 1.1), combined with their low densities compared with metals, makes them attractive candidate nanomaterials for use in nanofluids. Choi et al. (2001) were the first to disperse MWNTs into a host material, synthetic poly(α-olefin) oil by the two-step method and measured the effective thermal conductivity of nanotube-in-oil suspensions. They discovered that nanotubes yield an anomalously large increase in thermal conductivity (up to a 150% increase in the conductivity of oil at approximately 1 vol% nanotubes), which is by far the highest thermal conductivity enhancement ever achieved in a liquid. This measured increase in thermal conductivity of nanotube nanofluids is an order of magnitude higher than that predicted using existing theories (Maxwell, 1873; Hamilton and Crosser, 1962; Bonnecaze and Brady, 1990). In fact, all values calculated from these models are almost identical at low volume fractions. The results of Choi et al. show another anomaly. The measured thermal conductivity is nonlinear with nanotube loadings, while all theoretical predictions clearly show a linear relationship. This nonlinear behavior is not expected in conventional fluid suspensions of micrometer-sized particles at such low concentrations. Interestingly, similar results have been reported for polymer–nanotube composites (Devpura et al., 2001; Biercuk et al. 2002). Thus, there could be some common enhancement mechanism (such as percolation) between these two dispersions of carbon nanotubes, one in liquids and the other in polymers.

Xie et al. (2003) dispersed MWNTs is in water and ethylene glycol without any surfactant for the first time. The as-received nanotubes were treated with concentrated nitric acid, and their surface was made hydrophilic using a oxygen-containing functional group. Yang et al. (2006) studied the dispersing energy effect on the thermal conductivity of CNT nanofluids and showed that the aspect ratio of the nanotubes decreased significantly with increased sonication time or dispersing energy, confirming the proposition of Assael et al. (2005). Ding et al. (2006) were the first to study temperature-dependent conductivity of CNT–water nanofluids.

It should be noted that nonlinear relationship between thermal conductivity and concentration has been found with Fe–ethylene glycol nanofluids (Hong et al., 2005). It is interesting to note that the enhancement they got was higher than that obtained by Eastman et al. (2001) with Cu nanoparticles. Murshed et al. (2005) also discovered the nonlinear behavior of water-based nanofluids containing spherical and rod-shaped T_iO_2 nanoparticles. The $Al_{70}Cu_{30}$ nanofluids produced by Chopkar et al. (2006) also show strong nonlinear behavior and thus more than 200% enhancement in thermal conductivity with less than 2.0 vol% of the nanoparticles, which is probably due to uniformly dispersed $Al_{70}Cu_{30}$ nanoparticles in ethylene glycol.

Strongly Temperature-Dependent Thermal Conductivity Das et al. (2003b) discovered that nanofluids have strongly temperature-dependent conductivity compared to base fluids. Their data for water-based nanofluids containing Al_2O_3 or CuO nanoparticles show a two- to fourfold increase in thermal conductivity enhancement over a small temperature range between 20 and 50°C. This work opens up the possibility that nanofluids could be employed as "smart fluids," "sensing" local hot spots, spontaneously increasing their thermal conductivity, and providing more rapid cooling in those regions. This unique feature would make nanofluids very attractive coolants for high–heat-flux devices or applications at elevated temperatures. Das et al. suggested that the strong temperature dependence of thermal conductivity is due to the motion of nanoparticles.

Strongly Size-Dependent Thermal Conductivity The size of suspended nanoparticles is critical to the thermal properties of nanofluids. Chon et al. (2005) measured the temperature and nanoparticle size dependency of nanofluid thermal conductivity. Recently, Chopkar et al. (2006) studied the effect of particle size on the thermal conductivity of ethylene glycol–based nanofluids containing $Al_{70}Cu_{30}$ nanoparticles and showed a strongly size-dependent thermal conductivity.

1.4.2. Milestones in Convection Heat Transfer

Although increases in effective thermal conductivity are important in improving the heat transfer behavior of fluids, a number of other variables also play key roles. For example, the heat transfer coefficient for forced convection in tubes depends on many physical quantities related to the fluid or the geometry of the system through which the fluid is flowing. These quantities include intrinsic properties of the fluid such as its thermal conductivity, specific heat, density, and viscosity, along with extrinsic system parameters such as tube diameter and length and average fluid velocity. Therefore, it is essential to measure the heat transfer performance of nanofluids directly under flow conditions.

Experimentalists have shown that nanofluids have not only better heat conductivity but also greater convective heat transfer capability than that of base fluids. Experiments show unexpectedly that the heat transfer coefficients of nanofluids are much better than expected from enhanced thermal conductivity alone in both laminar and turbulent flow. However, for natural convection, nanofluids have lower heat transfer than that of base fluids.

Two- to 3.5-fold Increase in the Laminar Heat Transfer Coefficient Faulkner et al. (2004) conducted fully developed laminar convection heat transfer tests and made the startling discovery that water-based nanofluids containing CNTs provide significant enhancements to the overall heat transfer. First, the heat transfer coefficients of the nanofluids increase with Reynolds number. The heat transfer coefficients of the nanofluid were roughly twice those of plain water at the upper end of the Reynolds number range tested, and it appears that this enhancement will continue to increase with larger Reynolds numbers. Second, nanofluids outperform water, but nanofluids with low particle concentrations (1.1 vol%) perform better than those with higher concentrations (2.2 and 4.4 vol%). This is an unexpected and, indeed, counterintuitive result. This negative concentration dependence of the heat transfer enhancement could be due partially to the interaction between particles. Faulkner et al. proposed that the pseudoturbulence induced by rolling and tumbling CNT agglomerates in a microchannel results in microscale mixing, which enhances the laminar heat transfer coefficient. Since heat transfer applications operate over a wide range of Reynolds numbers and heat fluxes, additional work is needed to develop nanofluids that can provide the most significant benefit to specific heat transfer applications.

In contrast to the work of Faulkner et al., Yang et al. (2005) measured the convective heat transfer coefficients of several nanofluids under laminar flow in a horizontal tube heat exchanger. The average diameter of the disk-shaped graphite nanoparticles used in this research is about 1 to 2μm, with a thickness of around 20 to 40 nm. Their results indicate that the increase in the heat transfer coefficient of the nanofluids is much less than that predicted from a conventional correlation. Near-wall particle depletion in laminar shear flow is one possible reason for the phenomenon. However, there is a doubt whether this work falls in the category of nanofluids at all because the particle diameter is too large for the particles to be called nanoparticles.

Wen and Ding (2004) were first to study the laminar entry flow of nanofluids and showed a substantial increase in the heat transfer coefficient of water-based nanofluids containing γ-Al_2O_3 nanoparticles in the entrance region and a longer entry length for the nanofluids than for water. Ding et al. (2006) were first to study the laminar entry flow of water-based nanofluids containing multiwalled carbon nanotubes (CNT nanofluids). For nanofluids containing only 0.5 wt% CNTs, the maximum enhancement in the convection heat transfer coefficient reaches over 350% at $Re = 800$. Such a high level of enhancement could not be attributed purely to enhanced thermal conductivity. They proposed possible mechanisms such as particle rearrangement, reduction of thermal boundary layer thickness due to the presence of nanotubes, and the very high aspect ratio of CNTs.

Significant Increase in the Turbulent Heat Transfer Coefficient Xuan and Li (2003) were first to show a significant increase in the turbulent heat transfer coefficient. They found that at fixed velocities, the heat transfer coefficient of nanofluids containing 2.0 vol% Cu nanoparticles was improved by as much as 40% compared to that of water. The Dittus–Boelter correlation failed to predict the improved experimental heat transfer behavior of nanofluids. Recent unpublished work shows that the effect of particle size and shape and dispersion becomes predominant in enhancing heat transfer in nanofluids. Even greater heat transfer effects are expected for nanofluids produced by the one-step process. Therefore, there is great potential to "engineer" ultra-energy-efficient heat transfer fluids by choosing the nanoparticle material as well as by controlling particle size, shape, and dispersion.

Decrease in the Natural Convection Heat Transfer Coefficient Putra et al. (2003) were first to study natural convection in nanofluids. Using water with 130-nm Al_2O_3 and 90-nm CuO particles, they showed that the natural convective heat transfer is lower in nanofluids than in pure water and that this decrease in natural convection heat transfer coefficient increases with particle concentration. Interestingly, they attributed this deterioration to the slip between fluid and particle because the denser CuO particles show more deterioration. Wen and Ding (2005a) studied natural convection in water-based nanofluids containing TiO_2 particles and confirmed the deterioration of heat transfer discovered by Putra et al. (2003). However, they attributed this deterioration to convection driven by concentration gradient, particle–surface and particle–particle interaction, and modification of dispersion properties.

1.4.3. Milestones in Boiling Heat Transfer in Nanofluids

Most investigators observed deterioration of pool boiling in nanofluids. However, some experiments with nanofluids have shown a completely different picture by yielding up to a 40% increase in boiling heat transfer coefficient. The ability to greatly increase the critical heat flux (CHF), the heat flux limit in boiling systems, is of paramount importance to ultrahigh–heat-flux devices such as high-powered lasers and reactor components. The enhancement of CHF in nanofluids has been reported by all investigators.

Boiling Heat Transfer Coefficient Das et al. (2003) were first to study the pool boiling characteristics of water-based nanofluids containing 1, 2, and 4 vol% Al_2O_3 nanoparticles and unexpectedly, showed a deterioration of the boiling performance with particle concentration. Later, the same authors (Das et al., 2003) showed that that the deterioration of pool boiling heat transfer in nanofluids is less in small tubes than in large industrial tubes. Bang and Chang (2005) confirmed the deterioration of pool boiling in nanofluids. Furthermore, they observed that the Rohsenow correlation with effective nanofluid properties alone fails to predict their experimental data but with a combination of nanofluid properties and surface characteristics shows good agreement with their data.

In contrast to work by Das et al. (2003, 2003a) and Bang and Chang (2005), Wen and Ding (2005b) reported enhanced boiling heat transfer with nanoparticle concentration and heat flux in nanofluids. Their data show an increase as high as 40% in heat transfer coefficient at about 0.3 vol%, which cannot be explained by conductivity enhancement alone. This could be because Wen and Ding (2005b) conducted pool boiling experiments with nanofluids containing fewer nanoparticles than were used in previous studies. Liu and Qiu (2007) investigated the boiling of an impinging jet of CuO–water nanofluids on a flat surface and showed that nanofluids in jet impingement have poorer boiling characteristics than those of to pure water.

Threefold Increase in CHF You et al. (2003) measured the CHF in pool boiling of Al_2O_3-in-water nanofluids for the first time and discovered an unprecedented phenomenon: a threefold increase in CHF over that of pure water at the mass fraction O (10^{-5}). The enhancement of CHF was confirmed further by Vassallo et al. (2004) with SiO_2 nanoparticles in water despite some differences in nanoparticle materials and concentration range and heater geometry (silica nanoparticles between 2 and 9 vol%, in contrast to Al_2O_3 nanoparticles between ~0.001 and 0.013 vol%, and a horizontal 18-gauge NiCr wire versus the heating surface used by You et al.). Vassallo et al. (2004) also observed a thin coating of silica particles on the wire after boiling but concluded that the increase in roughness alone cannot explain such as unusual rise in CHF.

1.5. MECHANISMS AND MODELS FOR ENHANCED THERMAL TRANSPORT

The marvelous experimental discoveries described in Section 1.4 clearly offer theoretical challenges because they show the fundamental limits of conventional heat conduction, convection, or boiling models for solid–liquid suspensions. Most of the thermal properties of nanofluids measured greatly exceed the values predicted by classical macroscopic theories and models. For example, classical conductivity theories of solid–liquid suspensions used for traditional solid–liquid suspensions (Maxwell, 1873; Hashin and Shtrikman, 1962; Jeffrey, 1973; Jackson, 1975; Davis, 1986; Bonnecaze and Brady, 1990, 1991; Lu and Lin, 1996) cannot explain why low concentrations of nanoparticles can enhance the thermal conductivity of base fluids significantly larger than the theoretical prediction. The big gap between conductivity data measured and model predictions, particularly for copper and CNT nanofluids, which conduct heat 10 times faster than predicted possible, clearly suggests that conventional heat conduction models, developed for fluids containing relatively large particles (three to six orders of magnitude larger than nanoparticles), are inadequate for nanofluids. Other important thermal phenomena in nanofluids, such as a threefold increase in CHF and a twofold increase in convection heat transfer, cannot be explained by conventional convection or boiling theories. In trying to understand the unexpected discoveries and so to overcome the limitations of the classical models, a number of investigators have proposed new physical concepts and mechanisms and developed new models for the enhanced thermal conductivity of nanofluids.

Although there is a substantial number of mechanisms proposed and modeling work related to enhanced conductivity, other important thermal phenomena, such as anomalous increases in CHF and the convection heat transfer coefficient, have not yet led to new mechanisms or models. These unexpected thermal phenomena in nanofluids also necessitate new physical concepts, mechanisms, and models. Therefore, when we realize that nanofluids contain a small quantity of tiny nanoparticles and yet show interesting but unexpected properties, nanofluid is still a mystery calling for new and comprehensive theories to explain these unexpected thermal features.

1.5.1. Milestones in Mechanisms and Models for Enhanced Thermal Conductivity

Conventional solid–liquid suspensions can be described as macroscopic continuum systems. Therefore, existing continuum models of the thermal conductivity of solid–liquid suspensions, all of which are based on the central assumption that the heat transport in each phase is governed by the diffusion equation, adequately represent conventional suspensions of micrometer or larger particles. In these models the particle volume fraction, shape, and orientation and the thermal conductivities of particle and liquid are the important factors controlling the thermal conductivity of conventional suspensions.

For nanofluids the existing continuum model predictions begin to diverge from the experimental data at low volume fractions (Lee et al., 1999; Eastman et al., 2001). As a result, continuum models developed for suspensions of millimeter- or micrometer-sized particles can no longer describe the enhanced thermal conductivity of nanofluids observed in most thermal conductivity measurements. Therefore, it appears that the thermal behavior of nanofluids with nanoscale solid–liquid interface structures or nanoscale particle motion is more complex than that of conventional solid–liquid suspensions and so cannot be explained by the diffusive heat transport mechanism alone. It is expected that energy transport mechanisms at the nanoscale would differ from macroscale mechanisms.

What intrigued nanofluids researchers most in the early days of nanofluids was the experimental discovery that nanofluids can conduct heat much faster than scientists had predicted possible at the low volume fractions of nanoparticles. In addition to this big gap between measured conductivity data and model predictions, the strongly temperature- and size-dependent conductivities of nanofluids have created a great need to understand the thermal transport mechanisms in nanofluids. The expectation that the traditional understanding of how heat is conducted based on the Fourier law of heat conduction could be refined by these discoveries has motivated a number of nanofluids researchers to move to the frontiers of intense search for new mechanisms behind such dramatic property enhancement.

Wang et al. (1999) were first to propose new mechanisms behind enhanced thermal transport in nanofluids, such as particle motion, surface action, and electrokinetic effects. They suggested for the first time that nanoparticle size is important in enhancing the thermal conductivity of nanofluids. Xuan and Li (2000) suggested several possible mechanisms for enhanced thermal conductivity of nanofluids, such as the increased surface area of nanoparticles, particle–particle collisions, and the dispersion of nanoparticles. Years later, Keblinski et al. (2002) proposed four possible microscopic mechanisms for the anomalous increase in the thermal conductivity of nanofluids, which include Brownian motion of the particles, molecular-level layering of the liquid at the liquid–particle interface, the ballistic rather than diffusive nature of heat conduction in the nanoparticles, and the effects of nanoparticle clustering.

Modeling for the thermal conductivity of nanofluids typically falls into two broad categories: extension of existing conduction models and development of new models. Briefly, structural models such as nanolayer, fractal, or percolation structures and dynamic models such as Brownian motion-based collision of nanoparticles belong to the first category. Nanoconvection induced by Brownian motion of nanoparticles and near-field radiation belong to the second category. A number of investigators have proposed both static (or structural) and dynamic mechanisms and models in both categories to account for the anomalously high thermal conductivity enhancements reported in recent measurements. It is interesting to see that the shape of nanoparticles is critical in determining the key mechanism of heat transport in nanofluids. For example, it seems that dynamic mechanisms such as Brownian motion play a key role in nanofluids containing spherical nanoparticles, but structural mechanisms such as percolation are dominant in nanofluids containing CNTs. In some nanofluids there may be a synergistic effect of static and dynamic mechanisms.

Structure-Based Mechanisms and Models Major static or structural models are based on the concepts of nanolayers acting as thermal bridge, fractal structure of agglomerates, percolation structure of high-aspect-ratio nanotubes, cubic arrangement of spherical nanoparticles, interfacial thermal resistance, and surface charge state of nanoparticles. Although liquid molecules close to a solid surface are known to form a solidlike nanolayer, little is known about the connection between this nanolayer and the thermal properties of solid–liquid suspensions. Yu and Choi (2003) proposed for the first time a new mechanism in which, unlike that normally found in solid–solid composite materials, the nanolayer acts as a thermal bridge between a solid nanoparticle and a bulk liquid, so is a key structure-based mechanism of enhancing thermal conductivity of nanofluids. They then developed a renovated Maxwell model for the effective thermal conductivity of solid–liquid suspensions to include the effect of this ordered nanolayer. They extended this simple nanostructural model to nonspherical particles and renovated the Hamilton–Crosser model (Yu and Choi, 2004). The two nanostructural models developed by Yu and Choi are not able to predict the nonlinear behavior of nanofluid thermal conductivity. Xue (2003) was the first researcher to model the nonlinear behavior of nanofluid thermal conductivity. He developed a structural model for nanofluid thermal conductivity based on the liquid layering mechanism and the average polarization theory.

Wang et al. (2003) were first to study the effect of particle clusters and cluster distribution and developed a fractal model for thermal conductivity of nanofluids. Xie et al. (2002a) were first to report the effects of the shape (spherical and cylindrical) of nanoparticles on the enhancement of the thermal conductivity of SiC nanofluids. Because carbon nanotubes have extremely high aspect ratios (or high values of shape factor n in the Hamilton–Crosser model), they have more potential for thermal conductivity enhancement than do spherical nanoparticles. Nan et al. (2003) presented a simple model for thermal conductivity enhancement in CNT composites, taking the effective-medium approach. Nan et al. (2004) have developed a new model by incorporating interface thermal resistance with an effective-medium approach. Recently, Ju and Li (2006) and Xue (2006) considered the interfacial thermal resistance effect in their models for the effective thermal conductivities of carbon nanotube–based mixtures.

Xie et al. (2002b) showed first that the effective thermal conductivity of aqueous Al_2O_3 nanofluids increases with the difference between the pH value and the isoelectric point of nanofluids. Lee et al. (2006) studied the effect on thermal conductivity of the surface charge state of nanoparticles and showed strongly pH-dependent thermal conductivity. Yu and Choi (2005) were first to model the effective thermal conductivity of nanofluids with a cubic arrangement of spherical nanoparticles with shells and to show a nonlinear dependence on the particle-volume concentrations of the effective thermal conductivity of nanofluids containing spherical nanoparticles.

Dynamics-Based Mechanisms and Models The effective thermal conductivity of nanofluids depends not only on the nanostructures of the suspensions but also on the dynamics of nanoparticles in liquids. Nanofluids are dynamic systems, so the motion of nanoparticles and the interactions between dancing nanoparticles or between dancing nanoparticles and liquid molecules should be considered to develop more realistic models. Interestingly, the Brownian motion of nanoparticles was considered as a most probable mechanism. The studies of Wang et al. (1999) and Keblinski et al. (2002) clearly showed that Brownian motion is not a significant contributor to heat conduction based on the results of a time-scale study. However, it is important to understand that the heat transfer mechanism that Wang et al. (1999) and Keblinski et al. (2002) explored is heat conduction through particle–particle collisions caused by the Brownian motion of nanoparticles. Despite the work of Wang et al. (1999) and Keblinski et al. (2002), a few investigators did not drop the idea that Brownian motion of nanoparticles is a most probable mechanism. In fact, one of the key concepts used in most dynamic models is that nanoparticle motion is essential to enhanced energy transport in nanofluids. This is to address one of the most important thermal phenomena in nanofluids: the strongly temperature-dependent thermal conductivity of nanofluids.

Xie et al. (2002b) measured the thermal conductivity of aqueous Al_2O_3 nanofluids with varying particle sizes and showed for the first time that the thermal conductivity of nanofluids depends strongly on particle size. Xuan et al. (2003) were first to develop a dynamic model that takes into account the effects of Brownian motion of nanoparticles and fractals. However, their model has not correctly predicted the strongly temperature-dependent thermal conductivity data obtained by Das et al. (2003b).

Even though it had been stated earlier that Brownian motion is not a significant contributor to enhanced heat conduction (Wang et al., 1999; Keblinski et al., 2002), three dynamic models, all of which show the key role of Brownian motion in nanoparticles in enhancing the thermal conductivity of nanofluids, have been published (Bhattacharya et al., 2004; Jang and Choi, 2004; Hemanth et al., 2004). However, they show large discrepancies among themselves, and the validity of these competing theoretical models is hotly debated.

Yu et al. (2003) were first to develop a simplified one-dimensional drift velocity model of a nanofluid thermal conductivity. They assumed that in the presence of a temperature gradient, the thermophoretically drifting nanoparticles superimposed on their Brownian motion drag a modest amount of the surrounding fluid with them. However, this type of convection model failed to show the effect of nanoparticle size. Jang and Choi (2004) proposed the new concept that nanoscale convection induced by purely Brownian motion of nanoparticles without thermophoretically drifting velocity can enhance the thermal conductivity of nanofluids. Their new dynamic model, which accounts for the fundamental role of nanoconvection, predict strongly temperature- and size-dependent conductivity. Prasher et al. (2005) extended the concept of nanoconvection by considering the effect of multiparticle convection and developed a semiempirical Brownian model to show that nanoconvection caused by the Brownian movement of nanoparticles is primarily responsible for the enhanced conductivity of nanofluids. Recently, Patel et al. (2005) developed a microconvection model for evaluation of thermal conductivity of the nanofluid by taking into account nanoconvection induced by Brownian nanoparticles and the specific surface area of nanoparticles. Ren et al. (2005) considered kinetic theory–based microconvection and liquid layering in addition to liquid and particle conduction.

Koo and Kleinstreuer (2004) extended the convection model of Yu et al. (2003) to consider fluids dragged by a pair of nanoparticles. Furthermore, Koo and Kleinstreuer (2005) show that the role of Brownian motion is much more important than that of thermophoretic and osmophoretic motion and that particle interaction can be neglected when the nanofluid concentration is low ($< 0.5\%$).

Near-Field Radiation Recently, Domingues et al. (2005) proposed a new physical mechanism based on near-field heat transfer. When the volume fraction exceeds a few percent, the mean distance between particles in nanofluids is on the order of the particle diameter. This distance is much lower than the dominant wavelength of far-field radiation (i.e., when photons are emitted or absorbed), and near-field radiation (i.e., Coulomb interaction) may become important.

1.5.2. Milestones in Mechanisms and Models for Convection Heat Transfer

Experimental investigations have demonstrated a remarkable heat transfer enhancement when using nanofluids in forced convection: a 40% increase in turbulent convection heat transfer with the addition of 2.0 vol% of Cu nanoparticles in water and roughly a twofold increase in laminar convection heat transfer by the addition of 1.1 vol% CNTs in water (Xuan and Li, 2003; Faulkner et al., 2004). The enhancement of heat transfer coefficient measured is much higher than that of predictions based on enhanced effective thermal conductivity of nanofluids alone. Such dramatic enhancement of convective heat transfer has inspired several investigators to propose new mechanisms of enhanced convection heat transfer coefficient under both laminar and turbulent flow. In the flow of a nanofluid, thermal dispersion, particle migration, and Brownian diffusion may be some mechanisms of enhanced convection in nanofluids.

Xuan and Roetzel (2000) were first to employ the concept of thermal dispersion for modeling enhanced convection in nanofluids. This concept adds a fictitious conductivity called the *thermal dispersion coefficient* to the effective thermal conductivity of nanofluids by assuming that there is velocity slip between nanoparticle and liquid and that the nanoparticles induce a velocity and temperature perturbation. Xuan and Li (2000) advanced the concept of dispersion further by solving the energy equation under the assumption that axial dispersion is negligible. Khaled and Vafai (2005) investigated the effect of thermal dispersion on heat transfer enhancement of nanofluids and provided thermal dispersion as a possible explanation of the increased thermal conductivity of nanofluids.

Faulkner et al. (2004) were first to propose the concept that pseudoturbulence induced by the rolling and tumbling of CNT agglomerates results in microscale mixing, which nearly doubles the laminar heat transfer coefficient of CNT nanofluids flowing in a microchannel. Ding and Wen (2005) were first to develop a theoretical model to predict particle migration in pressure-driven laminar pipe flows of relatively dilute nanofluids. They showed that shear-induced, viscosity gradient–induced, and concentration gradient–induced particle migration results in the large radial variation of particle distribution, viscosity, and thermal conductivity. The results suggest the existence of an optimal particle size for enhanced thermal conductivity with little penalty on pressure drop.

Buongiorno (2006) considered seven possible mechanisms of fluid particle slip during the convection of nanofluids and showed that Brownian diffusion and thermophoresis are important mechanisms in laminar flow and in the viscous sublayer of turbulent flow, but are negligible in the turbulent region, where the nanoparticles are carried by turbulent eddies. Kim et al. (2004) studied convective instability in nanofluids and predicted enhanced heat transfer in natural convection of nanofluids where the Soret effect is significant. Later, Kim et al. (2007) considered both the Soret and Dufour effects in their study of convective instabilities in binary nanofluids for absorption application and derived the linear stability equation. They calculated the stability parameters for copper and silver nanofluids and showed that the Dufour and Soret effects make nanofluids unstable, but the Soret effect is more important for heat transfer.

Gosselin and da Silva (2004) were first to show that there are optimum particle loadings for the highest heat transfer in laminar and turbulent flow in nanofluids. Mansour et al. (2007) studied the effect of uncertainties in physical properties of nanofluids on forced convection heat transfer in nanofluids and showed that the estimated performance of nanofluids such as pumping power or heat exchanger sizing depends on the models of nanofluid properties. This work shows the importance of developing accurate models of nanofluid properties for practical applications.

1.6. FUTURE RESEARCH

Despite recent advances in the field of nanofluids, the mysteries of nanofluids are unsolved, presenting new opportunities and challenges for thermal scientists and engineers. Nanofluid research could lead to a major breakthrough in developing next-generation coolants for numerous engineering applications. Better ability to manage thermal properties translates into greater energy efficiency, smaller and lighter thermal systems, lower operating costs, and a cleaner environment.

Future research on nanofluids can be classified in two broad categories: basic research, and applied research including development and demonstration. However, basic research and applied research in nanofluids are not separate but go hand in hand. Therefore, a high level of interaction and integration between basic and applied research is required to advance not only nanofluid science but also nanofluid development and demonstration. Because the fundamental mechanisms for energy transport in nanofluids underlie heat transfer processes involving nanofluids, developing a new understanding of energy transport in nanofluids is vitally important for potential cooling applications of nanofluids in multibillion-dollar industries, including electronics, photonics, transportation, MEMS/NEMS, biological and chemical sensors, and biomedical applications. Basic nanofluid research would greatly enable creative development and application of future nanofluid technologies. For example, nanoscale phenomena and nanoscale transport mechanisms discovered or to be discovered in basic research would be very useful in the design of next-generation liquid coolants for a wide range of applications. In short, basic scientists will be able to explain the anomalous behavior of nanofluids, and application engineers will be able to design ultra-energy-efficient nanofluids. In this section we illustrate some challenges in basic and applied nanofluids research and give research directions for basic and applied research in order to create new understanding about the nanofluids and develop commercial nanofluids.

1.6.1. Future Basic Research on Nanofluids

Key Energy Transport Mechanisms The goal of future basic research on nanofluids is to gain a fundamental understanding of the static and dynamic mechanisms of enhanced heat transfer in nanofluids. At present, understanding the fundamental mechanisms of the enhanced thermal conductivity of nanofluids remains a key challenge in nanofluid research. The three main categories of new mechanisms proposed for enhanced thermal conductivity of nanofluids are conduction, nanoscale convection, and near-field radiation. Although these mechanisms have been proposed the validity of most of them remains a subject of debate, and there is no agreement in the nanofluids community about their use. Furthermore, there are few experimental data at the nanoscale level with which to test proposed nanoscale mechanisms such as nanoscale structures and dynamics. The true contribution of the proposed and potential new mechanisms can only be validated by highly sophisticated systematic experiments. Therefore, in the future, such experiments are needed to explore, for example, structure-enhanced energy transport and nanoparticle-mobility-enhanced energy transport. These future studies will reveal key energy transport mechanisms that are missing in existing theories and add to the understanding of the fundamental mechanisms of the thermal conductivity enhancement behind nanofluids. Understanding the fundamentals of energy transport in nanofluids is important not only for advancing basic nanofluid research, but also for validating competing theoretical models and ultimately for developing extremely energy-efficient nanofluids for a range of heat transfer applications.

In conjuction with experimental studies of fundamental energy transport mechanisms, we need to develop tools with high spatial and temporal resolution, for example, to measure the dynamic behavior of a single nanoparticle in suspension or to measure the thermal conductance between two nanoparticles that are suspended in liquid less than 50 nm apart. Development of a technique for temperature measurement at nanometer or subnanometer resolution and the application of x-ray methods to the determination of interface nanostructures would be very useful in advancing thermal physics of nanofluids. New tools and techniques are essential to better understand the physics and chemistry responsible for the anomalous increase in conductivity and to validate new mechanisms, such as nanoconvection or near-field radiation. If we can understand the mechanisms of enhanced thermal conductivity in nanofluids, we can control the thermal properties of nanofluids for nanoengineering of smart coolants.

When the size of an object or device is reduced down to nanometer scale, its surfaces and interfaces are very important. Understanding the thermal characteristics of interfacial nanolayers is important for the growing realm of nanotechnology in general and nanofluids in particular. To understand the laws of physics and chemistry that govern the interface structure and thermal properties, we need to measure the thickness and thermal conductivity of the interface nanolayer. Currently, very little information is available on the structure or chemical and physical properties of nanoparticle–liquid interfaces, and additional experimental studies are needed in this area. It has been observed that the modification of nanoparticle surfaces with surface-modifying additives such as surfactants has a strong influence on the thermal conductivity of nanofluids. For example, copper nanoparticle surfaces modified with thioglycolic acid can significantly increase the effective thermal conductivity of nanofluids (Eastman et al., 2001). Therefore, particle size, and hence large surface area, is not the only important parameter controlling the thermal conductivity of nanofluids. There is growing evidence that particle surface charge, surface chemistry, and interface thermal resistance are important. The development of nanoparticle surface modification methods and materials for improved thermal interfaces as well as the stability of nanofluids would provide great opportunities for the design of next-generation liquid coolants.

Validity of Thermal Conductivity Data and Expansion of Properties and the Cooling Performance Database A number of experimental nanofluid groups have shown that when uniformly dispersed and stably suspended in host liquids, nanoparticles can significantly increase the thermal conductivity of the host liquids. In almost all cases, a transient hot-wire method was used to measure the thermal conductivity of nanofluids. However, few groups have not observed any significant effect of suspended nanoparticles on thermal conductivity. For example, one group used a microscale beam deflection technique to measure the thermal conductivity of extremely dilute nanofluids and did not observe any significantly larger conductivity enhancement than the prediction of effective medium theory (Putnam et al., 2006). Thus, there is a new issue in nanofluids research regarding the validity of the conflicting experimental data. The structural characteristics of nanoparticles, such as the mean particle size, particle size distribution, and shape, depend on the synthesis method. At present it is not clear how many of the conflicting data are due to differences among the nanofluid samples produced by different synthesis techniques and how many are due to thermal conductivity measurement techniques used by the various groups. Therefore, this new issue would require use of at least two different methods to measure the thermal conductivity of the same nanofluid samples and check if data are different due to different methods. It would be vital to characterize and compare the thermal properties of a number of nanofluids accurately using new experimental techniques as well as the commonly used transient hot-wire technique.

In conjunction with this issue, test methods for measurement of thermal conductivities of nanofluids need to be standardized to provide nanofluid researchers with high-quality sample preparation and testing procedures for evaluating the thermal properties of nanofluids so that others can repeat the experiments, produce reliable results, and verify published data. When standardized test methods are established, the thermal properties database should be expanded for nanofluid applications. Furthermore, basic studies on single- and two-phase nanofluid flow and heat transfer in minichannels and microchannels should be conducted for cooling applications. In the future, nanofluid properties and cooling performance should be tested under potential service conditions.

Comprehensive Thermal Conductivity Models In addition to basic experimental study of new mechanisms, we need integrated experimental, modeling, simulation, and theoretical studies. Classical models for the effective properties of solid–liquid suspensions account for the particle concentration, shape, orientation, and distribution, as well as the thermal conductivity of the particle and liquid.

These conventional continuum models should be modified based on a number of nanoscale transport mechanisms, such as interface structures, nanolayer chemistry, and nanoparticle dynamics related to temperature and nanoparticle size. New and comprehensive models of energy transport in nanofluids should then link microscopic parameters such as particle size, shape, polydispersity index, zeta potential, surface chemistry, particle motion, interface structure and properties, and other parameters to the macroscopic properties of nanofluids.

Theory of Nanofluids One of the goals of theoretical research on nanofluids is to develop a theory of nanofluids to explain how nanoparticles change the thermal properties of nanofluids. A theory of nanofluids would also provide a theoretical foundation for physics- and chemistry-based predictive models. There are several reasons that a theory of thermal conductivity of nanofluids has not yet emerged. First, the thermal behavior of nanofluids is quite different from that of solid–solid composites or conventional solid–liquid suspensions. For example, the thermal conductivity of solid-solid composites is reduced when the grain size is reduced. In contrast, the effective thermal conductivity of nanofluids is increased when the nanoparticle size is reduced. Second, nanofluids and conventional solid–liquid suspensions are quite different not only in the magnitude of the thermal conductivity, but also in the dependence of thermal conductivity on temperature and particle concentration and size. Third, nanofluids comprise an emerging, highly interdisciplinary field combining some aspects of such traditional fields as materials science, colloidal science, physics, chemistry, and engineering. So a full understanding of nanofluids requires some knowledge of each field. Therefore, developing a theory of nanofluids is very challenging.

There are two major theoretical approaches to the thermal conductivities of materials: (1) first-principles atomistic simulations, such as equilibrium and nonequilibrium dynamic simulations, and (2) continuum kinetics, such as the Boltzmann transport equation. Atomistic simulations have been employed to determine diffusion coefficients, viscosities, and thermal conductivities for fluids. The Boltzmann equation has been used for various solids. However, there is still no satisfactory extension of the Boltzmann equation to fluids with collisions of more than two bodies.

The theories of thermal conductivity of pure liquids are not well developed. Some old models are based on the assumption that liquid molecules are arranged in a cubic lattice and that energy is carried by phonons from one lattice plane to another with the speed of sound (Bridgman, 1923; Horrocks, 1960). Predictions of the thermal conductivity of liquids based on old theoretical liquid models do not agree well with experimental data for pure liquids. So predictions would get worse when nanoparticles are suspended in a liquid because they would interact with each other or with lattice to allow electromagnetic or particle–lattice heat transfer on top of the lattice vibrational heat transfer of the liquid models.

Therefore, a theory of thermal conductivity of nanofluids may be developed initially by considering two distinctive parts: one that is given in terms of static mechanisms such as the nature of interface layering and thermal resistance, and a second that is given in terms of dynamic mechanisms such as nanoparticle motion and nanoconvection. Later, other mechanisms may be considered. For example, near-field radiation in nanofluids appears to be a really attractive and interesting hypothesis at this stage.

The theoretical result obtained for thermal conductivity should be tested against experimental data available on nanofluids in the literature and from future nanoscale experiments. Theoretical predictions should be in good agreement with experiments with regard to concentration, particle size, and temperature dependence. One proposed theory or model of nanofluids may not be able to explain all experimental data, and only realistic theories can guide the formulation of optimized nanofluids. However, it should be noted that the subject of nanofluids is a continuing study, and it is likely that several generations of theories will be required to arrive at a model that can explain all the data satisfactorily. This is how we advance scientific knowledge. No theory or model is perfect. Each time we take one small new step in developing a theory or model, we move it closer to reality.

New Mechanisms and Models of Enhanced Convection and Critical Heat Flux
It seems that investigators are having difficulty in understanding the anomalous behavior of nanofluids in regard to the enhanced convection heat transfer coefficient and critical heat flux since little work on the mechanisms and models of enhanced convection and CHF has appeared in the literature. In fact, such a large enhancement in heat transfer and CHF of nanofluids cannot be explained by the classical theories and models currently used for traditional solid–liquid suspensions. Therefore, we need to understand the underlying fundamentals of the role of nanoparticles in convection heat transfer and CHF, such as nanoscale mixing, bubble growth, and bubble dynamics by discovering missing heat transfer and CHF enhancement mechanisms at the nanoscale.

1.6.2. Future Applied Research on Nanofluids

Experiments have shown that a number of nanofluids provide extremely desirable thermal properties, such as higher thermal conductivity, convection heat transfer coefficients, and CHF compared to their base liquids without dispersed nanoparticles. These key thermal features of nanofluids, together with excellent nanoparticle suspension stability, would open the door to a wide range of engineering applications, such as engine cooling and microelectronics cooling, and biomedical applications, such as cancer therapy. Nanofluid research presents us with very promising opportunities for applications, but there are still a number of technical issues on the road to commercialization. In this connection, in this section we identify some technical barriers facing the development of commercially available nanofluid technology and suggest research needed to overcome the barriers and to achieve cost-effective, high-volume production of nanofluids.

Volume Production of Nanofluids Production of nanofluids is currently limited to laboratory-scale research. Therefore, high-volume low-cost production of well-dispersed nanofluids is one of the most serious technical barriers to the development and commercial use of nanofluids.

Barriers and Challenges in the Two-step Process An advantage of the two-step technique in terms of eventual commercialization of nanofluids is that the inert-gas condensation technique has already been scaled up to produce tonnage quantities of nanoparticles economically (Romano et al., 1997). Therefore, nanopowders produced in bulk at low prices can be used to make nanofluids by the two-step method. Because these nanoparticles are commercially available in volume orders and relatively cheap, the two-step method is very attractive for industrial applications of nanofluids. However, nanofluids produced by the two-step process contain large aggregates and require high-volume concentrations of oxide nanoparticles (approximately 10 times those of metallic nanoparticles produced and dispersed by the one-step process) to achieve comparable thermal conductivity enhancement. Although the problem of aggregation of nanoparticles is particularly severe at particle concentrations greater than 20 vol%, it often occurs in nanofluids, depending on the characteristics of nanoparticles and the liquid environment. Therefore, it is important to minimize aggregation in nanofluids. Some surface-treated nanoparticles show excellent dispersion and thermal properties. The challenge is to develop innovative ways to improve the two-step process to produce well-dispersed nanofluids in volume. In fact, some nanoparticles are available commercially in the form of liquid suspensions. Ceramic suspensions are available in large quantities in the market. Therefore, the real challenge appears to be significant cost reduction in nanofluid production using the two-step process.

Barriers and Challenges in the One-Step Process Although the two-step technique works well for oxide nanoparticles, it is not as effective for metal nanoparticles such as copper. For nanofluids containing high-conductivity metals, it is clear that the single-step technique is preferable to two-step processing. However, although the one-step physical method developed by Argonne is excellent for research, it is not likely to become the mainstay of nanofluid production because the process would be hard to scale up, for two reasons: Processes that require a vacuum slow the production of nanoparticles and nanofluids significantly, and the production of nanofluids by this one-step physical process is expensive.

Recently, an alternative one-step chemical method for making copper nanofluids has been reported (Zhu et al., 2004). Nearly monodisperse copper nanoparticles less than 20 nm in diameter were produced and dispersed in ethylene glycol by the reduction of a copper salt by sodium hypophosphite. Poly(vinylpyrrolidone) was added as a protective polymer and stabilizer that inhibited particle aggregation. Copper nanofluids produced by this one-step chemical method show nearly the same thermal conductivity enhancement as the nanofluids produced by the one-step physical method. Although this new one-step chemical process was used only to produce small quantities of nanofluids in a laboratory, with some development it appears that it has the potential to produce large quantities of nanofluids faster than the one-step physical process. Therefore, it is needed to study the potential of the new one-step chemical method of making stable nanofluids and scaling up to commercial production. Since a one-step chemical method can minimize nanoparticle agglomeration, it can produce well-dispersed nanofluids containing monosized nanoparticles. However, a significant limitation to the application of this technique is that the volume fractions of nanoparticles and quantities of nanofluids that can be produced are much more limited than with the two-step technique. Unlike the two-step process, these one-step processes are not yet available commercially. Therefore, the challenge is to develop innovative ways to improve the one-step chemical process to produce large quantities of nanofluids economically. It should be noted that the current one-step physical or chemical production systems run in batch mode with limited control over a number of important parameters, including those that control nanoparticle size. The one-step physical and chemical processes are likely to have commercial potential if they allow making nanofluids in a continuous process.

Future focus should be on identifying promising methods that do not require a vacuum and that provide continuous fluid feed and extraction capabilities in a production system. New technologies for making stable nanofluids which do not require a vacuum and utilize a semicontinuous or continuous process will probably replace current methods of producing nanofluids. In the future, these methods could lead to the ability to make nanofluids much faster and cheaper than can be accomplished using current methods. The critical technical breakthroughs in industrial-scale production of nanofluids to bring nanofluids to commercialization are expected to be achieved through continued support of nanofluids R&D and collaboration with industrial partners.

Long-Term Stability In addition to the production-scale-up issue, we need to address a number of concerns related to the use of nanofluids, including clogging, fouling, corrosion, abrasion, compatibility, and long-term stability. Making stable nanofluids containing monosized nanoparticles is challenging in lab-scale research. But long-term stability of the nanofluids could be a practical issue in the commercialization of nanofluids. Long-term stability of nanoparticle suspensions, by making small (1- to 10-nm) nanoparticles and dispersing them without agglomeration using special mechanical dispersing techniques and the creative use of chemical dispersants, is critical to fully appreciate the benefits of nanofluids.

Green Nanofluids Nanotechnology is a compelling solution to our urgent need for the more efficient use of energy in general and for the faster cooling of devices and systems in particular. However, we now face public concerns and challenges to make sure that nanotechnology is safe. We need to address public concerns about potential health and environmental hazards of nanotechnology.

Nanoparticles are very different from micro- or macro-sized materials. Because it is not known yet if nanoparticles of certain materials and size would have undesirable effects on the environment and health, we should care about the potential negative impact of nanoparticles on humans or the environment.

Systematic research into potential risks and benefits of nanofluids would require the development of methods for evaluating the health and environmental impact of nanofluids and models for predicting the potential health and environmental impact of nanofluids. The public needs to be informed of research findings on nanofluid risks and benefits. Looking forward, it seems prudent for nanofluid engineers to think about and develop green nanofluids by choosing nontoxic nanoparticles that would pose no environmental, safety, and health danger so that nanofluids could be produced in large quantities and used widely in industrial and consumer thermal management applications. Biodegradable nanoparticles could be used in making nanofluids for biomedical applications. Low-cost, high-volume production of green nanofluids would be one of the most challenging future research directions.

REFERENCES

Akoh, H., Y. Tsukasaki, S. Yatsuya, and A. Tasaki (1978). Magnetic properties of ferromagnetic ultrafine particles prepared by a vacuum evaporation on running oil substrate, *J. Cryst. Growth*, 45: 495–500.

Assael, M. J., I. N. Metaxa, J. Arvanitidis, D. Christophilos, and C. Lioutas (2005). Thermal conductivity enhancement in aqueous suspensions of carbon multi-walled and double-walled nanotubes in the presence of two different dispersants, *Int. J. Thermophys.*, 26: 647–664.

Bang, I. C., and S. H. Chang (2005). Boiling heat transfer performance and phenomena of Al_2O_3–water nano-fluids from a plain surface in a pool, *Int. J. Heat Mass Transfer*, 48: 2407–2419.

Bhattacharya, P., S. K. Saha, A. Yadav, P. E. Phelan, and R. S. Prasher (2004). Brownian dynamics simulation to determine the effective thermal conductivity of nanofluids, *J. of App. Phys.*, 95: 6492–6494.

Biercuk, B. J., M. C. Llaguno, M. Radosavljevic, J. K. Hyun, and A. T. Johnson (2002). Carbon nanotube composites for thermal management, *Appl. Phys. Lett.*, 80: 2767–2772.

Bonnecaze, R. R., and J. F. Brady (1990). A method for determining the effective conductivity of dispersions of particles, *Proc. R. Soc. London. A*, 430: 285–313.

Bonnecaze, R. R., and J. F. Brady (1991). The effective conductivity of random suspensions of spherical particles, *Proc. R. Soc. London. A*, 432: 445–465.

Bridgman, P. W. (1923). *Proc. Natl. Acad. Sci.*, 9: 341.

Buongiorno, J. (2006). Convective transport in nanofluids, *J. Heat Transfer*, 128: 240.

Choi, S. U. S. (1995). Enhancing thermal conductivity of fluids with nanoparticles, in *Developments and Applications of Non-Newtonian Flows*, D. A. Singer and H. P. Wang, Eds., American Society of Mechanical Engineers, New York, FED–231/MD-66: 99–105.

Choi, S. U. S., C. S. Rogers, and D. M. Mills (1992). High-performance microchannel heat exchanger for cooling high-heat-load x-ray optical elements, in *Micromechanical Systems*, D. Cho, J. P. Peterson, A. P. Pisano, and C. Friedrich, Eds., American Society of Mechanical Engineers, New York, DSC–40: 83–89.

Choi, S. U. S., Z.G. Zhang, W. Yu, F.E. Lockwood, and E. A. Grulke (2001). Anomalous thermal conductivity enhancement in nano-tube suspensions, *Appl. Phys. Lett.*, 79: 2252–2254.

Chon, C. H., K. D., Kihm, S. P. Lee, and S. U. S. Choi (2005). Empirical correlation finding the role of temperature and particle size for nanofluid (Al_2O_3) thermal conductivity enhancement. *Appl. Phys. Lett.*, 87: 153107.

Chopkar, M, P. K. Das, and I. Manna (2006). Synthesis and characterization of nanofluid for advanced heat transfer applications, *Scr. Mater*. 55: 549–552.

Das, S. K., N. Putra, and W. Roetzel (2003). Pool boiling characteristics of nano-fluids. *Int. J. Heat Mass Transfer,* 46: 851–862.

Das, S. K., N. Putra, and W. Roetzel (2003a). Pool boiling of nano-fluids on horizontal narrow tubes. *Int. J. Multiphase Flow,* 29: 1237–1247.

Das, S. K., N. Putra, P. Thiesen, and W. Roetzel (2003b). Temperature dependence of thermal conductivity enhancement for nanofluids, *J. Heat Transfer*, 125: 567–574.

Davis, R. H. (1986). The effective thermal conductivity of a composite material with spherical inclusions, *Int. J. Thermophys*. 7: 609–620.

Devpura A., P. E. Phelan and R. S. Prasher (2001). Size effect on the thermal conductivity of polymers laden with highly conductive filler particles, *Microscale Thermophys. Eng.*, 5: 177–189.

Ding, Y., and D. Wen (2005). Particle migration in a flow of nanoparticle suspensions. *Powder Technol.*, 149 (2–3): 84–92.

Ding, Y., H. Alias, D. Wen, and R.A. Williams (2006). Heat transfer of aqueous suspensions of carbon nanotubes (CNT nanofluids)., *Int. J. Heat Mass Transfer,* 49: 240–250.

Domingues, G., S. K. Volz, Joulain, and J.-J. Greffet (2005). Heat transfer between two nanoparticles through near-field interaction, *Phys. Rev. Lett.*, 94: 085901.

Duncan, A. B., and G. P. Peterson (1994). Review of microscale heat transfer, *Appl. Mech. Rev.*, 47: 397–428.

Duncan, M. A., and D. H. Rouvray (1989). Microclusters, *Sci. Am.*, Dec., pp. 110–115.

Eastman, J. A., S. U. S. Choi, S. Li, L. J. Thompson, and S. Lee (1997). Enhanced thermal conductivity through the development of nanofluids, *Proc. Symposium Nanophase and Nanocomposite Materials II*, Materials Research Society, Boston, MA, 457: 3–11.

Eastman, J. A., S. U. S. Choi, S. Li, W. Yu, and L. J. Thomson (2001). Anomalously increased effective thermal conductivities of ethylene glycol based nanofluids containing copper nanoparticles, *Appl. Phys. Lett.*, 78: 718–720.

Faulkner, D. J., D. R. Rector, J. J. Davidson, and R. Shekarriz (2004). Enhanced heat transfer through the use of nanofluids in forced convection, Paper IMECE2004-62147, presented at the 2004 ASME International Mechanical Engineering Congress and RD&D Expo, Anaheim, CA, Nov. 13–19.

Fissan, H. J., and J. Schoonman (1998). Vapor-phase synthesis and processing of nanoparticle materials (nano):-a European Science Foundation (ESF) program, *J. Aerosol Sci.*, 29: 755–757.

Gleiter, H. (1989). Nanocrystalline materials, *Prog. Mater. Sci.*, 33: 223–315.

Gosselin, L., and A. K. da Silva (2004). Combined heat transfer and power dissipation optimization of nanofluid flows, *Appl. Phys. Lett.*, 85: 4160–4162.

Granqvist, C. G., and R. A. Buhrman (1976). Ultrafine metal particles, *J. Appl. Phys.*, 47: 2200.

Grimm, A. (1993). Powdered aluminum-containing heat transfer fluids, German patent DE 4131516 A1.

Hamilton, R. L., and O. K. Crosser (1962). Thermal conductivity of heterogeneous two-component systems, *Ind. Eng. Chem. Fundam*. 1: 187–191.

Hashin, Z., and S. Shtrikman (1962). A variational approach to the theory of the effective magnetic permeability of multiphase materials, *J. Appl. Phys.*, 33: 3125–3131.

Hayashi, C., and M. Oda (1998). Research and applications of nano-particles in Japan, *J. Aerosol Sci.*, 29: 757–760.

Hemanth, K. D., H. E. Patel, K. V. R. Rajeev, T. Sundararajan, T. Pradeep, and S. K. Das (2004). Model for heat conduction in nanofluids, *Phys. Rev. Lett.*, 93: 144301.

Hong T. K., H. S. Yang and C. J. Choi (2005). Study of the enhanced thermal conductivity of Fe nanofluids, *J. Appl. Phys*., 97: 064311.

Horrocks, J. K. (1960). *Trans. Faraday Soc.*, 56: 206.

Jackson, D. J. (1975). *Classical Electrodynamics*, 2nd ed., Wiley, London.

Jang, S. P., and S. U. S. Choi (2004). Role of Brownian motion in the enhanced thermal conductivity of nanofluids, *Appl. Phys. Lett.*, 84: 4316–4318.

Jeffrey, D. J. (1973). Conduction through a random suspension of spheres, *Proc. R. Soc. London A*, 335: 355–367.

Ju, S., and Z. Y. Li (2006). Theory of thermal conductance in carbon nanotube composites, *Phys. Lett. A*, 353: 194–197.

Keblinski, P., S. R. Phillpot, S. U. S. Choi, and J. A. Eastman (2002). Mechanisms of heat flow in suspensions of nano-sized particles (nanofluids), *Int. J. Heat and Mass Transfer*, 45: 855–863.

Khaled, A. R. A., and K. Vafai (2005). Heat transfer enhancement through control of thermal dispersion effects, *Int. J. Heat and Mass Transfer*, 48: 2172.

Kim, J., Y. T., Kang, and C. K. Choi (2004). Analysis of convective instability and heat transfer characteristics of nanofluids, *Phys. Fluids*, 16: 2395–2401.

Kim, J., Kang, Y. T., and C. K. Choi (2007). Soret and Dufour effects on convective instabilities in binary nanofluids for absorption application, *Int. J. Refrig.*, 30: 323–328.

Kimoto, K., Y. Kamilaya, M. Nonoyama, and R. Uyeda (1963). An electron microscope study on fine metal particles prepared by evaporation in argon gas at low pressure, *Jpn. J. Appl. Phys.*, 2: 702.

Koo, J., and C. Kleinstreuer (2004). A new thermal conductivity model for nanofluids, *J. Nanopart. Res.* 6 (6): 577–588.

Koo, J., and C. Kleinstreuer (2005). Impact analysis of nanoparticle motion mechanisms on the thermal conductivity of nanofluids, *Int. Commun. Heat Mass Transfer*, 32 (9): 1111–1118.

Lee, S., S. U. S. Choi, S. Li, and J. A. Eastman (1999). Measuring thermal conductivity of fluids containing oxide nanoparticles, *J. Heat Transfer*, 121: 280–289.

Lee, D., J.-W. Kim, and B. G. Kim (2006). A new parameter to control heat transport in nanofluids: surface charge state of the particle in suspension, *J. Phys. Chem. B*, 110: 4323–4328.

Li, B. C. (1998). Nanotechnology in China, *J. Aerosol Sci.*, 29: 751–755.

Liu, Z. and Y. Qiu (2007). Boiling heat transfer characteristics of nanofluids jet impingement on a plate surface. *Heat Mass Transfer*, 43: 699–706.

Lu, S., and H. Lin (1996). Effective conductivity of composites containing aligned spherical inclusions of finite conductivity, *J. Appl. Phys.*, 79: 6761–6769.

Mansour, R. B., N. Galanis, and C.T. Nguyen (2007). Effect of uncertainties in physical properties on forced convection heat transfer with nanofluids, *Appl. Therm. Eng.*, 27: 240–249.

Masuda, H., A. Ebata, K. Teramae, and N. Hishinuma (1993). Alteration of thermal conductivity and viscosity of liquid by dispersing ultra-fine particles (dispersion of r-Al_2O_3, SiO_2, and TiO_2 ultra-fine particles), *Netsu Bussei (Japan)*, 4: 227–233.

Maxwell, J. C. (1873). *Treatise on Electricity and Magnetism*. Clarendon Press, Oxford.

Murshed, S. M. S., K. C. Leong, and C. Yang (2005). Enhanced thermal conductivity of TiO_2–water based nanofluids, *Int. J. Therm. Sci*, 44: 367–373.

Nan, C.-W., Z. Shi, and Y. Lin (2003). A simple model for thermal conductivity of carbon nanotube-based composites, *Chem. Phys. Lett.*, 375: 666–669.

Nan, C.-W., G. Liu, Y. Lin, and M. Li (2004). Interface effect on thermal conductivity of carbon nanotube composites, *Appl. Phys. Lett.* 85: 3549–3551.

Patel, H. E., T. Sundararajan, T. Pradeep, A. Dasgupta, N. Dasgupta, and S. K. Das (2005). A micro-convection model for thermal conductivity of nanofluid, *Pramana J. Phys.*, 65: 863–869.

Prasher, R., P. Bhattacharya, and P. E. Phelan (2005). Thermal conductivity of nanoscale colloidal solutions (nanofluids), *Phys. Rev. Lett.*, 94: 025901.

Putnam, S. A., D. G. Cahill, P. V. Braun, Z. Ge, and R. G. Shimmin (2006). Thermal conductivity of nanoparticle suspensions, *J. Appl. Phys.*, 99: 084308.

Putra, N., W. Roetzel, and S. K. Das (2003). Natural convection of nano-fluids, *Heat Mass Transfer*, 39: 775–784.

Ren, Y., H. Xie, and A. Cai (2005). Effective thermal conductivity of nanofluids containing spherical nanoparticles, *J. Phys. D Appl. Phys.*, 38: 3958–3961.

Roco, M. C. (1998). Nanoparticle and nanotechnology research in the U. S. A., *J. Aerosol Sci.*, 29: 749–751.

Rohrer, H. (1996). The nanoworld: chances and challenges, *Microelectron. Eng.*, 32: 5–14.

Romano, J. M., J. C. Parker, and Q. B. Ford (1997). *Adv. Powder Metall. Partic. Mater.*, 2: 12–3.

Siegel, R. W., and J. A. Eastman (1993). A small revolution creates materials one atomic building block at a time, *Logos*, 11: 2–7.

Suslick, K. S., M. Fang, and T. Hyeon (1996). *J. Am. Chem. Soc.*, 118, 11960.

Touloukian, Y. S., R. W. Powell, C. Y. Ho, and P. G. Klemens (1970). *Thermophysical Properties of Matter*, Vol. 2, Plenum Press, New York.

Tuckerman, D. B., and R. F. W. Peace (1981). High-performance heat sinking for VLSI, *IEEE Electron. Devices. Lett.*, EDL-2: 126–129.

Vassallo, P., R. Kumar, and S. D'Amico (2004). Pool boiling heat transfer experiments in silica–water nano-fluids, *Int. J. Heat Mass Transfer*, 47: 407–411.

Wagener, M., B. S. Murty, and B. Günther (1997). Preparation of metal nanosuspensions by high-pressure dc-sputtering on running liquids, in *Nanocrystalline and Nanocomposite Materials II*, S. Komarnenl, J. C. Parker, and H. J. Wollenberger, Eds., Materials Research Society, Pittsburgh, PA, 457: 149–154.

Wang, B.-X., L.-P. Zhou, and X.-F. Peng (2003). A fractal model for predicting the effective thermal conductivity of liquid with suspension of nanoparticles, *Int. J. Heat Mass Transfer*, 46: 2665–2672.

Wang, X., X. Xu, and S. U. S. Choi (1999). Thermal conductivity of nanoparticle–fluid mixture, *J. Thermophys. Heat Transfer*, 13: 474–480.

Wen, D., and Y. Ding (2004). Experimental investigation into convective heat transfer of nanofluids at the entrance region under laminar flow conditions, *Int. J. Heat Mass Transfer*, 47: 5181–5188.

Wen, D., and Y. Ding (2005a). Formulation of nanofluids for natural convective heat transfer applications, *Int. J. Heat Fluid Flow*, 26: 855–864.

Wen, D., and Y. Ding (2005b). Experimental investigation into the pool boiling heat transfer of aqueous based alumina nanofluids, *J. Nanopart. Res*. 7: 265–274.

Xie, H. Q., J. C. Wang, T. G. Xi, and Y. Liu (2002a). Thermal conductivity of suspensions containing nanosized SiC particles, *Int. J. Thermophys.*, 23: 571–580.

Xie, H. Q., Wang, J. C., Xi, T. G., Liu, Y., Ai, F., and Wu, Q. R. (2002b). Thermal conductivity enhancement of suspensions containing nanosized alumina particles, *J. Appl. Phys.*, 91: 4568–4572.

Xie, H., H. Lee, W. Youn, and M. Choi (2003). Nanofluids containing multiwalled carbon nanotubes and their enhanced thermal conductivities, *J. Appl. Phys.*, 94: 4967–4971.

Xuan, Y., and Q. Li (2000). Heat transfer enhancement of nano-fluids, *Int. J. Heat Fluid Flow*, 21: 58–64.

Xuan, Y., and Q. Li (2003). Investigation on convective heat transfer and flow features of nanofluids, *J. Heat Transfer*, 125: 151–155.

Xuan, Y., and W. Roetzel (2000). Conceptions for heat transfer correlation of nanofluids, *Int. J. Heat Mass Transfer*, 43: 3701–3707.

Xuan, Y., Q. Li, and W. Hu (2003). Aggregation structure and thermal conductivity of nanofluids, *AIChE J.*, 49: 1038–1043.

Xue, Q.-Z. (2003). Model for effective thermal conductivity of nanofluids, *Phys. Lett. A*, 307: 313–317.

Xue, Q. Z. (2006). Model for the effective thermal conductivity of carbon nanotube composites, *Nanotechnology*, 17: 1655–1660.

Yang, Y. Z., Z. G. Zhang, E. A. Grulke, W. B. Anderson, and G. Wu (2005). Heat transfer properties of nanoparticle-in-fluid dispersions (nanofluids) in laminar flow, *Int. J. Heat Mass Transfer*, 48: 1107–1116.

Yang, Y. Z., E. A. Grulke, Z. G. Zhang, and G. Wu (2006). Thermal and rheological properties of carbon nanotube-in-oil dispersions, *J. Appl. Phys.*, 99: 114307.

You, S. M., J. H. Kim, and K. M. Kim (2003). Effect of nanoparticles on critical heat flux of water in pool boiling of heat transfer, *Appl. Phys. Lett.*, 83: 3374–3376.

Yu, W., and S. U. S. Choi (2003). The role of interfacial layers in the enhanced thermal conductivity of nanofluids: a renovated Maxwell model, *J. Nanopar Res.* 5: 167–171.

Yu, W., and S. U. S. Choi (2004). The role of interfacial layers in the enhanced thermal conductivity of nanofluids: a renovated Hamilton–Crosser model, *J. Nanopart Res.* 6: 355–361.

Yu, W., and S. U. S. Choi (2005). An effective thermal conductivity model of nanofluids with a cubic arrangement of spherical particles, *J. Nanosci. Nanotechnol.*, 5: 580–586.

Yu, W., J. H. Hull, and S. U. S. Choi (2003). Stable and highly conductive nanofluids: experimental and theoretical studies, Paper TED-AJ03-384, *Proc. 6th ASME-JSME Thermal Engineering Joint Conference*, Hawaiian Islands, Mar. 16–23, 2003, ASME, NewYork.

Zhu, H., Y. Lin, and Y. Yin (2004). A novel one-step chemical method for preparation of copper nanofluids, *J. Colloid Interface Sci.*, 277: 100–103.

2 Synthesis of Nanofluids

In this chapter we are concerned with the synthesis of a diverse variety of nanofluids. As this book is concerned with nanofluids designed for specific applications, we limit our discussion to stable nanofluids and free-standing nanosystems which may be made in the form of fluids. Although the first category of fluids may not be separable into the constituent phases (i.e., solid and liquid), the other is in the separated phases to begin with. Nanoparticles formed in porous media such as zeolites, and embedded particles in glasses are not dealt with, although in a larger sense, particles in solids may be considered as solutions.

From a general perspective, a two-phase colloidal system can be classified in terms of a dispersed phase and a dispersion medium. The dispersed phase and dispersion medium can be any one of the three phases (i.e., gas, liquid, or solid) except that the first category (i.e., gas in gas) is unknown. From this, a solid nanoparticle dispersed in an amorphous solid may be considered as a colloidal system and consequently, a nanofluid. In our descriptions, fluids will be liquids at ordinary conditions of temperature and pressure, and for that reason, supercritical fluids and gases as the dispersion phase are not considered. It may be noted that the synthesis of nanoparticles in these media (i.e., solid matrices[1] and supercritical fluids[2]) is a large and advanced area of science. From a historical perspective, it is also important to remember that some of the early applications of nanoparticles were in the form of embedded particles in glasses.[3] Supercritical fluids are a recent area of development in nanoparticle science.

2.1. GENERAL ISSUES OF CONCERN

There are several factors of interest when considering a given synthetic approach: (1) thermal stability, (2) dispersability in diverse media, and (3) chemical compatibility and ease of chemical manipulation. Each of these parameters is discussed in some detail below. It should be noted that several are intimately connected to each other.

Nanofluids: Science and Technology, By Sarit K. Das, Stephen U. S. Choi, Wenhua Yu, and T. Pradeep

2.1.1. Thermal Stability

Nanoparticles are metastable systems. This means that they will be transformed to stable materials that have global energy minima in the free-energy landscape. If one plots the free energy of a system for a fixed amount of material (as free energy is an extensive property), nanoparticles exist at a higher energy than the bulk materials, although a given nanostructure may have a local minimum compared to other structures. In the limit of extremely small particles, these local minima correspond to magic numbers with unusual structural–electronic stability. Such particles, called clusters, are discussed in a later section. In the case of isolated atoms and molecules, the total energy is much larger than that for bulk materials. Nanoparticles constitute a regime in which the energy is in between that of bulk materials and molecules/atoms. It is possible to convert one form to the other by physical or chemical means.

In the smaller size regime of less than 1 nm, nanoparticles possess distinct structures and may be regarded as *nanocluster molecules*. Each of these clusters may have isomeric structures, and one may be more stable than the other. Various structures of a given size may have large differences in properties. In the larger size regime, numerous structural forms can exist, but it is difficult to distinguish them. When it comes to large changes in the geometry, as in the case of a nanoparticle and a nanorod, distinctions can be made on the basis of their electronic properties. Synthetic methodologies are available to make some of these structures in the case of a few metals and ceramics.

As is evident from Figure 2.1, transformation from one form to another is possible. Although the transformation from atoms, molecules, and nanoparticles to bulk is spontaneous, stabilizing the nanoparticle regime requires careful control. The formation of nanoparticles from constituent atoms is referred to as the *bottom-up approach*, while the synthesis of nanostructures from bulk is referred to as the *top-down* approach. In our discussion, the principal focus is on bottom-up cases, although top-down is mentioned occasionally. Because nanoparticles are metastable, over an infinite period of time they would revert to bulk. In some cases, time has no practical consequence: Faraday's colloids, made in 1857, remain stable today.[4] Although bulk metal is more stable, it is analogous to the stability of graphite compared to diamond. Diamond, although metastable, does not become graphite at normal conditions of temperature and pressure even if kept for a millennium. The kinetics of diamond-to-graphite transformation is very slow and therefore, is insignificant under normal conditions. The case of nanoparticles is analogous and is referred to as *kinetic stability*.

If held at a close distance, nanoparticles will lead to interparticle interaction and that will cause aggregation or coalescence. In *aggregation*, the particles retain their individuality, but part of their surface area is lost due to the interaction. Such interactions occur because the surface of the nanoparticles contains groups or molecules that aid in aggregation. Familiar cases are hydrogen bonding or ionic interactions between surface groups. Aggregation is also referred to as *coagulation*, which is especially important in cases where particles are stabilized by an electrical double layer (see below). Interparticle interaction can also result

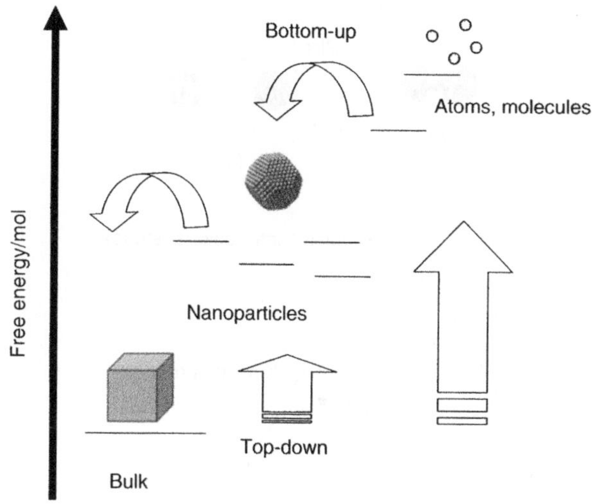

Fig. 2.1 Nanoparticles shown as a metastable system; their energies are between atoms, molecules, and bulk. There can be several different types of nanoparticles with distinct energy, and each nanoparticle can be converted to bulk.

in *coalescence* of particles, the irreversible fusion of particles that results in larger particles.

The situation can be better understood in terms of two isolated particles. Nanoparticles generally contain an overlayer of stabilizing groups or molecules. When they are at a finite distance d, interactions between particles will result in an energy minimum. The interactions can be electrostatic or van der Waals (due to the type of protecting group). Most often, these two types of interactions dominate, especially in a dielectric solvent such as water. Whereas electrostatic interactions try to keep the particles away from each other, van der Waals interactions between the particle cores bring them together. This results in a net energy minimum, as shown in Figure 2.2. These van der Waals interactions are strong at short distances and the particles coalesce in the absence of a shell that imparts repulsion. The repulsion can, additionally, be due to steric forces, as in the case of a molecule covering a particle. The nature of repulsive interactions change depending on the type of the shell. For the van der Waals forces to be effective, the distance has to be short and there is a barrier that prevents this interaction from being dominating. If the height of this barrier is greater than the thermal energy kT, the system is kinetically stable. As can be seen, the particles possess greater energy to overcome this barrier at higher temperatures, and the colloidal system aggregates beyond a critical value called the *critical flocculation temperature*. It may be noted that the stability of the shell over the nanoparticle is also temperature dependent. The discussion here is analogous to that of colloids.

In the limit of a covalently bound shell on a nanoparticle surface, the shell is stable under normal conditions of temperature and pressure encountered in a

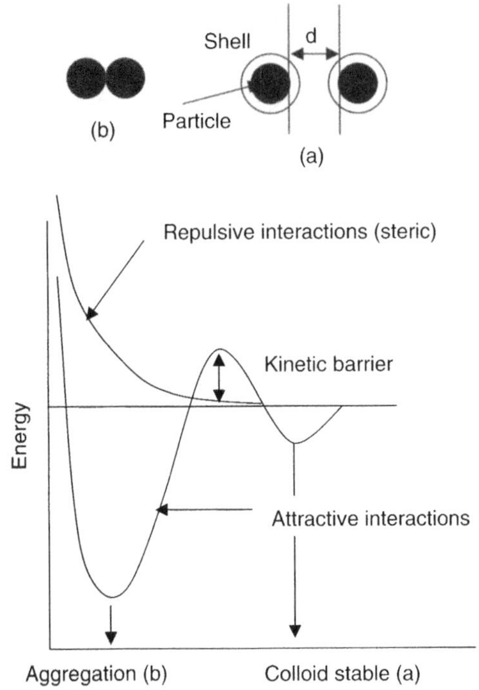

Fig. 2.2 Kinetic stability of a nanoparticle system. As the barrier for van der Waals interaction is greater than thermal energy, the isolated nanoparticles are stable. While attractive interactions bring down the energy, repulsive interactions increase the energy. In the absence of a shell over the nanoparticle, kinetic stability does not exist and the particles coalesce or aggregate.

nanofluid application. Thus, the system can be infinitely stable. Nevertheless, it is important to note that the nanoparticle core may undergo irreversible changes in temperature cycling. These changes correspond to structural, electrical, or magnetic phase transitions. In molecular detail, the structure of the shell may also undergo a transformation, such as *conformational ordering*, changes in the relative orientation of the bonds in the molecule.

In most of the cases described, stability is related to kinetic stability. Nevertheless, there are also thermodynamically stable nanoparticles, particles of specific shapes and structures, as in the case of molecular nanoparticles. Here a given structure is *thermodynamically stable* under the conditions of temperature, pressure, and concentration of the species under question. For example, a nanocluster of Au_{55} is a thermodynamically stable entity, with a specific number of ligands, in a medium or in the solid state. Varying conditions will lead to collapse of the structure. This may be said about particles of specific shapes, such as nanorods, which will be transformed to other shapes under appropriate conditions. Micelles correspond to another example, which may be transformed to lamellar or liquid-crystalline phases upon variation in conditions. The various phases are thermodynamically stable.

2.1.2. Dispersability in Diverse Media

A nanoparticle is composed of two entities: the *core*, often ceramic, metallic, or polymeric, and a thin *shell*, which may be ionic, molecular, polymeric, ceramic, or metallic. In most cases we encounter a ceramic or metallic core and a molecular shell (Figure 2.3). The properties of a nanoparticle are due principally to the core, and the shell is used to provide a protective layer. Often, the nature of the shell is extremely significant in a number of applications, such as luminescence of the particles. The shell and the core may have underlying structures and may be composed of more than one entity. The solubility of a nanoparticle is determined by the chemical nature of the shell. *Solubility* is not the appropriate term, as the solution or fluid formed is in effect a dispersion, which may be separated by physical means such as centrifugation.

A molecular shell has a characteristic chemical affinity to the nanoparticle core, due to its specific atoms or groups. For example, in the case of an oxide nanoparticle, the metal at the surface can link with an alkoxide (–OR, where R is alkyl). In the case of a metal nanoparticle such as gold, the metal can link with a sulfur atom of the thiolate (–SR). Such a link present throughout the surface of the nanoparticle is called a *protective monolayer* or *capping layer*. A nanoparticle so produced is called a *protected* or *capped nanoparticle*. The sulfur at the end (–SR) is a surface-active head group, as it links with a nanoparticle surface due to its specific chemical affinity. The chemical bond so formed gives thermal stability to the nanoparticle system. The weaker it is, the easier it desorbs from the surface, and a nanoparticle will be less stable. Au–S bond has a bond

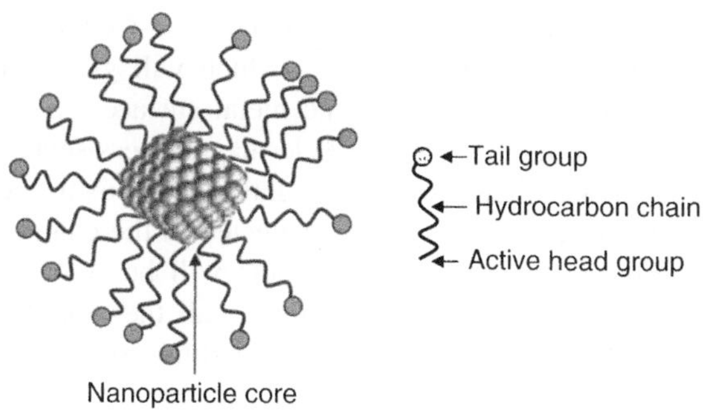

Fig. 2.3 Schematic of a nanoparticle, showing the core and shell. Unlike the one shown here, the core need not be a crystalline assembly with a regular arrangement of like atoms but may be a mixture of different types of atoms. The molecular shell has three distinct regions, although one or more of these may be absent in a specific case. A hydrocarbon chain may be long, as in a polymer, or completely absent, as in an ion protecting the nanoparticle. The shell may also be an extended solid, such as SiO_2.

strength on the order of 50 kcal mol^{-1}, and the thiolate desorbs from the metal surface only above 270°C.[5] Thus, the metal-active head group binding is an important parameter in determining the thermal stability of the nanosystem.

In our discussion we are concerned primarily with large nanoparticles whose typical size is on the order of nanometers. In such cases the nanoparticle core contains several thousands of atoms. For example, a 3-nm gold particle has approximately 1100 atoms, considering the nanoparticle to be a sphere. In reality, these particles are faceted or the outer surfaces terminate at specific crystallographic planes (assume a cube with the corners cut off). On these planes the head groups of the protecting molecules occupy specific locations, decided by the available space, packing density, and the van der Waals diameter of the molecules. If a larger number of monolayers are present on each crystallographic plane, the alkyl chains are arranged closeby and their interchain van der Waals interaction becomes important (as these interactions act at short distances). This gives additional stability to the system. In addition to breaking the nanoparticle–head group interaction, the van der Waals interaction also needs to be broken to destabilize the nanoparticle. In most cases the interchain interaction is generally weaker than the head group–nanoparticle interaction. As the monolayer assembly gets organized, as in the case of long monolayers, the core becomes inaccessible for ions and molecules in the medium. This leads to increased chemical stability for the core. The strength of the van der Waals interaction increases with increased chain length. In the limiting case of a polymer or a ceramic shell, chemical bonds in the shell are comparable or stronger than those in the nanoparticle core.

The tail group is the part that interacts with the solvent or medium. As a result of a favorable interaction, the nanoparticle gets dispersed in the medium. Thus, to make the nanoparticle disperse in water, a hydrophilic cover is required, whereas a hydrophobic cover makes the nanoparticle nondispensible in organic media such as toluene. By varying the polarity of the tail group, it is possible to get the system dispersed in solvents of varying dielectric constants. In the case of a hydrophilic monolayer, the shell has groups such as –COOH or $-NH_2$, which may be ionized to yield $-COO^-$ or NH_3^+, which will give a net negative or positive charge per monolayer chain to the metal surface. As the nanoparticle contains several such monolayers, the particle may possess several charges. The particle may be such that there are both negative and positive charges on the same particle, the net effect of these being reflected in the charge of the system. At a specific pH, the net charge on the particle will be zero. For example, in the case of an amine ($-NH_2$)-terminated surface, all the monolayers will be in the form $-NH_2$, and not $-NH_3^+$. This pH is called the *isoelectric point*. This is generally encountered in the case of proteins, where each molecule can exist as a *zwitterion* (having both negative and positive charges on the same molecule). In the case of proteins, this occurs as a result of the existence of $-COO^-$ and NH_3^+ on the same molecule. Only at the isoelectric point is the molecule not ionized.

As mentioned earlier, the shell present on a nanosurface need not be a molecule. In several cases the shell itself is an inherent part of the core. For example, in the case of silica nanoparticles, the surface is often a layer of hydroxyl groups and the particles can easily be suspended in water. In contrast, a hydrocarbon monolayer will make the particles disperse in organics. Gold nanoparticles can be made hydrophilic or hydrophobic in a similar fashion. In the case of reactive nanoparticles such as copper, the shell can get oxidized easily and there is always a layer of oxide over the surface, especially when the particles are exposed to air.

The tail group can change its character depending on the medium. This is particularly significant in cases where the group is $-COOH$, $-NH_2$, or $-OH$, for example, where the pH of the medium can greatly affect the nature of the group. In the case of a $-COOH$-terminated monolayer for example, in acidic media we get $-COOH$, and in alkaline media we get $-COO^-$. The pH values of a nanoparticle dispersion (without the additional base or acid) depend on the pK_a value of the acid in question. The change makes a large difference in the charge on the nanoparticle surface. This changes the zeta potential of the particle (see Section 2.3.5) and may have an effect on the properties. During changes in conditions such as pH, it is possible that the core is also affected.

In early thermal conductivity studies,[6] nanoparticles prepared by diverse routes were stabilized by dispersants and activators[7] such as laurate salts $[CH_3(CH_2)_{10}COO{-}X]$ and oleic acid $[CH_3(CH_2)_7CH{=}CH(CH_2)_7COOH]$. The purpose of this approach was to stabilize nanoparticles in diverse media, such as transformer oil, water, and ethylene glycol. The general approach used is appropriate surface functionalization, so that the nanoparticle surface is friendly to the medium.

The core–shell structure of a nanoparticle system is not limited to spherical particles. The very same general structure may be considered for nanorods, nanotubes, and nanoshells, where a chemically compatible shell is put around the nanosystem to make it go into the solution, biological environment, and so on.

2.1.3. Chemical Compatibility and Ease of Chemical Manipulation

The size, shape, and properties of nanoparticles, which depend on the synthetic conditions, are significant if the same core size has to be used in diverse applications. This is to be expected for a method that makes a metastable system. A given nanoparticle is kinetically trapped in a local minimum of free energy, and the synthetic parameters are crucial in deciding the final result. Thus, to preserve the core size, it is important to follow the same methodology. Often, this causes limitations in the adaptability of the system to various chemicals and conditions. For example, if a system is sensitive to a given chemical due to its core or shell, the shell can be suitably modified so that the chemical has no access to the shell and the shell does not react to the chemical. This means that the shell has to be manipulated after the nanoparticle synthesis, which is possible if a suitable shell were to be chosen that has distinct chemical features, allowing it to be functionalized.

Solvent compatibility has been brought about by changing the entire monolayer in a postsynthetic operation referred to as *ligand exchange*, in which the monolayer ligand molecules are exchanged with another one in the medium. This exchange process leads to equilibrium between molecules in the adsorbed and free states, and by repeating this process a few times, complete exchange can be achieved in several cases.

Chemical manipulation of monolayers can be done just as in the case of solution chemistry with simple molecules. The chemistry of the monolayer is utilized, as in the case of free molecules, to make suitable postsynthetic changes. For example, a given monolayer may be polymerized or may be included in a polymeric matrix by utilizing functional group chemistry. Chemical, thermal, and photochemical processes may be utilized to achieve this. A nanosystem can be manipulated to trap it in a cavity of a large molecule so that the system can be shipped in a suitable medium. Examples include the use of dendrimers and cyclodextrins.

2.2. SYNTHETIC METHODS: COMMON ISSUES OF CONCERN

Nanoparticles in general, and metal nanoparticles in particular, are investigated in the context of diverse research perspectives. Among these, catalysis, biology, drug delivery, materials science, photophysics, and novel phenomena are most important. Each of these areas has a specific emphasis, although the synthetic methodologies have some overlap. The particles may have to be presented in different forms, and for that, specific modifications in the synthetic approach are necessary. Numerous books of nanoparticles are available and may be consulted for specific details on the adaptability of a given technique for specific application. Some synthetic methods are totally unrelated; for example, the synthesis of nanoparticles and carbon nanotubes will have more differences than similarities, although both are nanosystems.

Any synthetic methodology produces particles of a specific size distribution. In the simplest case of spherical particles, one linear dimension, the diameter, is adequate to describe the size of the particles. It is best to describe the particle size in terms of statistical analysis. Here we take a collection of N particles. The particles are first sorted in terms of classes, with a class mark, a_i, with narrower size distributions. The class has a midpoint and a distribution. The distribution of particles among various classes can be plotted as a histogram (Figure 2.4). The number of classes increase as the width of the interval decreases, and we finally get a smooth curve. The distribution is characterized by an average and a standard deviation. The average,

$$\bar{a} = \sum_i \frac{n_i}{\sum_i n_i} a_i \tag{2.1}$$

where the first term, the fraction of the number of particles having the class mark, is denoted $f_{n,i}$. Note that this is a number-averaged diameter of particles. The standard deviation,

$$\sigma = \left[\sum_i \frac{n_i}{\sum_i n_i} (a_i - \bar{a})^2 \right]^{1/2} = \left[\sum_i n_i \frac{(a_i - \bar{a})^2}{\sum_i n_i} \right]^{1/2} \tag{2.2}$$

Deviation of a value from the mean is given by $a_i - \bar{a}$. This can be positive or negative. It is clear that σ^2 is the number average of the square of standard

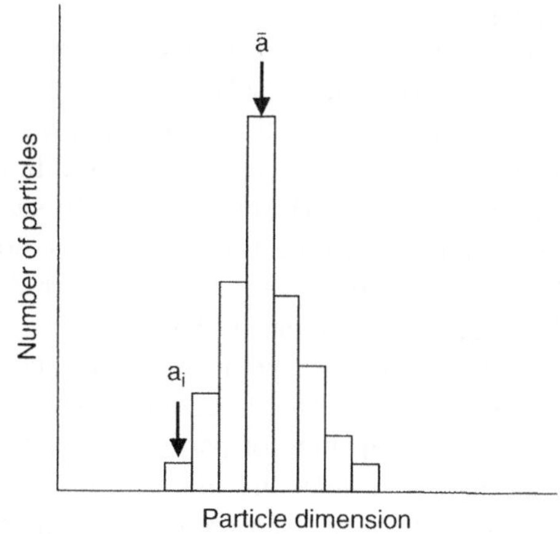

Fig. 2.4 Distribution of particles among various classes in a given synthetic approach. The number-averaged particle size is shown. The class width for one class is also indicated. As the width decreases, the distribution becomes a smooth line.

deviations, $(a_i - \overline{a})^2$. The square root of this square is therefore the spread of the data. Because of this, σ is called the *root mean square* (rms) *deviation*. For ease of computation, σ is given as $(\overline{a^2} - \overline{a}^2)^{1/2}$.

Synthetic methodology becomes important if it can produce particles of a given size distribution in a simple process. Often, the interest is to get as narrow a size distribution as possible. In case the methodology fails to generate particles of narrower size distribution, postsynthetic processes are utilized to select particles of interest or to convert one to the other. These processes are often laborious and lead to poor yields. Thus, a one-pot methodology is desirable.

In a generalized approach to nanoparticle synthesis in solution, the precursor species (i.e., metal ions, organometallics, complexes, etc.) are reduced, decomposed, or hydrolyzed, as the case may be, in the presence of an appropriate stabilizer. The conditions of the medium are adjusted such that the nucleation of the particles is fast and the surfaces of the particles thus formed are protected by the stabilizer. Depending on the type of reaction, the synthetic conditions vary; temperature, pH, and medium are the most common variables used. In the case of thermal decomposition of precursors or more complex reactions, conditions are more cumbersome and the procedure may be conducted in inert atmospheres. The synthesized particles are precipitated out of the medium by varying the solvent polarity or by solvent evaporation at reduced pressure. The material is purified by repeated solvent washing or dialysis or re-precipitation, depending on the case.

2.2.1. Size Control

A synthetic methodology may not be capable of giving size exclusivity which means that a variety of sizes are possible in the as-synthesized particles. Selection of a given size requires post-synthetic approaches. There are several such processes, the first being size exclusion chromatography. In this, a mixture of nanoparticles is passed through a size-selective stationary phase such as a gel, which has definite pore sizes. The eluant (solvent medium) used elutes the material as a function of size. Agarose and Sephadex are the two common media used. The other method involves solvent-selective precipitation. In this, the polarity of the medium is changed progressively (from low to high) such that larger particles precipitate from the mixture. By repeating this process, proper size control is possible, although the stability of the material may vary in different media. The other approach used is digestive ripening, in which the nanoparticle is digested with the protecting agent used in the synthesis, at elevated temperatures in a selected series of temperature steps. The process consumes particles of smaller sizes. The approach of Ostwald ripening or particle coarsening is a similar process in which the as-prepared particles are allowed to age for a finite period, during which large particles grow at the expense of smaller particles, narrowing the particle size distribution. This may be achieved along with temperature cycling.

2.3. HOW WE STUDY NANOPARTICLES

Although there are numerous ways to study nanoparticles, we discuss next the tools most commonly used.

2.3.1. Transmission Electron Microscopy

The formation of nanoparticles is best studied by transmission electron microscopy (TEM), which gives two types of information in routine examination. The first is the particle size distribution, which is normally represented in terms of a mean diameter and a standard deviation. Both are not calculated rigorously in most studies; instead, a histogram of size distribution is presented (Figure 2.4) along with the TE micrograph. The second type of information is the crystallinity of a sample, obtained through electron diffraction or nanodiffraction. More detailed information on particle shape, phase transitions, two- and three-dimensional ordering, in-situ nanomeasurements, and evaluation of other properties are possible using TEM.[8]

2.3.2. Optical Spectroscopies

When it comes to metal nanoparticles in general and gold particles in particular, optical absorption spectroscopy is a powerful tool. The optical properties of nanoparticles have been investigated extensively in recent years. When an electromagnetic wave passes through a metal particle, the electronic and vibrational states get excited. The optical interaction induces a dipole moment that oscillates coherently at the frequency of the incident wave. The frequency of this oscillation depends on the electron density, its effective mass, and the shape and size of the charge undergoing oscillation. There can also be other influences, such as those due to other electrons in the system. The restoring force arises from the displacement of the electron cloud relative to the nuclei, which results in oscillation of the electron cloud relative to the nuclear framework. The collective oscillation of the free conduction electrons is called the *plasmon resonance* or *dipole plasmon resonance* of the particle.[9] In this resonance, the total electron cloud moves with the field applied. There can be higher modes of plasmon resonance as well. In the quadrupole mode, half the electron cloud is parallel while the other half is antiparallel to the field.

The dipole plasmon frequency is related to the dielectric constant of the metal. The frequency-dependent dielectric constant of a bulk metal [$\varepsilon(\omega)$] is measurable. To simplify matters, we consider a spherical particle whose diameter is much smaller than the wavelength of the electromagnetic radiation. Under such conditions, the electric field of light felt by the particles can be regarded as a constant. This reduces the interaction to be treated by electrostatics rather than electrodynamics. This treatment is referred to as the *quasistatic approximation*— "quasi" because we consider the wavelength-dependent dielectric constant. In electrostatic theory, when the incident electric field of the radiation interacts with the electrons, we get a net field due to the applied field and its induced field. This field in reality is radiating and contributes to extinction and Rayleigh scattering by the particle. The strength of extinction (note: extinction = absorption + scattering) and scattering can be given in terms of their efficiencies:

$$\text{extinction efficiency, } Q_{\text{ext}} = 4x \cdot \text{Im}(g_d) \tag{2.3}$$

$$\text{scattering efficiency, } Q_{scat} = \frac{8}{3}x^4|g_d|^2 \tag{2.4}$$

where $x = 2\pi R\varepsilon_m/\lambda$, $g_d = (\varepsilon_c - \varepsilon_m)/(\varepsilon_c + 2\varepsilon_m)$, ε_c and ε_m are the dielectric constants of the metal and the medium, respectively, and R is the particle radius. Dielectric functions are complex quantities, Im refers to the imaginary part, and the efficiency = cross section/area (πR^2).

In particles less than 10 nm in diameter, light scattering does not make a significant contribution.

$$Q_{\text{ext}} \sim Q_{\text{abs}} = \frac{4(2\pi R\varepsilon_0^{1/2})}{\lambda}\text{Im}\left(\frac{\varepsilon_c - \varepsilon_m}{\varepsilon_c + 2\varepsilon_m}\right) \tag{2.5}$$

When $\varepsilon_c = -2\varepsilon_m$, we get the resonance condition and Q_{abs} goes to a maximum. Since the dielectric function is a complex quantity, this equation can be given in terms of the real and imaginary parts of the metals dielectric function, ε' and ε'', respectively. There are two distinct size regimes of the particles; in both, the plasmon resonance depends on size. For particles larger than 10 nm in diameter, the dielectric function itself is independent of size. The shape and size dependence of plasmon resonance in this regime is due to the dependence of electrodynamics on size and shape. This is called the *extrinsic size regime*. In the *intrinsic regime*, for particles less than 10 nm in diameter, the dielectric function itself changes with size. For metals, the absorption characteristics depend, to a large extent, on the conduction band electrons. The spatial confinement of the free conduction band electrons results in plasmon excitations that are restricted to a small range of frequencies, usually in the ultraviolet (UV)–visible region. Bulk metals absorb very strongly in the infrared (IR) or near-IR region, but colloidal metals are transparent.

The optical absorption spectrum of a nanoparticle solution of gold is shown in Figure 2.5, made by the citrate route (see Section 2.4.1). The peak at 520 nm is due to plasmon absorption, and the position and shape of the absorption peak are characteristic features of the particle size. This feature is not shown by bulk gold. A smaller particle size, as in the case of a thiol-protected gold nanoparticle in toluene, shows considerably broader plasmon absorption. Particles smaller than 2 nm will show no plasmon absorption, as is the case with Au_{25} particles. These molecular clusters of gold show distinct features due to their molecular energy levels. These spectra show that plasmon resonance exhibited by a metal nanoparticle is a good indicator of size. Plasmon resonance is also sensitive to the molecular shell on the surface, and its thickness is reflected in the spectrum. For a given molecular shell, the properties of the medium are reflected in the spectrum when the shell thickness is small. Both the interactions of nanoparticles with ions or molecules in the medium and interparticle interactions are manifested in

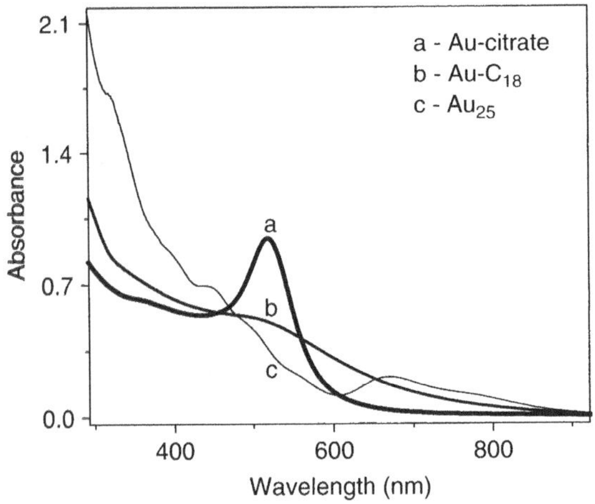

Fig. 2.5 Optical absorption (extinction) spectra of (a) 15-nm gold particles in aqueous solution (labeled Au-citrate), (b) 3-nm particles in toluene protected with octadecane thiolate monolayers, labeled Au-C_{18}(note the broadening of the plasmon feature), and (c) Au_{25} in water, where there is no plasmon excitation and all the features are due to molecular absorptions of the cluster.

the absorption features. An interested reader may consult an article by Link and El-sayed[10] for a detailed discussion of plasmon resonance of gold particles and its applications.

In the case of quantum dots, one gets discrete energy levels. A complete discussion of this topic is beyond the scope of this chapter, but it is important to mention that optical spectroscopy can be used to understand these energy levels, and therefore, quantum dots. The simplest model to represent the energy states of a nanocrystal is a spherical quantum well with an infinite potential barrier. Although the model is simple, if we include the coulombic interaction between the charge carriers (electron, e and hole, h), analytic solutions for the Schrödinger equation are not possible. Disregarding e–h interaction is possible in the strong confinement regime, as confinement energies scale with d^{-2} (as energy goes as n^2/d^2) while Coulomb interaction scales with d^{-1} (d being the diameter of the particle). This yields states with distinct n, l, and m quantum numbers, referring to symmetry, orbital angular momentum, and its projection, respectively (similar to electrons in the orbitals of an atom). The wavefunctions are represented as products of several terms. The energies of the states can be given as

$$E_{n,l}^{e,h} = \frac{h^2 n^2}{8\pi^2 m_{e,h} d^2} \tag{2.6}$$

where n is a quantum number. The exact nature of the wavefunction and quantum number are not discussed here. The wavefunctions correspond to the S, P, D,... states, depending on the orbital angular momentum, ℓ. There is one more quantum number, m, which decides the degeneracy of the states. The energy states are shown in Figure 2.6. The energies are measured from the bottom of the conduction (valence) band for electrons (holes). The energy increases with higher quantum numbers. Since the electron mass is much smaller than that of the hole ($m_h/m_e \sim 6$ in CdSe), the electron levels are separated more widely than the hole levels.

Electronic transitions are possible between various energy levels. However, the wavefunctions corresponding to different n and/or ℓ are orthogonal and therefore it is not possible to observe all of these transitions. Optical transitions between states of the same symmetry are observable. The intensity of the transition is related to the degeneracy of the states in question. The transitions observed are far more complex than can be described by the spherical quantum well model. The scheme provided here is inadequate to describe the hole states. Spin–orbit and Coulomb e–h interactions must be considered to improve the energy-level picture. These transitions can be observed using both optical and fluorescence spectroscopy. One of the important aspects to be considered in interpreting experimental spectra is the size range of the particles prepared in a typical synthesis. Spectroscopic size selection is possible by techniques such as fluorescence line narrowing, spectral hole burning, and photoluminescence excitation. In these techniques, a narrow energy window is used for excitation (first two) or detection (last). This makes the technique sensitive only to a specific particle size.

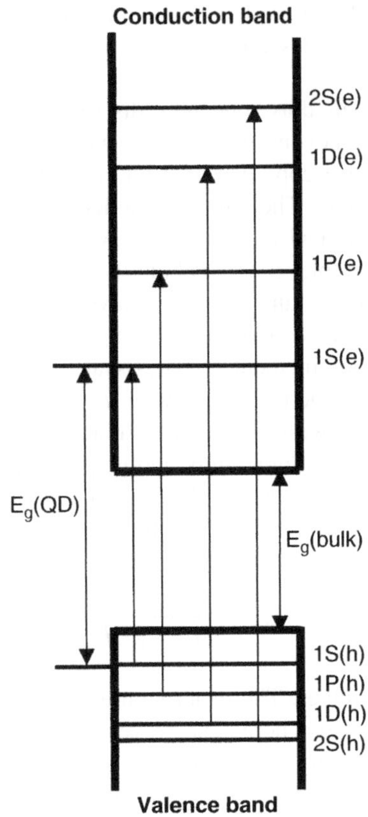

Fig. 2.6 Electronic states of a nanocrystal. The optical transitions allowed are marked.

Size selection with the red region of the spectrum is better, as it selects the particles of the largest size in the ensemble. An interested reader may consult the references cited in Ref. 11 to understand the details of the application of optical spectroscopies to quantum dots.

2.3.3. X-ray Diffraction

X-ray diffraction is another important tool used to understand the properties of synthesized materials. Metals have simple crystal structures and consequently fewer peaks in the diffraction pattern. The normal diffraction line is of finite width, due to several factors. These include the finite line width of the excitation source and the imperfections in the focusing geometry. The Bragg condition ($n\lambda = 2d \sin \theta$) occurs when each plane in a crystal diffracts exactly one wavelength later than the previous plane. Constructive interference occurs due to this condition. When the incident ray strikes at a larger angle, θ_1 than the diffraction angle θ, the phase lag will be greater than the wavelength λ and it becomes

$\lambda + \delta\lambda$. As the number of planes becomes $j+1$, the cumulative phase lag, $\Sigma\, \delta\lambda$, could increase to become $\lambda/2$, (i.e., $j\ \delta\lambda = \lambda/2$). For the ray incident at the larger angle θ_1, the diffracted rays from plane 1 and plane $j+1$ are 180° out of phase. As a result, there is no net intensity for the ray diffracted at this angle. Note that we have several planes in the crystallite, and the rays diffracted from the set of planes 1 through j are exactly canceled by planes $j+1$ through $2j$, if $2j$ planes are present in the crystallite. Thus, the intensity of the diffracted beam will fall to zero at a finite angle, with a peak maximum, as a result of this effect. One should note that there is also a phase difference, $\lambda - \delta\lambda$, which occurs for an angle θ_2 smaller than θ. The width of the diffraction peak is therefore determined by the number of planes present in the crystallite. For large crystallite, j is large, $\delta\lambda$ is small, and the width is negligible. The particle size effects, seen as broadening of the diffracted lines, are given by the *Scherrer formula*, $t = 0.9\lambda/(B \cos\theta)$, where t is the thickness of the crystallite (in angstroms) and θ is the Bragg angle. B is the line broadening, indicating the extra peak width of the sample compared to the standard, derived using the *Warren formula*, $B^2 = B_M^2 - B_S^2$, where M and S refer to specimen and standard. B's are measured in radians at half-height. The peaks of the sample and the standard should be close to each other.

Particle sizes up to 200 nm can be measured using the Scherrer formula. In the range 5 to 50 nm, the broadening is easy to determine. In Figure 2.7 we show the x-ray diffractograms of several gold particles taken with CuKα radiation. The spectrum of a standard bulk gold powder sample is shown for comparison. The conclusion is that at larger particle sizes the difference between the sample and the standard is small, and at small particle sizes the peak is difficult to distinguish from the background. For smaller particle sizes low-angle peaks are used for size determination, as they are less broad than large-angle peaks.

It should be noted that the powder pattern may be shifted or broadened as a result of stresses present in the material. Due to uniform compressive stress, the d spacing may decrease and the peaks may shift to larger angles. If the stress is nonuniform throughout the crystallite, the peaks will broaden. A composite of these size- and stress-induced effects are observed generally.

2.3.4. Infrared, Raman, and Other Spectroscopies

Infrared spectroscopy is an ideal tool to use to understand molecular vibrations in a nanomaterial. It is best suited to the study of monolayers on the surface of a nanoparticle. The nature of binding, the molecular nature of the ligand and its organization, the extent of order, and other details can be determined from infrared spectroscopy. The infrared spectrum of silver nanoparticles protected with octadecanethiol (ODT) $C_{18}H_{37}$–SH shows no peak due to the S–H frequency, and it can be concluded that the thiol is absorbed in the form of thiolate (RS^-) on the surface of the metal particle (Figure 2.8). At a characteristic temperature, the C–H vibrations of methylenes (–CH_2) shift to a higher frequency, in a variable temperature experiment indicating a phase transition. This corresponds to melting of the alkyl chain order on the monolayer assembly. This is observable

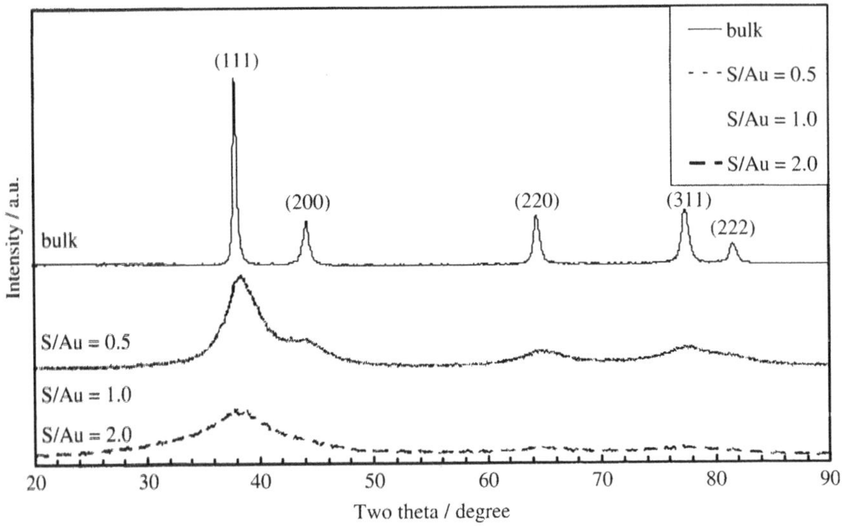

Fig. 2.7 Variation in the x-ray diffractograms of gold nanoparticles as a function of particle dimension compared with a bulk gold powder. The gold nanoparticle samples are protected with mercaptosuccinic acid and the S/Au ratio used in the synthesis determines the particle dimensions. Increasing the S/Au ratio decreases the metal core size. As the core size is reduced, the peaks broaden. The (220) and (311) reflections merge with the baseline in the last sample (S/Au = 2.0). The approximate dimensions of the nanoparticles are 4.0 ± 1.0, 3.0 ± 0.5, and 2.0 ± 0.5 nm. (Data courtesy of Tsugo Oonishi and Keisaku Kimura, Hyogo University, Japan.)

in differential scanning calorimetric analysis of the materials. The phase transition leads to increased freedom, and the monolayers possess orientation freedom and undergo rotational dynamics above this temperature. Part of this rotational freedom is evident in the infrared spectrum. At low temperatures, the r^- mode corresponding to the methyl ($-CH_3$) is split into two, as the chains have no rotational freedom and the modes are nondegenerate. When the temperature increases beyond the phase transition point, the methyl groups acquire rotational freedom and the vibrations become degenerate. The chains, to begin with, have a distinct all-*trans* conformation, as the gauche bonds are less prominent in the spectrum. The spectrum shows characteristic features due to the progression bands, suggesting an alkyl chain order. However, as temperature increases, the intensity of the C–S gauche increases. Also, at the phase transition point, the progression bands disappear. The infrared spectrum helps us in a number of ways to understand the structure of the monolayer chain.[12] If the functionality is modified, it is reflected in the spectrum.

In the case of a metal nanoparticle, infrared spectroscopy does not provide any information on the core. However, for a semiconducting or insulating nanoparticle, it gives invaluable information on the structure and phase transitions. Most

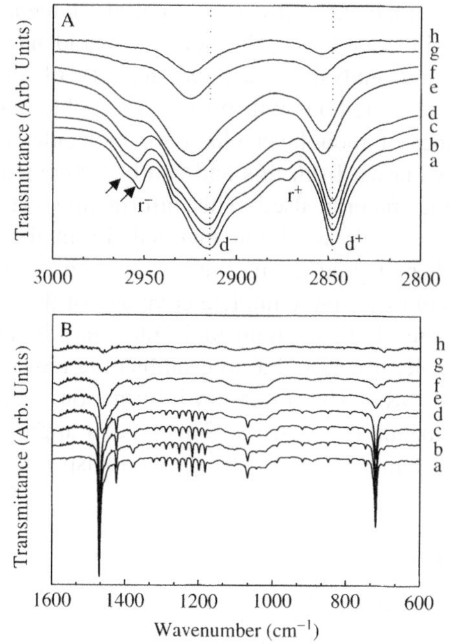

Fig. 2.8 Variable-temperature FT–IR spectra of Ag nanoparticles protected with ODT. A and B correspond to the C–H stretching and low-frequency (fingerprint) regions, respectively. The spectra were measured in KBr matrices. The temperatures are a, 298; b, 323; c, 348; d, 373; e, 398; f, 423; g, 448; and h, 473 K. The d^- and d^+ modes are the asymmetric and symmetric modes of CH_2 stretching. Note that in the spectrum at 398 K these peaks are shifted to higher values, indicating a more disordered methylene units, implying phase transition. Several peaks in the 1400–800 cm^{-1} window are the progression bands, which disappear at the phase transition temperature. The C–S gauche mode at ~700 cm^{-1} is retained while C–S trans at 720 cm^{-1} disappears.

of these studies are conducted with Raman spectroscopy and its variations, such as those at high pressures and high temperatures.

X-ray photoelectron spectroscopy is another valuable tool in the study of nanomaterials. For gold nanoparticles, only the Au^0 state has been observed, and the sulfur is in S^-. The nature of S has been a question of considerable debate, as the H_2 liberated upon thiol binding has not been detected, although there have been mass spectrometric reports. The presence of thiol itself on the surface of gold has been reported. Desorption spectroscopy has shown that it is always the disulfide that escapes the surface, not the sulfide. This suggests the presence of an S–S bond on the surface of the nanoparticle. In general, the core energy levels shift in the quantum dots due to the fact that there are fewer electrons in the valence bands to screen the charge created by photoemission. Thus, systematic shifts are seen in the core-level spectra as a function of dimension. The shift in the valence band can be seen in ultraviolet photoelectron spectroscopy.[12]

To understand the composition of a nanomaterial, elemental analysis is often useful. This may be illustrated with the help of a gold nanoparticle protected by thiols. The elemental analysis of a fixed quantity of nanomaterial by wet chemistry or instrumental methods will provide the gold content. From the sulfur content, the number of monolayer chains or molecules can be understood. The gold/thiol ratio is known from this and can be compared with a theoretical estimate of the starting material used. This information may also be available from thermogravimetric analysis of the material. From the diameter obtained using TEM, the number of gold atoms present may be determined considering spherical geometry, and assuming complete coverage of the gold surface with S, an approximate estimate of the capping molecules may be arrived at. From all the information available, it is possible to establish the molecular formula of the nanoparticle to be, Au_nSR_m. The minimum information necessary includes, the core diameter and elemental composition. This type of approach can be extended to any nanoparticle system, although extreme monodispersity is assumed.

2.3.5. Zeta Potential

Due to dipolar characteristics and ionic attributes, the colloidal particles (including nanoparticles) suspended in solvents are charged electrically. For example, the surface groups of a colloid may be ionized. This leads to a net electric charge at the surface that causes the accumulation of opposite charges (counterions) around them. This in turn results in an electrical double layer. The ion (with positive or negative charge) and a set of counterions form a fixed part of the double layer. The diffuse or mobile part of the double layer consists of ions of different polarities, which extend into the liquid phase. This double layer may also be considered to have two parts, an inner region that includes ions bound relatively strongly to the surface, and a diffuse region in which the ion distribution is determined by a balance of electrostatic forces and random thermal motion. When an electric field is applied, the particles are attracted to the electrodes, depending on their polarity. The potential at which the fixed part of the double layer along with a part of the mobile layer move toward an electrode is called the *Zeta* or *electrokinetic potential*. It can also be defined as the potential at the shear plane of the particle when it moves in the medium.

The zeta potential depends on a number of parameters, such as surface charges, ions adsorbed at the interface, and the nature and composition of the surrounding medium. The net charge in a specific medium depends on the particle charge and counterions. The zeta potential is an index of interaction between the particles. The zeta potential is calculated according to *Smoluchowski's formula*,

$$\zeta = \frac{4\pi\eta}{\varepsilon} \times U \times 300 \times 300 \times 1000 \tag{2.7}$$

where ζ is the zeta potential in mV, ε the dielectric constant of the medium, η the viscosity of solution, and U the electrophoretic mobility ($v/V/L$), where v is the velocity of the particles under an electric field in cm/s, V the applied voltage, and L the electrode distance.

Measure of the zeta potential throws light on the stability of colloidal and nanoparticle solutions. If all the particles in a suspension have large negative or positive zeta values, they will repel each other and there will be no tendency to flocculate. However, if the particles have low zeta potential values, there is no force to prevent the particles from coagulating. The threshold of stability of a colloidal–nanoparticle solution in terms of the zeta potential is $\pm 30\,\text{mV}$. The greater the zeta potential, the greater the stability will be. The value of the zeta potential is affected primarily by pH.

The zeta potential is measured traditionally using the *micro electrophoresis method*, which needs extreme dilutions and hence stringent sample-handling requirements. Microelectrophoresis is a technique based on light scattering by particles. In the case of nanoparticle solutions, however, microelectrophoresis is not ideal, due to the Doppler broadening of the light scattered from the fine particles. Modern methods used for zeta potential measurements are based on electroacoustic methods that rely on electrokinetic properties. In these methods, the application of a high-frequency electric field sets in motion electrophoretic movements of the particles. This generates an alternating acoustic wave due to the density difference between the particles and the medium. The velocity of the particles is measured using laser Doppler electrophoresis. The velocity of these particles or mobility is converted to the zeta potential using *Henry's equation*:

$$U = \frac{2\varepsilon z f(ka)}{3\eta} \tag{2.8}$$

where ε is the dielectric constant, z the zeta potential, η the viscosity, and $f(ka)$ is Henry's function. Zeta potential measurements in aqueous media and moderate electrolyte concentration generally employ an $f(ka)$ value of 1.5 (Smoluchowski's approximation). The $f(ka)$ value is generally taken as 1 for zeta potentials of small particles in nonaqueous media (Hückel approximation). The zeta potential measurement by microelectrophoresis is a passive technique, as it does not alter the chemical properties of systems.

2.4. VARIETY IN NANOMATERIALS

Next, we discuss various methods for each of the specific categories of nanoparticles. In his early review, Gleiter reviewed the various methods available for nanoparticle synthesis.[13] In the discussion below we discuss only those methods that can be useful in making nanofluids. As a result, several of the routes for purely ceramic powders are not discussed. However, the various methods available are noted. Although useful, mechanical attrition (high-energy ball milling) is not discussed.

2.4.1. Metals

History Metal colloids were the earliest nanoparticles to arouse scientific curiosity. Although nanoparticles of noble metals have been used to impart color to glass since the time of the early Romans, of which the Lycargus cup is famous, scientific study of these particles was not begun until the seventeenth century. The Lycargus cup, dating from the fourth century B.C., contains about 40 parts per million (ppm) gold and about 300 ppm silver, and the nanoparticles of these metals give the glass its distinct optical properties. The interest in that period was in solutions containing gold, due to their redness, as it was thought that the active principle of blood was its color, and blood itself was considered the essence of life. In the sixteenth century, Paracelsus described a method to make "Aurum portable" by reducing auric chloride by alcoholic plant extracts. Synthetic gold preparations have been used in the traditional Indian medical practice, Ayurveda. *Saraswatharishtam* uses gold particles and *Makaradhwaja* uses finely divided gold.[14] The use of gold metal in medicine and dentistry itself is much older. The first book on colloidal gold was published in 1618. A German chemist, Johan Kunckels, published a book in 1676[15] which described a drinkable gold that had curative properties. The presence of gold in the solution in invisible form was postulated in this book. In a detailed book published in 1718, Hans Heinrich Helcher wrote that starch enhances the stability of a gold preparation.[16] These preparations were used for dying in 1794. The difference in the color of various gold preparations was attributed to the size and nature of particle aggregation by Jeremias Benjamin Richters in 1818.[17] In 1857, Faraday[18] reported the synthesis of stable colloids by a two-phase reduction method in which gold chloride ($AuCl_4^-$) in water was reduced by red phosphorus in CS_2. The term *colloid* was coined by Graham in 1861 (from the French word *colle*, meaning "glue"). Numerous workers investigated colloidal gold in twentieth century. Methods have been reported using formaldehyde, hydrogen peroxide, hydroxyl amine, hydrazine, and gases such as CO and H_2. Various well-established synthetic methods such as the citrate reduction method[19] and the Brust method[20] increased the momentum in this field. Several subnanometer gold clusters were synthesized by Schmid[21] and Bartlett.[22]

The methodologies and the modifications introduced by later workers[23] have made colloidal gold science one of the most intensely pursued areas in science today. Although the subject area is vast, our focus is not on reviewing the research in any significant detail but in presenting various synthetic methodologies used to make stable, well-characterized nanofluids. Earlier work in the area of colloidal gold from the perspective of biology is compiled in a book by M. A. Hayat,[24] which also lists the various methodologies used for gold particles in a very large size range. Several reviews devoted to gold and metal particles in particular may be consulted for an exhaustive review of the literature.[25]

Solution-Phase Routes by Chemical Reduction Reduction of metal ions leads to the metal atom, which upon aggregation forms nanoparticles. The growth of the aggregate is arrested at some stage of its growth by stabilizing or protecting agents. The reduction reaction can be represented as

$$M^{n+} + ne^- \rightarrow M^0 \quad (2.9)$$

The electron is not supplied as an electron per se but as a reducing agent, which gets oxidized in the process:

$$Red^{m-} - ne^- \rightarrow Oxi^{(m-n)-} \quad (2.10)$$

where the reducing species (reductant) of finite charge gets oxidized, losing a certain charge. Note that both the metal and the reductant may not contain any distinct charge, and those mentioned are only nominal. The feasibility of the net reaction

$$M^{n+} + Red^{m-} \rightarrow M^o + Oxi^{m-n-} \quad (2.11)$$

depends on the thermodynamics of the process, which in turn is represented by the electrochemical potentials of the corresponding half-cell reactions called the *standard reduction potentials*. If the reduction potentials corresponding to reactions (2.9) and (2.10) are added (with their signs), and if we get a net positive value, the process is thermodynamically feasible. This corresponds to a net negative free-energy change $\Delta G = -nFE$, where ΔG is the free-energy change of reaction (2.11), n the number of electrons involved, F the Faraday constant, and E the electrochemical potential of reaction (2.11). Note that we have written E not E^0 as the potential has to be taken at the appropriate conditions. The process is thermodynamically feasible if ΔG is negative.

Let us illustrate this with some examples. The standard reduction potentials of common metal ions and reducing agents are presented in Table 2.1. If the potential is positive, it implies that the process can occur. Note that in the case of reducing agents, the reaction of importance is the reverse, oxidizing reaction, liberating electrons. These are the electrons that will be consumed by the metal ions as they are reduced. While considering these examples, it is clear that all the metal ions mentioned in Table 2.1 can be reduced by borohydride; that is, if one uses borohydride to reduce Ni^{2+} under standard conditions (i.e., at 25° C, 1 atm, and 1 M concentration of the ions), the electrochemical potential of the process is $(-0.257) - (-0.481) = +0.224$. Therefore, the reaction is feasible. However, the reduction of Ni^{2+} by hydrazine is not possible, as the potential is negative. The total ionic reduction of $AuCl_4^-$ by BH_4^- can be represented as

$$8AuCl_4^- + 3BH_4^- + 9H_2O \rightarrow 8Au + 3B(OH)_3 + 21H^+ + 32Cl^- \quad (2.12)$$

The discussion above suggests that the only point of concern is the reduction potential. It is important to emphasize that in several cases, the ions present in the solution are in complex form and reduction or oxidation is conducted on that ion. This changes the potentials substantially. As a result, although simple Au^{3+} in the

Table 2.1 Standard Reduction Potentials of Metal Ions and Reducing Agents

Reaction Process	Potential (V)
For metals:	
$AuCl_4^- + 3e^- \rightarrow Au + 4Cl^-$	+1.002
$Ni^{2+} + 2e^- \rightarrow Ni$	−0.257
$Co^{2+} + 2e^- \rightarrow Co$	−0.28
$Fe^{2+} + 2e^- \rightarrow Fe$	−0.447
For reducing agents:	
ABH_4 (A = alkali metal)	
Chemical reaction, $B(OH)_3 + 7H^+ + 8e^- \rightarrow BH_4^- + 3H_2O$	−0.481
Hydrazine (N_2H_4 forms $N_2H_5^+$ in water as it is basic)	
Chemical reaction, $N_2 + 5H^+ + 4e^- \rightarrow N_2H_5^+$	−0.23

Source: Ref. 26.

form of $AuCl_4^-$ can be reduced by mild reducing agents such as carboxylates or alcohols, this is not possible when the metal ion is in the presence of excess thiols. Here they form metal thiolates, and reduction of these complexes is possible only by strong reducing agents such as borohydride. If reduction is conducted along with mild reducing agents in the presence of gold metal particles it may occur on the surface of the gold, where it is easier (see later). On the other hand, some reductions that are not possible normally by consideration of electrochemical potentials can happen by varying conditions. For example, the reduction of Ni^{2+} by hydrazine hydrate is possible in ethylene glycol at 60° C in the presence of sufficient hydroxyl ions;

$$2Ni^{2+} + N_2H_5^+ + 5OH^- \rightarrow 2Ni + N_2 + 5H_2O \tag{2.13}$$

This particular method gives 9-nm particles.[27] It is important to emphasize that electrochemical potentials must be used only as guidelines in understanding the chemistry.

The metal ion can also be reduced by a molecule or ion, which itself can act as the stabilizing agent. This happens for a number of metals using citrate, amines, alcohols, and thiols. Reduction of $HAuCl_4^-$ by trisodium citrate is a classic example. Numerous such examples are known from the recent literature in which a variety of amines, alcohols, thiols, or complex ions are used. Alcohols by themselves are not good protecting agents, so polyols containing more hydroxyl groups per molecule are used, which effectively chelate (multiple coordination) the metal ions. Reduction processes employed are summarized in Table 2.2.

Strong Reducing Agents When a metal to be reduced has a large negative reduction potential, the reduction process is difficult and the reaction conditions require careful control. When the reducing agent is very strong, it can reduce the solvent and other reagents present in the medium. For example, if the reagent is

Table 2.2 Various Solution-Phase Reduction Processes Used to Make Metal Nanoparticles

Method	Summary	Example	Refs.
$NaBH_4$ route	Metal ion/BH_4^-	Au, Ag	28, 29
Citrate route	Metal ion/Cit_3^-	Au, Ag	19, 30, 31
Polyol route	Metal ion/ethylene glycol	Ag, Pd	32
Polyvinylpyrollidone (PVP) route	Metal ion/PVP	Pd	33
Amine route	Metal ion/APS, AES	Ag	34

too strong, water may have to be eliminated, so that the reaction

$$2H_2O + 2e^- \rightarrow H_2 + 2OH^- \tag{2.14}$$

is avoided. The standard reduction potential of this reaction is −0.828 V. The most powerful reducing agents are solvated electrons. In the laboratory, these are prepared by dissolving alkali metals in aprotic solvents such as diethyl ether or tetrahydrofuran in the presence of excess complexing agent, such as a crown ether. The resulting ionic compound is called an *electride*. The reaction can be written as

$$M + 2(15\text{-crown-}5) = M^+(15\text{-crown-}5)_2e^- \tag{2.15}$$

Crown ethers are macrocyclic polyethers and their common names have a prefix corresponding to the total number of atoms in the macrocycle and a suffix to indicate the number of oxygen atoms. 15-crown-5 has a total of 15 atoms in the macrocycle, of which 5 are oxygens.

If the concentration of the complexation agent is less, we can get alkalide (i.e., the alkali metal anion)[35]:

$$2M + (15\text{-crown-}5) = M^+(15\text{-crown-}5)M^- \tag{2.16}$$

Both of these reagents have low thermal stability, and as a result, it is important to conduct the reactions at low temperatures. Such synthesis has been done and several nanocrystalline metals and alloys have been prepared this way.[36] The other strong reducing agents are trialkylborohydrides ($ABEt_3H$, A = Li, Na, K). There have been other methods, such as the use of trialkyl aluminum. A variety of transition metal nanoparticles have been synthesized by these routes.[36,37]

Most Popular Methods for Synthesis We discuss next two of the most common methods used for the synthesis of gold nanoparticles: the citrate route and the Brust method.

Citrate Route This method, known as the *Turkevich method*[19] is the most convenient for the synthesis of colloidal gold nanoparticles of ~ 15 nm mean diameter. The synthesis involves the following steps. Make ~ 5.0×10^{-3} M $HAuCl_4$ in water. This is a stock solution. Take 1 mL and make it up to 19 mL using water. Heat the solution to boil and add 1 mL of 0.5% sodium citrate solution when the boiling begins. Continue heating until the color changes to pale purple. Remove the solution from the heating mantle and allow it to cool slowly. The colloidal solution prepared will have a net gold concentration of 2.5×10^{-4} M. A TEM image of the particles obtained is shown in Figure 2.9.

The characteristic feature of this nanoparticle solution is its color, which is due to the plasmon resonance of particles of this size range. As described above, the plasmon resonance, is due to collective electron oscillation of the nanoparticle. As the valence electrons in the metal particle are free, they contribute to the oscillation, being excited when photons of characteristic energy pass through the particles. In the case of gold particles of ~ 15 nm mean diameter, the oscillation occurs at 520 nm and the absorption is very strong, which results in a deep color for the nanoparticle solution, even if the concentration is weak. Thus the particles behave like dyes. These particles cannot be taken out of the solution, and if concentrated, the particles settle irreversibly. The citrate-protected particles can subsequently be covered with various molecules or ceramics such as silica and taken out of the solution and redispersed. These particles can be good starting points for a variety of investigations in biology and materials science, and therefore this methodology is practiced widely.[38]

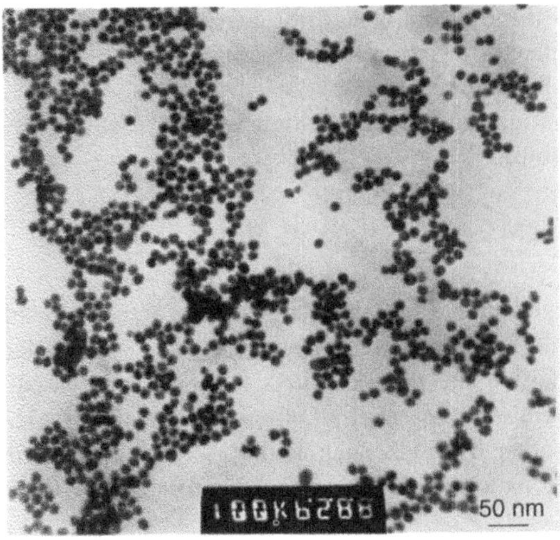

Fig. 2.9 A TEM image of Au–citrate prepared as described in the text with an average particle diameter of 15 nm. The samples were drop-cast from an aqueous solution onto the TEM grid.

Brust Reduction The *Brust method*[20] involves phase transfer of $AuCl_4^-$ from the aqueous phase to the organic phase by a phase-transfer reagent, tetraoctylammonium bromide, and its subsequent reduction at the interface by $NaBH_4$ in the presence of a thiol. The method produces a thiolate (RS^-)-protected gold nanoparticle with a core diameter in the range of 1 to 5 nm. The core dimension can be varied by varying the Au/thiol ratio used in the synthesis: The larger the thiol concentration, the smaller the particle formed. The nanoparticles can be taken out of the medium and dried. The powder can be stored for a long time and can be redispersed.

In a typical procedure, an aqueous solution of $HAuCl_4$ (30 mL, 30 mM) is mixed with a solution of tetraoctylammonium bromide in toluene (80 mL, 50 mM). The mixture is stirred vigorously until all the tetrachloroaurate is transferred completely into the organic layer (the aqueous phase becomes colorless). There is a visible change of color when gold gets phase transferred. Then the desired thiol (depending on the Au/S ratio) is added to the organic phase. A freshly prepared aqueous solution of sodium borohydride (25 mL, 0.4 M) is added slowly with vigorous stirring. The solution is kept stirring for several hours. The organic phase is separated and evaporated to 10 mL in a rotary evaporator. Solvents such as ethanol can be added to precipitate the particles. Washing with ethanol can be repeated to remove the free thiols. All the gold can be recovered. Depending on the dimension of the nanoparticle, the thiol content will vary, which decides the yield. A typical particle size distribution obtained in a 1:2 Au/S ratio synthesis is shown in (Figure 2.10).

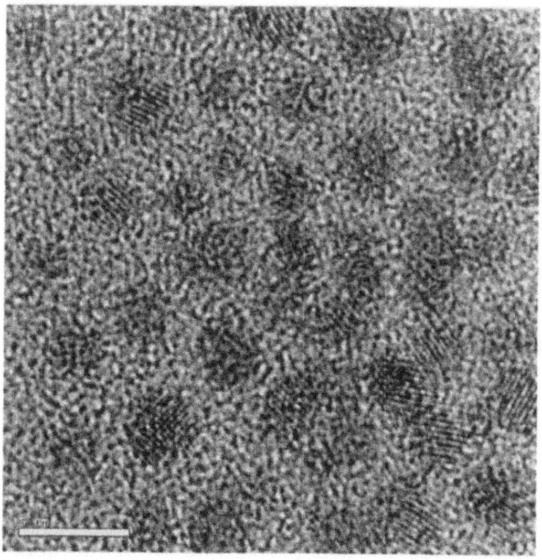

Fig. 2.10 A typical TEM image of Au–ODT nanoparticles prepared by the Brust method. Particles are 3 nm in diameter. The sample was drop-cast from a toluene solution onto the TEM grid.

The Brust method provides many advantages. Functionalization of the particles formed is possible using functionalized thiols in the synthesis. Another method of functionalization is the place exchange reaction, in which one type of thiol or another ligand is exchanged with that on the nanoparticle surface. These nanoparticles, also called *monolayer-protected clusters*, have been reviewed extensively.[12,39]

There are a variety of ways to conduct the synthesis. Phase transfer has been achieved using acid,[40] and this method avoids phase transfer catalyst impurity in the nanoparticles. The method produces monodisperse particles that order to form two-dimensional lattices on a TEM grid. For the normal Brust method, the particle size distribution can be reduced by digestive ripening,[41] a process in which prepared particles are heated in a temperature cycle in the presence of thiol. Particles with a narrow size distribution can be arranged to give two- and three-dimensional superstructures.[42] Synthesis can be achieved without phase transfer and a variety of reducing agents can be used instead of $NaBH_4$.

The monolayers on the nanoparticle surface are well ordered, and the structure of the monolayer assembly has been a subject of detailed examination, as it provides protection to the nano system.[43] The assembly can be investigated by a variety of techniques, such as NMR, IR, Raman, and fluorescence spectroscopies, and the phase behavior of the assembly can be probed by differential scanning calorimetry.[44] A range of techniques have been used to study such systems. From all of these studies it is clear that the structure is well organized with a distinct phase transition temperature,[45] which is a function of the monolayer chain length. The alkyl chain assembly is rotationally disordered in shorter chains, but no orientational freedom exists at room temperature in longer-chain monolayers.[46] The interaction between monolayers on adjacent clusters leads to superlattices, which melt to form a liquid in a first-order transition.[47]

The studies noted above have been performed principally on water-insoluble nanoparticles. From these studies it is clear that when the monolayer chain is on the order of eight or more carbon atoms long, the core of the nanoparticle is not experiencing the solvent. The monolayer assembly is rigid even in the solution phase. The first few outermost carbon atoms are flexible, and solvent penetration does occur to that extent. However, in the case of shorter monolayers, the entire monolayer itself can be exchanged with other suitable ligand molecules, and such ligand exchange chemistry[48] can be used effectively to change the surface properties of the nanoparticles.

A number of water-soluble nanoparticles have been synthesized using water-soluble thiols. Among these, glutathione[49] and mercaptosuccinic acid (MSA)[50] need to be mentioned, as both of them produce extremely water soluble clusters. The method involves reducing the Au–thiolate complex in methanol by $NaBH_4$ in water. The nanoparticles formed precipitate from the solution, as they are insoluble in methanol. The material can be washed in methanol repeatedly and dissolved in water. The MSA clusters form well-organized superlattices.[51] The glutathione clusters consist of a variety of molecular clusters, ranging from Au_8

to Au_{39}, and various fractions of these have been separated by polyacrylamide gel electrophoresis (PAGE).[52]

Electrochemical reduction Although chemical reduction is the most extensively investigated method used to make nanomaterials, various other methods have been studied in specific cases. One of these is electrochemical reduction, where the metal is dissolved at the anode and the metal ion formed is reduced at the cathode. The process is done in the presence of a stabilizer, so that the particles do not deposit at the cathode and lead to electroplating. Palladium nanoparticles have been made this way by passing a current of $0.1\ mA{\cdot}cm^2$ at 1 V in a 0.1 M TOAB solution in a 4 : 1 acetonitrile/THF mixture.[53] Particles of 4.8 nm diameter were precipitated in the process and could be redispersed in THF or toluene. The methodology can also be used for other metals, such as silver,[54] and also for the formation of gold nanorods.[55] In a variation of the method, an aqueous solution of Sr^{2+} and Fe^{2+} produced strontium ferrites.[56]

Radiation Radiation-assisted reduction is another method used to synthesize a variety of nanoparticles. Visible and ultraviolet light, x-rays, and γ-rays have been used to achieve this task. Typically, this method involves the use of a stabilizing agent while irradiating the metal salt solution. In the extreme case of photoreduction involving γ-rays, the species produced in the medium depend on the photon energy absorbed. Typically, in an aqueous solution, radiolysis of water produces H_2, $H^{\cdot}$, H_2O_2, $OH^{\cdot}$, and e^-. The electron is scavenged by nitrous oxide, used in the medium, generating OH^- and $OH^{\cdot}$ in the process[57]:

$$N_2O + e^-(aq) + H_2O \rightarrow N_2 + OH^- + OH^{\cdot}$$

The radicals produced are consumed by the alcohols used following the reactions

$$CH_3OH + OH^{\cdot} \rightarrow H_2O + {}^{\cdot}CH_2OH \qquad (2.17)$$

$$CH_3OH + H^{\cdot} \rightarrow H_2 + {}^{\cdot}CH_2OH \qquad (2.18)$$

The reducing agent in a reaction is the $^{\cdot}CH_2OH$ radical:

$$M^{n+} + n\,{}^{\cdot}CH_2OH \rightarrow M + nCH_2O + nH^+ \qquad (2.19)$$

The products formed are the metal and formaldehyde.

The reducing power would be greater if the solvated electron itself could be used. Au,[58] Ag,[59] Cu,[60] and Co[61] particles have been made this way. The radiolytic method is very useful for making complex structures such as core–shell particles, in which a shell of another metal is coated on an already prepared nanoparticle. Au–Ag,[62] Au–Pt,[63] Pt–Au,[63] Au–Pb,[64] and similar shells have been prepared. The metal nanoparticle is mixed with an aqueous metal ion and radiolyzed using a ^{60}Co source. The radicals produced transfer the electrons to the metal nanoparticle, charging it, and the particle subsequently reduces the metal ion present. The metal atom gets deposited at the nanoparticle, establishing the core–shell geometry. The approach can be used for the controlled increase of nanoparticle size. This has been demonstrated for Au nanoparticles using repeated radiolysis.[65]

Thermal Decomposition of Organometallics One method used to make metal nanoparticles is decomposition of carbonyls by heating in an inert solvent at elevated temperatures in the presence of a suitable stabilizing agent. Co nanoparticles have been made this way by heating $Co_2(CO)_8$ in decalin at 130 to 170° C.[66] The stabilizers used, often nitrogen-containing polymers, were found to form metal cluster macromolecules, in which the stabilizer acted as a complexation agent. By controlling the functionality of the polymer, particle size can be varied. Using different polymers, Fe[67], Ni, Cr, Mo, and W nanoparticles[68,69] and alloy nanoparticles have been prepared. One of the important aspects is that the method allows the use of ligands stable at high temperatures as capping agents in the synthesis. In one such approach a new metastable Co phase (ε-Co) has been formed,[70] stabilized by TOPO. Without TOPO, this phase was not formed. Details of the formation of various kinetically stabilized shapes have been investigated.[71,72] Various types of organometallic reagents and mixtures of those with carbonyls have resulted in FePt[73] and CoPt[74] alloy and core–shell nanoparticles. Thermal decomposition of metalloalkenes is another route for nanoparticle formation.[75,76] The ligands 1,5-cyclooctadiene, 1,3,5-cyclooctatriene, dibenzylidene, and cyclooctenyl (C_8H_{13}-) have been used for this purpose. Co, Ni, Ru, Pd, Pt nanoparticles, Co and Ni nanorods, and CoPt, CoRu, CoRh, and RuPt nanoalloys have been prepared.

Microwave-Assisted Synthesis This is a well-established methodology for the synthesis of a variety of organic and inorganic materials. It has both synthesis and processing aspects,[77] and a variety of materials are synthesized and processed this way. Processing usually refers to inorganic solid state materials such as ceramic oxides. One important aspect of the synthesis is the fast time scale involved as heating is achieved from within. Typical methodology uses a domestic microwave oven working at a frequency of 2450 MHz, and the mixture to be irradiated is placed in the oven with an appropriate stirring mechanism. In the simplest case of metals, metal ions and reducing agents, in a suitable medium, are placed in the oven. In the case of simple metals such as Au and Ag nanoparticles, the methodology is known to produce narrower size distribution than that produced by thermal reduction, using the same reducing agent.[78] Polyols can effectively reduce metal ions by microwave irradiation, and the approach is referred to as the *microwave polyol process*.[79] Irradiation of an aqueous solution of H_2PtCl_6, poly(vinylpyrollidone), ethylene glycol, and NaOH produced 2 to 4 nm Pt nanoparticles,[80] and 6-nm Ni particles were produced similarly.[81] Synthesis can also be adapted to continuous-flowreactors so that production can be automated.[82] Microwave-based methods have been reviewed.[83]

Sonolysis In this method, the reaction mixture is irradiated with ultrasound, typically of 20 kHz. The process of nanoparticle formation is called *cavitation*. This is a process of implosion of cavities of very small dimensions on a nanosecond time scale, leading to local hot spots of very high temperature (5000 K). Precursor species such as organometallics trapped in this atmosphere get decomposed due to high temperature, and the products instantaneously get quenched as a result of the solvent medium. This produces amorphous nanoparticles. Several transition metal nanoparticles have been made this way.[84,85,86] For example, sonolysis of $Fe(CO)_5$ in decane produces 8 nm Fe particles protected by oleic acid.[82] A number of alloy nanoparticles have also been prepared. Sonolysis route for the synthesis of nanomaterials has recently been reviewed.[87]

All nanometals synthesized prior to September 2006 are listed in the Appendix at the end of the book.

2.4.2. Oxides

Aqueous Route If the variety in metals is large, it is even more diverse in metal oxides. As a result, many more nanoparticle systems are investigated in this category. Metal oxides investigated are binary, ternary, and quaternary, and the complexity of the structure, as well as the synthetic methodologies and properties increases in this order. In general, the methodology employed involves precipitation of the oxide or its precursor species, such as hydroxide, carbonate, or oxalate, and subsequent heat treatment of the product. In both cases it is necessary to use protecting agents to prevent aggregation. The precursor species formed is often complex and difficult to characterize completely, as it is likely to be amorphous as a result of low-temperature processing. The complexity increases if multiple metals are involved. The particles produced in the case of oxides are much more polydisperse than metal nanoparticles, although there are several recent examples of monodisperse particles. The other issue is that as the temperature increases, the extent of aggregation increases, increasing the particle size. However, most precursors decompose at lower temperatures, minimizing crystal growth. Another aspect of importance is that just as in the case of metal particles, in certain cases high-energy metastable phases are stabilized at low temperatures.

Certain simple oxides are prepared in aqueous solutions at low temperatures without sintering. For example, 4-nm rutile TiO_2 particles can be made by precipitating aqueous $TiCl_3$ by NH_4OH.[88] Poly(methyl methacrylate) is used as the stabilizer. For an oxide ion-conducting electrolyte, $Ce_{0.8}Y_{0.2}O_{1.9}$, aqueous $Ce(NO_3)_3$, and $Y(NO_3)_3$ were precipitated by oxalic acid and the product was sintered to produce nanoparticles of the composition noted above.[89] By varying the sintering temperature, particles of different mean diameters were produced.

Ternary oxides have also been produced. When the structures are stable, such as spinals, the hydroxides can be converted to oxides by conducting the precipitation near the boiling temperature. This has been achieved in the case of Fe_3O_4,[90] $MnFe_2O_4$,[91] $CoFe_2O_4$,[92] and $Pr_xCe_{1-x}CeO_2$.[93] In the method used by Li et al., the ferromagnetic $CoFe_2O_4$ was stabilized by dilute HNO_3.[94] The product is a ferrofluid of considerable interest. Stable ferrofluids of Fe_3O_4 were also made by steric stabilization of poly(vinyl alcohol), starch, and other substances.[95,96]

Nonaqueous Route Several metal oxides have been prepared by the nonaqueous path. The methodology involved is useful for metals where precipitation is difficult in aqueous media. It is also helpful if more than one metal needs to be precipitated simultaneously, which requires widely different pH conditions if done in water. In the case of $LiCoO_2$, LiOH and $Co(OH)_2$ were precipitated simultaneously by dripping an ethanolic solution of $LiNO_3$ and $Co(NO_3)_2$ into 3 M ethanolic KOH.[97] A mixture of hydroxides obtained was heated to obtain nanoparticles of $LiCoO_2$. Oxides such as γ-Fe_2O_3,[98] $BaTiO_3$,[99] and MFe_2O_4 (M = Mn, Fe, Co, Ni, Zn) were prepared by using various precursors in different solvent systems.[100] Synthetic routes used for oxide nanoparticles are included in the Appendix.

2.4.3. Chalcogenides

One of the most extensively studied group of nanomaterials are the chalcogenides (sulfides, selenides, and tellurides). These are semiconducting quantum dots in which the dramatic effects of size quantization are manifested. The range of properties that these systems exhibit is also large, and as a result, there have been extensive investigations on establishing proper synthetic routes for particles of narrow size distribution. This is important, as the properties vary with size, just as in the case of any other nanomaterial. The chemical purity and surface functionalization are very important for some of the properties investigated, especially fluorescence. The surface states can destroy the photophysical properties completely. From several of the properties investigated, it is now very clear that synthesis plays a key role in understanding and utilizing the properties of these systems. Various synthetic methods available for chalcogenide particles have been reviwed.[101,102,103]

The simplest approach to making chalcogenides is to mix a chalcogenide ionic salt with a metal salt in aqueous solution. This leads to immediate precipitation of the metal chalcogenide in most cases, but the process is extremely rapid and control becomes difficult. The alternative approach is to use covalent chalcogenides and organometallics so that reaction at a higher temperature in an organic solvent can produce nanoparticles. Suitable passivating agents may be used, and the reaction is controlled kinetically (reaction conditions are controlled). Optimization of the parameters has produced several semiconductors of the II–VI and II–V categories.[104–109] In the typical approach, the Cd precursor used is $Cd(CH_3)_2$ and the chalcogenide precursors are $[(CH_3)_3Si]_2S$, $[(CH_3)_3Si]_2Se$, R_3PSe, and R_3PTe (R = C_4 to C_8 n-alkyl). A suitable solvent stable at high temperatures such as trioctylphosphine (TOP) or trioctylphosphineoxide (TOPO), is held at 340 to 360° C and a room-temperature solution containing the precursors is added, resulting in nucleation of the particles. Growth is controlled by reducing the temperature, and the reactions are allowed to continue, resulting in a particle size increase.

Further addition of reagents increases the size of the particle. The crystallinity can be controlled by annealing the mixture for extended periods. Size control in the range 1.2 to 12 nm is achieved.[110–113] The methodology has been used for In-group V semiconductors.[114–116] The surface of the particles is protected by TOP or TOPO, but other surface-passivating agents have also been used. In addition to the hexagonal and cubic phases, the wurtzite phase has also been stabilized by controlling synthetic parameters.[117–119] When used at higher precursor concentrations, the methodology gets nanorods of CdSe.[119] An aspect ratio of 30 has been achieved. Other morphologies have also been observed.[118]

Dimethylcadmium is pyrophoric, and conducting experiments at elevated temperatures poses a high risk. Therefore, various other precursors have been employed. CdO has been used for this purpose, and the methodology is similar to that described. Alkyl (C_6 or C_{14}) phosphonic acid is used as another reagent in the chemistry, along with TOPO as precursor.[120,121] Other precursors, such as cadmium carboxylates, have also been used.[121]

Microwave-assisted synthesis has been carried out for CdSe, PbSe, and other quantum dots.[122,123] Several other nanoparticles of oxides and chalcogenides have been prepared by the microwave route.[124–127] Sonochemical methods have also been used for the synthesis. By using an anionic surfactant template of SDS, hollow nanoshells of CdSe have been synthesized using the sonolysis route.[128] The use of such materials in biological or other environments where toxicity is a problem requires additional control. The surface can be protected with various inert oxides such as SiO_2. The synthetic approach is similar to making a nanoparticle: ligand exchange with a suitable molecule and using it to grow an oxide layer.[129]

2.5. MICROEMULSION-BASED METHODS FOR NANOFLUIDS

Microemulsions are micellar solutions. *Micelles* are self-organized spherical structures of amphiphilic molecules in a suitable medium. The micellar solutions contain an amphiphlic molecule or a surfactant, an organic medium, and water. There may also be a cosurfactant. In a typical micellar arrangement, water is outside the spherical structure. When the medium is organic (in large excess), we get reverse micelles in which the self-assembled structure encloses water as shown in Figure 2.11. Micelles change shape and several types of structures (e.g., vescicles, lamellar and cylindrical phases) exist, depending on the thermodynamic conditions. As they contain water within, reverse micelles can be loaded with metal ions or reducing agents. The micellar dynamics is such that two micelles in contact can fuse and exchange their contents. Due to Brownian motion, such collisions, leading to coalescene and decoalescence take place. This makes it possible to conduct reactions in confined spaces, and as a result, nanomaterials can be formed. Such materials can be purified, processed, and redispersed to make nanofluids. The important aspect is that since the concentration of the micelles is low, the synthesis does not lead to large quantities of materials as

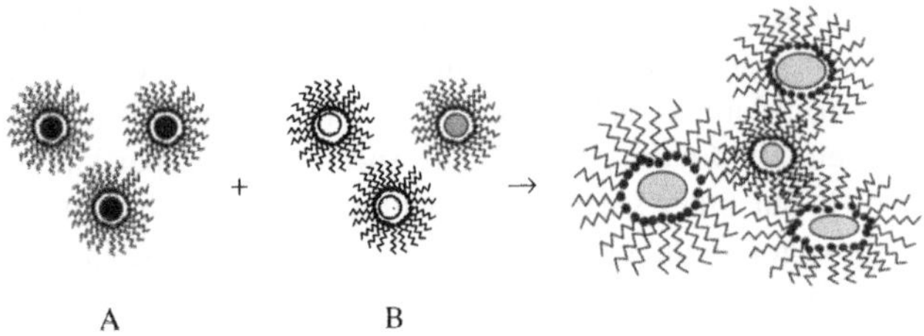

Fig. 2.11 Microemulsion-based synthetic approach. The reverse micelles shown here are water-in-oil structures (with a majority oil phase). The two solutions, with metal ions (A) and reducing agent (B), for example, can be mixed to make a nanoparticle. The micelles coalesce and a dynamic equilibrium is reached by which the contents of all get mixed.

with the chemical method. This may also be categorized as template-mediated synthesis.[130]

2.5.1. Metals

Water-soluble metal ions and reducing agents are used. $NaBH_4$ and $N_2H_4.H_2O$ are the two reducing agents commonly employed. Reduction by gases such as H_2 is slow, but this method is also used. Two types of surfactants are used, anionic and cationic. The most common and oldest cationic surfactant is cetyltrimetrylammonium bromide, $(C_{16}H_{33})(CH_3)_3NBr$ (CTAB), a quaternary ammonium salt. The most common anionic surfactant is sodium bis(2-ethylhexyl)sulfosuccinate, generally referred to by its trade name, Aerosol OT or AOT. Nonionic surfactants include polyethylene ethers [e.g., pentaethylene glycol dodecyl ether (PEGDE), $CH_3(CH_2)_{11}-O-(CH_2-CH_2-O-)_5-H$ (Triton-X)]. The structures of these surfactants are shown in Figure 2.12. The synthetic methodology employed is simple and easily adaptable in a number of cases. As the surfactants contain ions, it is important to use suitable metal ions which will not precipitate in the presence of these ions, an example being Ag^+ in the presence of CTAB.

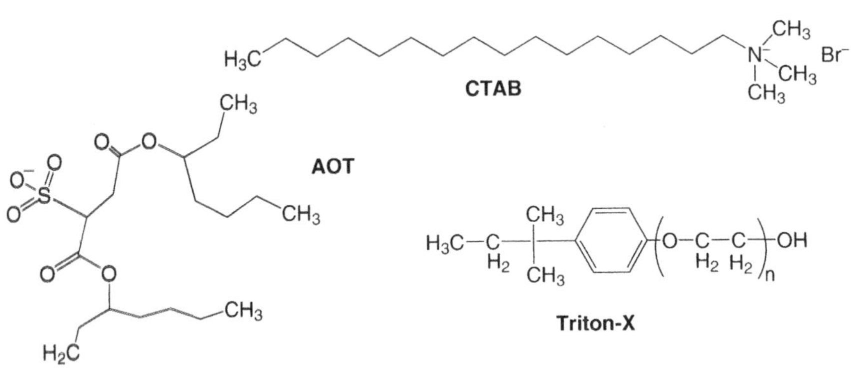

Fig. 2.12 Various surfactants used in microemulsion-based synthesis.

The chemistry is the same as discussed earlier, but the process now occurs in reverse micelles. In addition to metals, several alloys have also been made. To achieve this, two separate microemulsions (of the two metals) may be mixed with another mocroemulsion of the reducing agent, or a microemulsion of the two metal ions may be mixed with that of the reducing agent. A brief survey of the metal particles made by microemulsions is given in the Appendix. Various conditions, such as pH, and temperature, can be tuned to control the particle size and shape.

The synthesis of metal oxides can proceed in the same fashion. Here, metal oxides or hydroxides are precipitated by the addition of reagents such as NH_4OH or NH_4OH taken in a reverse micelle, with the hydroxide or oxide precipitated collected using centrifugation. Subsequently it is calcined at the temperature required to yield the oxide. For ions not soluble or stable in aqueous solutions, suitable precursors are used. Metal chalcogenides are prepared similar to metal oxides. In this case, Na_2S, or Na_2Se, or similar salts are typically used to precipitate metal ions such as Pb^{2+}. Several of these examples of microemulsion-based synthesis are presented in the Appendix.

2.5.2. Core–Shell Structures

Several complex structures can be prepared by the microemulsion route. One form of complexity is to produce core–shell structures. The shell itself can have several layers, increasing the complexity further with dissimilar materials. In the simplest metal-on-metal core and shell, the approach is to add the metal ion in a reverse micelle into a solution in which the particle has already been synthesized. The solution contains excess reducing agent from the particle synthesis and the addition of the metal ion in the reverse micelle produces reduction, thus coating the metal particles with a shell. It is likely that the metal ion added may be reduced before encountering the metal nanoparticle already synthesized. As a result of this, free metal nanoparticles of the second kind as well as uncoated nanoparticles of the first kind, can be present, although the objective was to prepare core–shell particles. A typical example is an Fe–Au[131,132] system, in which the synthesis also gives pure Fe particles.[133] The approach can be extended to get Au–Fe–Au systems.[134,135] The difference between this and the simultaneous reduction approaches is that it is very difficult to get conditions by which a less noble metal becomes the core, such as Ag–Au in simultaneous reduction, but such possibilities exist in the microemulsion route.

The core and shell can be made of oxides. Fe_3O_4–MnO has been synthesized. The core is 10 nm in diameter and the shell is 2.5 nm thick.[136] SiO_2–Fe_3O_4 has been prepared.[137] The complexity and variety of the core–shell system can be numerous. One could coat the surface of the nanoparticle with a polymer prepared in situ in the reaction mixture. The polymerization can occur over the nanoparticle surface, which is already coated with an appropriate group which aids polymerization. This type of approach can also be used to grow inorganic shells. This is the approach used by Liz-Marzán to prepare SiO_2 shells over gold nanoparticles.[138] Here the approach is to coat the surface of gold with a monolayer of aminopropyltrimethoxysilane. The amino group binds onto the gold surface and the trimethoxy group can be hydrolyzed in the presence of tetraethoxyorthosilane, resulting in the growth of a thin silica shell. The shell thickness can be increased further by using Stöber's method.[139] This synthetic approach can also be used for other nanoparticles.

The synthesis of core–shell materials can also be achieved by a single step, in which all materials and reagents are added into one pot. This approach has produced Au–TiO_2 and Ag–TiO_2 particles.[140,141] Here the reduction is achieved by dimethylformamide (DMF), and the precursor species are $HAuCl_4$, $AgNO_3$, and titanium isopropoxide. A similar method can be used for Au–ZrO_2 and Ag–ZrO_2, and the Zr precursor used is zirconium isopropoxide.[141] Corresponding alloys can also be prepared. All these core–shell materials (i.e., namely Au–SiO_2, Au–TiO_2, and Au–ZrO_2 and their Ag analog) are freely dispersible in organic media. To increase the dispersibility, the oxide surfaces can be coated with long-chain carboxylates. In this case the material obtained can be taken out of the solution in powder form and stored for extended periods, then redispersed.[142] Colloidal core–shell particles can be used to control the optical absorption of nanoparticles. The assemblies of these structures in various forms are also important ways of controlling optical absortion. Various aspects of such control are discussed in an article by Liz-Marzán.[143]

2.6. SOLVOTHERMAL SYNTHESIS

In the solvothermal method, the reaction is conducted in a closed vessel in which the solvent can achieve a supercritical state. The temperatures and pressures are high (e.g., above 374° C and 218 atm for water). Although many reactions are not conducted in the supercritical state, the reactions are done in pressure bombs typically lined with Teflon. In the supercritical state the liquid–vapor boundary disappears and the fluid acquires the properties of both the liquid and the vapor. Increased solubility and a higher reaction rate favor many reactions and processes are therefore better done at the supercritical state. However, for many reactions, increased reactivity at higher temperatures under controlled conditions is the sole aspect of interest. If the reaction is carried out in water it is called *hydrothermal*; if carried out in other solvents it is referred to as *solvothermal*. Several reviews are available on this topic and interested readers may consult them for details such as pressure conditions, and experimental apparatus, and various materials synthesized.[144–151]

One of the aspects of particular interest in this method is the significant reduction in reaction temperatures, even for many ceramic materials. The material thus synthesized is highly crystalline, and often, postsynthetic annealing operations are not necessary. The materials are monodisperse, and often they are suspendable in a suitable medium. In addition, the methodology can be adapted to suit several synthetic conditions and can also be used for large-scale synthesis. The heating process itself can be conducted in a microwave, making it energy efficient in addition to the distinct advantages that the technique offers.[152] The process can also be adopted to continuous flow conditions.[153] All of these make it an interesting methodology for synthesis. Besides, the synthetic procedures have the advantages of being relatively inexpensive in terms of the solvents used, and arguably, green (when water is the solvent).[146]

In a typical method, the precursor species are mixed with suitable reagents for reduction, precipitation, and so on, in a suitable medium. Often, stabilization and complexation agents are added so that the process happens under controlled conditions. The synthesis of anatase TiO_2 is achieved by controlled hydrolysis of $Ti(OEt)_4$ in ethanol.[154] Monodisperse particles were prepared by this route.[155] A TEM image of such TiO_2 particles, prepared starting from titanium isopropoxide, is shown in Figure 2.13. As the nanoparticles are not protected by strongly binding protecting agents, they are not as separated as in Figure 2.10. The particles are highly crystalline. The starting material can be $TiCl_4$[156,157] and the synthesis can also use stabilizers such as citric acid.[158] Microemulsion-based synthesis often includes a solvothermal step, and TiO_2 has also been prepared that way.[159] Large-scale synthesis of TiO_2 is significant in view of its use in photocatalysis, and solar cells. CeO_2 synthesis has been achieved with[160] and without[161] stabilizers. Complex oxides such as γ-Fe_2O_3, $CoFe_2O_4$,[162] and $ZnFe_2O_4$[147] were prepared by the hydrothermal route. The microwave–hydrothermal route is very useful for many complex oxides, such as the ferrites (MFe_2O_4, where M = Mn, Co, Ni, or Zn),[163] $BaTiO_3$,[152] and α-Fe_2O_3.[164] The methodology is also useful in making several other materials, especially mesoporous solids. Applications for these materials are numerous in catalysis and are covered extensively in other

Fig. 2.13 Lattice-resolved TEM image of TiO_2 particles synthesized by the hydrothermal route. (From the author's laboratory.)

places.[165–168] Hydrothermal processes are used to make metal chalcogenides.[169] CdSe particles were made from elemental Cd and Se at 180° C in a hydrothermal route.[170] The products were aggregated in the absence of suitable stabilizers. In the presence of TOPO, the synthesis of CdSe produces ~ 3-nm particles.[171] Many other chalcogenides (e.g., SnS_2, NiS_2, CoS_2, FeS_2, $NiSe_2$) have been synthesized.[172,173]

2.7. SYNTHESIS USING SUPPORTS

Various synthetic approaches are available to use supports for the preparation of nanoparticles. These vary from the use of nanoparticles themselves to materials with nanocavities. Particles prepared by one route can be used for subsequent growth. This results in the controlled growth of particles at nanoparticle surfaces. If the growth of smaller particles is avoided, one can achieve a programmed increase in size. This is possible by the use of suitable reactants. For example, hydroxyl amine will reduce Au^{3+} but will work better on Au particles. As a result, larger particles can be grown by adding Au^{3+} to a nanoparticle solution containing this reducing agent. This approach, called *electroless plating*,[174] is useful for making various kinds of structures, such as core–shell particles. An interested reader may consult reviews in this regard.[175,176] The templates and covers need not be metals alone. In the case of Au or Ag coated with ZrO_2, by removing the metal one can get oxide shells.[177] The cores and shells prepared can be redispersed, even in organic solvents, by suitable functionalization of the oxide surface. The approach is similar to that discussed earlier for metal particles.[142]

The support used for particle growth can be a polymer. In this case a metal-coated polymer can be obtained and the polymer can be removed to get hollow nanoshells. An approach of this kind was developed using polystyrene coated with a positively charged polymer such as poly(allylamine hydrochloride). This was coated with 4-(dimethylamino)pyridine–protected gold nanoparticles, and the shell cover can be increased using the hydroxylamine method. The polymer can be removed by heating or by washing with organics.[178]

Dendrimers as templates have been used for nanoparticle synthesis. This topic has been extensively reviewed.[179] Dendrimers are branched polymeric molecules with inner and peripheral functional groups which are known in terms of their generation. The polyamidoamine (PAMAM) dendrimers have been well studied. The interest in this approach is that the dendrimer is of nanometer dimensions. For example, a fourth-generation PAMAM is 4.5 nm in diameter and can encapsulate nanoparticles, those encapsulated in single dendrimers being different from those protected with multiple dendrimers. Depending on the functionality of the dendrimer, the nanoparticle formed can be redispersed in a variety of solvents.

In a typical approach, the metal ion is mixed with the dendrimer and stirred so that the metal ion complexes with the ligand locations. Then a reducing agent is added, creating a dendrimer- encapsulated nanoparticle. The particles, in general, are in water, and if necessary can be phase transferred to the organic phase by stirring with a thiol or suitable ligand taken in an organic phase. The resulting dendrimer will remain in the aqueous phase. Dendrimer-encapsulated particles show high monodispersity and high catalytic activity. Other dendrimers have also been used to prepare nanoparticles.[180]

Mesoporous solids are excellent templates. MCM-41 is one such material. The pores of these systems are of nanometer size, and by incorporating metal ions, with subsequent reduction, several nanostructures can be obtained. The approach can be used for chalcogenides by diffusing H_2S, for example, after metal ion incorporation. Various materials have been prepared in this way.[168,181–183] A similar approach can be used in the case of organic and inorganic membranes. NAFION membrane has been used to grow semiconductor nanoparticles[184] and the same approach has been used for a sulfonic acid membrane (eg., Dowex).[185] Alumina membranes have been used to grow particles this way.[186] Packing of nanoparticles leading to nanowires is possible in such membranes. Nanowires can be created in pores by reducing ions in them. The sheath can be dissolved later to get isolated nanowires.[187] A similar approach has been used with MCM-41 to make Pt nanowires, and such materials have been used for catalysis.[183] Templates of carbon, especially nanotubes, are very interesting, due to their one-dimensionality. Metals, alloys, semiconductors, and oxides have been filled in carbon nanotube templates. The nanotubes can be burned to create anisotropic nanoparticles.[188,189] A similar method utilizes inorganic nanotubes.[190] However, all these approaches, in general, do not produce stable suspensions. Short-term stability can be achieved with stabilizers.

2.8. USING BIOLOGY

Nanoparticles have been found in several organisms. Silver[191] and Fe_3O_4[192] nanoparticles have been found in bacteria. Therefore, it is natural to believe that the ions taken inside organisms may be subjected to reduction. Reduction of Au^{3+} by fungi has been achieved in this way.[193] Common *lactobacilli* reduce gold.[194] Such a reduction also happens in human cells.[195] An important aspect is that all the metal nanoparticles can be extracted and resuspended. This approach makes it possible to make large quantities of nanoparticles at less cost. It is also possible to achieve reduction extracellularly so that the organism is alive to continue the process and thus can be used for manufacture. This approach has been used for metal oxides, sulfides, and magnetic nanoparticles.[196,197] Biological templates such as DNAs can be used to organize nanoparticles.[198] Patterning of nanoparticles to two-dimensional arrays has been a subject of intense investigation. The subject matter is covered in reviews.[199–201]

2.9. MAGNETIC NANOFLUIDS

Magnetic particles are important in ferrofluidics, magnetic refrigeration, information storage, magnetic drug delivery, and as contrast enhancement agents in imaging and fundamental applications. These particles in the nanometer regime are superparamagnetic; that is they behave like paramagnetic atoms, with large magnetic moments in an isolated state. Most magnetic particles are prepared in water in oil micelles or oil-in-water reverse micelles. These approaches can be adapted very well for ferrite spinnels, MFe_2O_4 (M = Mg, Co, Mn, Ni, Zn, etc.). With suitable protection the particles can be redispersed in media.[202] The other approach is to inject precursor species such as carbonyls to a hot surfactant solution. It is also possible to mix the precursors at a low temperature followed by slow heating. Quick nucleation and slow particle growth in appropriate conditions are the keys to efficient synthesis. Nanoparticles of Fe, Co, Ni, and their alloys and ferrites have been prepared in this way. Details may be had in an article by Hyeon.[203] Organometallic precursors may also be used.[204]

Magnetic Fe–Au core shell particles have been synthesized by laser ablation. In this approach, Fe and Au particles protected by oleic acid and CTAB, respectively, were irradiated in hexane by a 532-nm laser. The 18-nm core, 3-nm shell particles formed were superparamagnetic and had a shelf life of four months.[205] This method is significant, as normally one would have got Fe over Au, as Au is noble. FePt nanoparticles are diverse systems for a variety of applications. Various kinds of such particles with shells of Fe_3O_4, CdS, and so on, for permanent magnet and biological applications are discussed in a review by Sun.[206] The approach for the synthesis adopted is thermal decomposition of $Fe(CO)_5$ and reduction of platinum acetylacetonate, $Pt(acac)_2$, in the presence of 1,2-alkanediol with suitable stablizing agents. $Fe(CO)_5$ is thermally unstable and makes Fe atoms. $Pt(acac)_2$ is reduced readily by the diol, and the atoms formed result in nuclei over which Fe and Pt condense to make the nanoparticles. Oleic acid and oleylamine (or other long-chain carboxylic acids or primary amines) may be used to stabilize the surfaces. The normal approach is to conduct the reactions in an inert atmosphere in the absence of moisture. The product composition is controlled by varying the starting mixture. Replacing $Fe(CO)_5$ with $Fe(acac)_2$ or $Fe(acac)_3$ and simultaneous reduction of the salts is possible. Details of these methodologies and original references may be found in the review cited above.

2.10. INERT GAS CONDENSATION

In this method, the nanoparticle is made in the gas phase by creating a condition of supersaturation. The condition in the preparation chamber is made such that the solid phase is more stable than the gas phase. At this point the cluster nucleates and deposits from the gas phase. The deposition will lead to aggregation and can be controlled suitably by surfactants or suitable protecting agents. This makes it possible for the materials to be redispersed with not much effort. The evaporation conditions can be varied such that the gas-phase species undergo reactions such as oxidation to produce oxides (forming CuO, Al_2O_3, etc.). The nanoparticles formed are transported and deposited by thermophoretic diffusion on a cold finger. The gas pressure in the chamber can be controlled to modify the sample morphology.[207] Evaporation can be achieved thermally or by the use of a laser or spark or discharge. It is also possible to use organic precursors which are decomposed in the gas phase to make appropriate species. The methods employed include spray pyrolysis, chemical vapor deposition, and flame pyrolysis. Short summaries of the synthetic tools are given in References 208 and 209.

A similar method is the direct condensation of gas-phase species on a flowing low-vapor-pressure liquid.[210] This method is also called *vacuum evaporation onto a running oil substrate*(VEROS). Materials prepared by a similar route were used for early thermal conductivity measurements.[211–213] An important advantage of this method is that it avoids agglomeration.

2.11. ANISOTROPIC NANOPARTICLES

Although nanoparticles often refer to spherical shapes, there are also various anisotropic shapes. These refer to all shapes other than spherical. Such shapes require more than one parameter to describe their shapes. The more common ones are nanorods and nanotriangles. Several other shapes, such as tripods, tetrapods, stars, flowers, and sheets, are known, and in several cases synthetic flexibility does not exist. Several of these shapes are used as starting points for the synthesis of more complex shapes. However, it is not possible to get most of these in the solution phase, so we will not discuss them. A discussion of rods and tringles follows.

2.11.1. Gold and Silver Nanorods

Nanorods of gold and silver are stable colloidal solutions. They are typically 10 to 30 nm in diameter and 50 to 1000 nm long. They show two plasmon absorptions, one, due to the long axis, called the *longitudinal plasmon*, and the other, due to the short axis, called the *transverse plasmon*. The nanrods can be synthesized principally in three different ways: by the electrochemical route, by the photochemical route, and by the seed-mediated route. In the electrochemical route, electrochemical reduction is employed, whereas in the photochemical route, light-induced reduction is used.[214] The seed-mediated method uses chemical reduction.[215–217] All methods use a structure-directing template for the growth of the rods. Nanorod synthesis has been reviewed.[218]

In the chemical method 4-nm seed particles of gold are prepared by the borohydride reduction route (at 0° C). The seed is allowed to grow in a growth solution that is prepared with Au^{3+}, CTAB, and ascorbic acid. Au^{3+} is reduced to Au^{+} by ascorbic acid, but it is not capable of reducing Au^{3+} to Au(0). However, Au(0) can be formed on the surface of Au particles by ascorbic acid. CTAB is a structure-directing agent that guides the growth of the rod on certain specific sites on the gold surface. The presence of about 5% silver is necessary to increase the yield of nanorods to nearly 100%. However, with silver the aspect ratio of

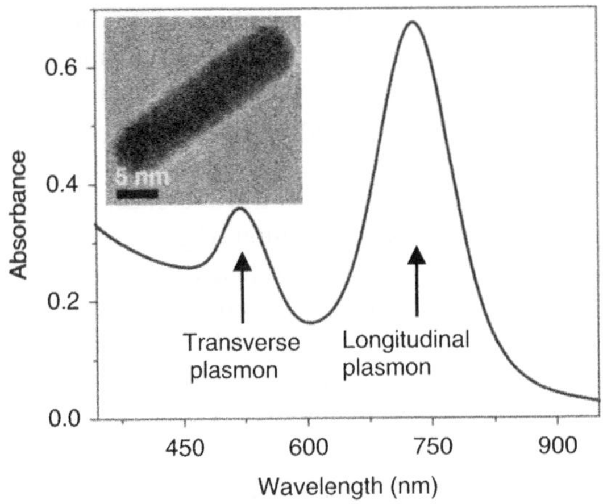

Fig. 2.14 UV/Vis absorption spectrum of gold nanorods along with a TEM of a single nanorod.

the rods is a maximum of 6, whereas in its absence, the maximum aspect ratio is about 25; nevertheless, the yield is less. This is the maximum length observed for gold. For silver rods, micrometer lengths are observed.

The CTAB used is of high concentration (0.2 M), far above the critical micelle concentration (CMC). It is also known that other counterions, such as Cl^-, do not help in nanorod growth. In fact, only particles are formed in this way. Iodide makes assorted particles. Br^- is believed to be important in making thin layers of AgBr (it is not found as a precipitate) on certain surfaces of nanoparticles, arresting growth along these planes.[218]

The absorption spectrum of a typical gold nanorod solution is shown in Figure 2.14. The rods shown are 18 nm in diameter and 60 nm long. The transverse plasmon appears at 500 nm and the longitudinal plasmon appears at 745 nm. The position of the longitudinal plasmon is a good measure of the length of the rod. The greater the length, the farther the position shifts to red, and ultimately it occurs in the infrared. This is advantageous for biological applications, as the skin penetration depth of infrared is higher than that of visible. The figure also shows a TEM image of a single purified nanorod. The rods are stable in the solution state, and with a thin layer of CTAB, are infinitely stable at ambient conditions. The solution can be heated to boiling without apparent change in the rod morphology. Short rods are far more reactive than longer ones and undergo radical, ion-mediated corrosion. In the presence of excess Au^{3+}, rods become particles.

2.11.2. Triangles

Nanoparticles of silver become nanotriangles or nanoprisms upon light irradiation. The silver nanoparticles were prepared by borohydride reduction and were protected with trisodium citrate. Irradiation with a fluorescent lamp was done in the presence of bis(p-sulfonatophenyl)phenylphosphine dehydrate dipotassium salt solution for extended periods (of about 70 hours). The nanotriangles are stable in the solution and self-organize when dried.[219] It has been shown that irradiation of silver nanoparticles (with citrate protection) by a sodium lamp produces triangles with no need for a surfactant.[215] Gold nanoplates were prepared through reduction of hydrogen tetrachloroaurate by a reduced amount of sodium citrate in the presence of poly(vinylpyrrolidone). The nanoplates differ in optical characteristics with strong infrared absorption and can be tuned depending on the dimension. These absorptions are due to quadrupolar plasmon resonances.[220] Gold triangles have also been prepared by a lemon grass extract.[221] A summary of the synthetic routes used for anisotropic nanoparticles is presented in the Appendix.

2.12. OTHER NANOFLUIDS

Nanofluids need not be made only from metal or inorganic nanoparticles. Organic nanoparticles and nanomolecules can also be used to make nanofluids. Several classes of nanomolecules belong in this category. Although there are numerous types of systems, the most important are (1) fullerenes, (2) nanotubes, (3) dendrimers, and (4) polymers. In this section we outline the synthetic aspects of first two types of nanofluids.

2.12.1. Fullerenes

Fullerenes are all carbon spheroidal molecules of the general formula C_{20+2n6} (where n_6 is the number of hexagonal faces). A closed cage requires that

$$12 = 3n_3 + 2n_4 + 1n_5 + 0n_6 - 1n_7 - 2n_8 - \cdots, \tag{2.20}$$

where n_k is the number of k-sided faces. For carbon, the only k values are 5 and 6, although 7 has been detected in carbon nanotubes. This means that in carbon one has to make closed-cage structures with pentagonal and hexagonal faces. So there should be 12 pentagonal faces in a given structure, and the number of hexagonal faces is arbitrary. In C_{60} there are 12 pentagonal faces and 20 hexagonal faces. The most intensely explored molecules in this category are C_{60} and C_{70}, although there are lesser known analogs such as C_{76} and C_{82}, which are more difficult to synthesize in larger quantities. Fullerenes were discovered in 1985 in the laser evaporation of carbon[222] and were prepared in the condensed phase in 1990 by thermal evaporation of graphite in an inert atmosphere.[223]

To synthesize fullerenes, all that one needs is a welding transformer, a chamber connected to a vacuum pump (even a single-stage oil-sealed rotary pump is adequate), and some graphite rods. The graphite electrodes are brought in close contact with each other and an arc is struck in an atmosphere of 100 to

200 torr of helium or argon. To sustain the arc, a voltage of 20 V (ac or dc) may be necessary. For a graphite rod 6 mm in diameter, a current of about 50 to 200 A may be consumed. Generally, spectroscopic purity graphite of high porosity is used so that the evaporation rate is high. The soot generated is collected on water-cooled surfaces, which could even be the inner walls of the vacuum chamber. After sustaining the arc for several minutes, the vacuum is broken and the soot is collected and soxhlet-extracted for about 5 to 6 hours in toluene or benzene, resulting in a dark reddish-brown solution which is a mixture of fullerenes. Some 20 to 30% of the soot collected is soluble, and this soluble material is then subjected to chromatographic separation.[224] Over the years, several simple methods have been discovered, including filtration over an activated charcoal–silica gel column. About 80% of the soluble material is C_{60}, which can be collected in one pass using toluene as the mobile phase. C_{70} can be separated using a toluene/*o*-dichlorobenzene mixture as the eluant. Repeated chromatography may be necessary to get pure C_{70}. C_{60} solution is violet in color, whereas that of C_{70} is reddish brown. Higher fullerenes (i.e., C_{76}, C_{78}, C_{82}, etc.) require high-performance liquid chromatography for purification. Spectroscopic properties of several of these less common fullerenes are now known. On the lab scale, preparation of a gram of C_{60} requires about 5 hours of work, starting from graphite. But 250 hours are required to make about 1 mg of many of the higher fullerenes. C_{60} and C_{70} are available commercially from several sources. Fullerenes crystallized from saturated solutions retain solvent molecules, and removing them may require long hours of vacuum drying. Crystal growth by vapor transport is an excellent way to grow millimeter-sized crystals devoid of solvent for sensitive measurements. For solid-state spectroscopic measurements, it is better to use evaporated fullerene films in high or ultrahigh vacuum to avoid solvent contamination. Evaporation is also used as a method of purification, as there are substantial differences in the onset of evaporation between C_{60} and C_{70}.

Arc evaporation is not a unique way to make C_{60}. Fullerenes have been found in flames, upon chemical vapor deposition used to produce diamonds, in a 1.85-billion-year-old bolide impact crater, and from spacecraft. It has also been made from diamond. No one has made it by chemical reaction, but such a possibility has excited many organic chemists. It has been synthesized from camphor. Mass spectrometry has shown that higher clusters of carbons could be formed by laser evaporation of polymers. Upon laser evaporation, highly unsaturated carbonaceous ring systems produce C_{60}. There are several other exotic means of producing C_{60}. However, total synthesis would indeed be a landmark in chemistry.

Solid C_{60} dissolves in a number of organic solvents, such as hexane, CH_2Cl_2, and toluene, yielding magenta solutions. These solutions may be stored for extended periods without degradation. However, C_{60} gets deposited on the walls of the vessel. The solubility of C_{60} is about 7.2 mg/mL at room temperature. The solutions are stable over a range of temperatures.

2.12.2. Nanotubes

Carbon nanotubes[225] are one-dimensional cylinders of carbon with single or multiple layers of carbon. The tube diameters are in the range of a few nanometers and the length is on the order of micrometers. This makes carbon nanotubes one of the longest-aspect-ratio materials. As the tubes exhibit many properties as a result of the confinement of electrons to one dimension, they are new types of model systems in which a number of quantum phenomena can be investigated. As a result, they are also one of the most extensively investigated nanosystems.

A single sheet of graphite is called *graphene*. Rolling the sheet produces a carbon nanotube. Planar carbon sheets can be rolled in a number of ways. This makes the carbon sheet helical around the tube axis, as shown in Figure 2.15. If we fold the graphene symmetrically as shown in Figure 2.15, the resulting tube will have the hexagons neatly arranged side to side, as shown by the arrow. Imagine that the graphene is folded differently, at an angle. This results in a tube in which the hexagons form a coil around the tube axis. One can see that there are infinite ways by which such folding can be done, resulting in tubes of different helicities. These are all different types of tubes. The extent of helicity varies, and as a result, numerous tube structures are possible, which provides variety that leads to diversity of properties. The electronic structure of the tube varies as the helicity changes.

The structure of a cylindrical tube is best described in terms of a tubule diameter d and a chiral or helical angle θ, as shown in Figure 2.16. The chiral vector $\mathbf{C} = n\mathbf{a}_1 + m\mathbf{a}_2$ and the two parameters d and θ define the tube. The unit vectors $\mathbf{a}_1$ and $\mathbf{a}_2$ define the graphene sheet. In a planar sheet of graphene, carbon atoms are arranged in a hexagonal structure, and each atom is connected to three neighbors. In the figure, each vertex corresponds to a carbon atom. The vector $\mathbf{C}$ connects two crystallographically equivalent points. The angle θ is with respect to the zigzag axis, and it is 30° for the armchair tube. If one rolls over one end

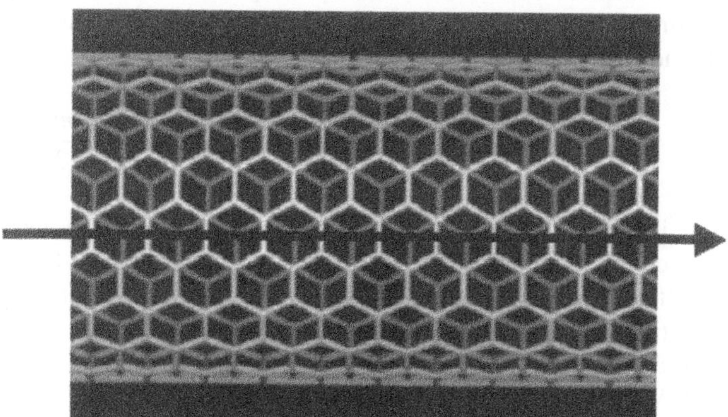

Fig. 2.15 Part of a nanotube. The tube is highly symmetrical and is made from a graphene sheet.

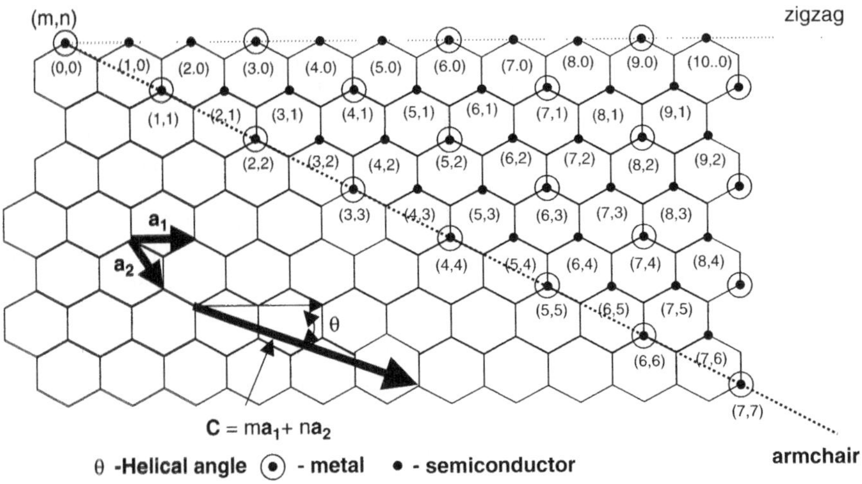

Fig. 2.16 Notation used to demonstrate carbon nanotubes.

of the tube to the other, we get a cylinder. The rolling can be done in several ways. The bond angles of the hexagons are not distorted in making the cylinder. The properties of the tube get modified depending on the chiral angle θ and the diameter d.

The tubes are characterized by the (n, m) notation. Assume the construction of a tube, (4,2). Here the vector $\mathbf{C} = 4\mathbf{a}_1 + 2\mathbf{a}_2$. It is made by making four translations along the zigzag direction and two translations at 120° from the zigzag axis. There are numerous ways in which tubes can be rolled. The $(n,0)$ tubes are called *zigzag* where θ is zero. The (n, n) tubes are called *armchair* where θ is 30°. These two types of tubes have high symmetry and have a plane of symmetry perpendicular to the tube axis. Any other tube (n, m) is a chiral tube, which can be left- or right-handed. The tubes will be optically active to circularly polarized light, circulating along the tube axis. The two important tube parameters, d and θ, can be found from n and m:

$$d = \frac{C}{\pi} = \frac{\sqrt{3}r_{\text{C-C}}(m^2 + mn + n^2)^{1/2}}{\pi} \quad \text{and} \quad \theta = \tan^{-1}\frac{\sqrt{3}m}{(m + 2n)} \tag{2.21}$$

where rC–C is the C–C distance of the graphene layer (1.421 Å) and C is the length of the chiral vector. Due to the symmetry of the graphene layer, several tubes with different (n, m) notations are the same. A tube of $(0, n)$ is the same as $(n,0)$. The tube diameter will increase with larger n and m.

So far the discussion has focused on single nanotubes, termed *single-walled nanotubes* (SWNTs). However, the first experimentally observed tubes were multiwalled (i.e., several tubes are stacked one within the other). One observes only the sidewalls of the tube in a two-dimensional projection of the tube in a TEM image. In nanotube assemblies of this kind, there is no three-dimensional order between the graphite layers as in the case of bulk graphite. This is due to the rotational freedom that exists between the tubes, called *turbostratic constraint*. This lack of three-dimensional order within a MWNT has been found in atomically resolved STM measurements. In a given tube, it is not possible to fit any other tube, as no space may be available. For a tube to fit into another, there must be a gap of at least 3.44 Å between the layers. We can fit a (10,0) tube in a (19,0) tube, but not in a (18,0) tube. This is because in order to insert a 7.94 Å diameter tube, the larger-diameter tube has to have a diameter of 14.82 Å or larger [(7.94 + 2(3.44) Å]. The diameters of (19,0) and (18,0) are 15.09 and 14.29 Å, respectively.

Synthesis and Purification Carbon nanotubes were first noticed in the graphitic soot deposited on the negatively charged electrodes used in the arc-discharge synthesis of fullerenes. In the Kratschmer–Huffman procedure,[223] the graphite rods are evaporated in a dynamic atmosphere of helium (helium is leaked in while the vacuum system is pumped). Typically, a pressure of 130 torr of helium is used and the arc is run at 30 V dc and current is maintained at ~ 180 A. Increasing the helium pressure to 500 torr increases the MWNT yield.[226] The carbon deposited on the cathode has a soft inner core and a hard outer cover. The core containing MWNTs is extracted and suspended in suitable solvents. The tubes were observed as empty cylinders lying perpendicular to the electron beam along with amorphous carbon material. Mutiwalled tubes were observed and the interlayer gap was 0.34 Å, close to the spacing found in graphite. The very first images taken by Iijima[225] showed two continuous lines in the TEM. The lines were as shown in Figure 2.17 along with the cross section of the MWNTs proposed. The tube's inner diameter, interlayer spacing, length, and chiral angle θ can be determined from the TEM images. Although the first three are straightforward from a high-resolution image, determination of the chiral angle requires measurement of the interference pattern of the parallel planes and is often not done in routine TEM examination of nanotubes.

Nanotubes are found with closed ends on either side, although open tubes are also seen. Thus, these are three-dimensional closed-cage objects and may be considered as elongated fullerenes. There must be at least 12 pentagons to make a closed-cage structure, according to Euler's theorem, considering only pentagons and hexagons. The hexagons make the elongated body of the tube and the ends contain both hexagons and pentagons, six pentagons on each face being a minimum. However, the tube body and the ends can have defects. Whereas pentagons result in positive curvature, heptagonal defects make negative curvature. Both of these have been observed. The former makes a larger tube smaller and the latter can remove this curvature. Various types of end-tube morphologies have been found.

Various modifications to the arc-discharge process are reported in the literature for the synthesis of nanotubes. In the process, a smaller-diameter (typically, 3 mm) anode evaporates on the face of a larger-diameter (6 mm) cathode in a

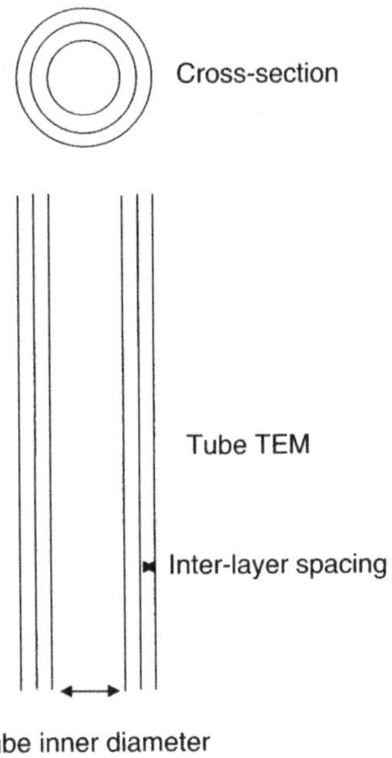

Fig. 2.17 Schematic of the observed multiwalled nanotube in a two-dimensional projection.

direct-current arc-discharge apparatus. The bowl that grows on the cathode contains multiwalled tubes. This can be broken, ground, and the nanotubes suspended in a suitable solvent and deposited on the TEM grids for examination.

Today, most extensive investigations are happening on SWNTs. Principally three methods are used for their synthesis.

1. Incorporating transition metals in catalytic amounts in the arc-discharge process results in the formation of SWNTs. The catalytic metal is added into the anode. The most common metals used are iron and nickel, but a mixture of transition metals is better. Several bimetallic systems (e.g., Co–Ni, Co–Pt, and Ni–Y) have been tried for this purpose. Weblike deposits are found around the cathode or the cooler regions of the reaction vessel. These materials contain significant quantities of single-walled nanotubes; they are seen in the form of ropes containing 5 to 100 individual SWNTs with amorphous carbon and nanoparticles of the metal/metal–carbon compounds. Optimized synthesis utilizes Ni–Y catalysts in a 4:1 atomic ratio, and several grams of SWNT-containing material can be prepared.[227,228]

2. Laser evaporation is another way to produce SWNTs in good yield.[229] By heating a mixture of graphite with Fe and Ni catalysts at a temperature of 1200° C and irradiating the material with a laser, it is possible to synthesize SWNTs. The yield of the nanotubes is about 50 to 70% of the product. Nanotubes thus synthesized are found to form ropes in which individual tubes are organized into

a hexagonal assembly. This very clearly shows the homogeneity of the tubes synthesized.

3. Chemical vapor deposition is another useful method for the synthesis of SWNTs and MWNTs.[230] Here, an organometallic precursor is mixed with a carbon-containing feed gas pyrolyzed in a quartz tube and the nanotubes are collected from the cooler end of the reaction vessel. The feed gas may contain several species and is often mixed with an inert gas. Nanotubes are also grown on solid catalytic substrates such as SiO_2 and quartz which contain transition metal precursors. Such approaches are important to make supported MWNT assemblies for specific applications. By feeding suitable precursor species it is possible to incorporate other atoms, such as nitrogen, into the nanotube structure, by substitution. It is also possible to change the morphology of the tubes by changing the precursors. Nanotube arrays can be produced using CVD on surfaces with patterning done by suitable catalysts.[231] One of the important aspects of CVD growth is the possibility of making nanotubes in batch processes. It is also possible that by using proper precursors, suitable properties such as elemental or ^{13}C doping can be achieved.

Both MWNTs and SWNTs are formed with significant quantities of carbonaceous material. One way to separate the tubes from the carbon mass is to heat-treat the product. Although all the carbon forms react with oxygen, they do so at different rates. All of the amorphous carbon materials can be burned off by heating the soot at 750° C for half an hour. By this process less than 1% of the original material is left, but the product is found to be essentially nanotubes. The presence of a large number of defects on amorphous carbon makes it react at a higher rate than nanotubes. There are acid-based cleaning procedures as well.

To make larger quantities of nanotubes, the HiPco (high-pressure CO disproportination) method is used.[232,233] This involves the high-pressure disproportination reaction of CO in the presence of catalysts. The feed gas contains organometallic precursors such as $Fe(CO)_5$, which is passed into an oven held at 1100° C. The carbonyl decomposes and produces metal atoms that form clusters. CO undergoes disproportination, $2CO \rightarrow C(SWNT) + CO_2$. About 97% of the material formed is nanotubes. Various methods are available to remove metal impurities from the SWNTs formed. This involve washing with HCl, liquid bromine, or other solvents.[234]

In a number of applications, it is important to have nanotubes oriented perpendicular to the surface. An approach that has received significant attention is the synthesis of aligned nanotube bundles on substrates.[235] Here a two-furnace approach is used along with metallocenes and organic precursors. Compact, aligned nanotube bundles could be obtained by introducing acetylene during the sublimation of ferrocene. Such assemblies grown on substrates, especially in a patterned fashion, can be important for uses such as field emission displays.

Using anionic, nonionic, and cationic surfactants, SWNTs can be suspended in solvents to various degrees. A common anionic surfactant is sodium dodecyl sulfate (SDS), which suspends nanotubes in water. An example of a cationic surfactant is cetyltrimethylammonium bromide (CTAB), which suspends SWNTs in organics. Typically, a 0.2% solution of CTAB can be used to suspend nanotubes.[236] There are several polymeric surfactants, such as poly(vinylpyrrolidone); depending on the nature of the surfactant, they can suspend nanoparticles in organic or inorganic media. An aqueous suspension is important for applications in biology. In a standard protocol, the nanotubes are suspended in dimethylformamide by repeated sonication. The dispersion is centrifuged at high speeds, typically in an ultracentrifuge at 60,000 to 100,000g. The free-standing nanotubes are used for investigations. Stabilization can be accomplished using surfactants. The solution is stable for extended periods.

2.13. SUMMARY

In this chapter we discussed methods of synthesizing nanofluids. A variety of nanosystems were discussed, including metal and semiconductor spherical nanoparticles, anisotropic nanostructures, fullerenes, and nanotubes. Chemical, physical, mechanical, and biological routes of synthesis were outlined. In the case of nanoparticles, principally solution chemistry approaches were discussed. The discussion focused on methods of making stable and redispersible nanoparticles. A concise summary of the various methods used for the synthesis of diverse nanosystems is presented. Methods for their characterization were outlined with specific examples in the entire range of particle sizes, from nanoparticles to clusters. The synthetic issues involved were discussed briefly. It is clear from the data presented that almost the entire periodic table could be converted to the nano form.

REFERENCES

1. K. S. Morley, P. C. Marr, P. B. Webb, A. R. Berry, F. J. Allison, G. Moldovan, P. D. Brown, and S. M. Howdle, Clean preparation of nanoparticulate metals in porous supports: a supercritical route, *J. Mater. Chem.*, 12 (2002), 1898–1905.
2. J. D. Swalen, D. L. Allara, J. D. Andrade, E. A. Chandross, S. Garoff, J. Israelachvili, T. J. McCarthy, R. Murray, R. F. Pease, J. F. Rabolt, K. J. Wynne, and H. Yu, Molecular monolayers and films, *Langmuir*, 3 (1987), 932–950.
3. http://www.nanovic.com.au/?a=education.history&p=30.
4. http://www.rigb.org/rimain/heritage/faradaypage.jsp.
5. N. Sandhyarani, M. R. Resmi, R. Unnikrishnan, et al., Monolayer-protected cluster superlattices: structural, spectroscopic, calorimetric, and conductivity studies, *Chem. Mater.*, 12 (2000), 104–113.
6. X. Wang, X. Xu, and S. U. S. Choi, Thermal conductivity of nanoparticle–fluid mixture, *J. Thermophys. Heat Transfer*, 13 (1999), 474.
7. Y. Xuan and Q. Li, Heat transfer enhancement of nanofluids, *Int. J. Heat Fluid Flow*, 21 (2000), 58–64.

8. N. Yao and Z. L. Wang (eds), *Handbook of Microscopy for Nanotechnology*, Kluwer Academic Publishers, Boston (2005).

9. K. L. Kelly, E. Coronado, L. L. Zhao, and G. C. Schatz, The optical properties of metal nanoparticles: the influence of size, shape, and dielectric environment, *J. Phys. Chem. B*, 107 (2003), 668–677.

10. S. Link and M. A. El-Sayed, Spectral properties and relaxation dynamics of surface plasmon electronic oscillations in gold and silver nanodots and nanorods, *J. Phys. Chem. B*, 103 (1999), 8410–8426.

11. D. Gammon, S. W. Brown, E. S. Snow, et al., Nuclear spectroscopy in single quantum dots: nanoscopic Raman scattering and nuclear magnetic resonance, *Science*, 277 (1997), 85–88.

12. N. Sandhyarani and T. Pradeep, Current understanding of the structure, phase transitions and dynamics of self-assembled monolayers on two- and three-dimensional surfaces, *Int. Rev. Phys. Chem.*, 22 (2003), 221–262.

13. H. Geliter, Nanocrystalline materials, *Prog. Mater. Sci.*, 33 (1989), 223–315.

14. S. Mahdihassan, Cinnabar-gold as the best alchemical drug of longevity, called *makaradhwaja* in India, *Am. J. Chin. Med.*, 13 (1985), 93.

15. J. Kunckels, *Nuetliche Observationes oder Anmerkungen von auro und argento Potabili*, Schutzens, Hamburg, 1676; G. Savage, *Glass and Glassware*, Octopus Books, London, 1975.

16. H. H. Helcher, *Aurum Potabile oder gold Tinstur*, J. Herbord Klossen, Breslau and Leipzig, 1718.

17. W. Ostwald, Zur Geschichte des colloiden Goldes, *Kolloid Z.*, 1909, 4, 5.

18. M. Faraday, Experimental relations in gold (and other metals) to light, *Philos. Trans. R. Soc. London*, 147 (1857), 145.

19. J. Turkevich, P. Stevenson, and J. Hillier, A study of the nucleation and growth processes in the synthesis of colloidal gold, *Discuss. Faraday Soc.*, 11 (1951), 55–75.

20. M. Brust, M. Walker, D. Bethel, D. J. Schriffin, and R. Whyman, Synthesis of thiol derivatised gold nanoparticles in a two-phase liquid/liquid system, *J. Chem. Soc. Chem. Commun*, 7 (1994), 801–802.

21. G. Schmid, R. Boese, R. Pfeil, F. Bandermann, S. Meyer, G. H. M. Calis, and J. W. A. van der Velden, $Au_{55}[P(C_6H_5)_3]_{12}Cl_6$ - ein Goldcluster ungewöhnlicher Größe $[Au_{55}[P(C_6H_5)3]_{12}Cl_6$:- a gold cluster of an exceptional size], *Chem. Ber.*, 114 (1981), 3634–3642.

22. P. A. Bartlett, B. Bauer, and S. Singer, Synthesis of water-soluble undecagold cluster compounds of potential importance in electron microscopic and other studies in biological systems, *J. Am. Chem. Soc.*, 100 (1978), 5085–5089.

23. T. Yonezawa, and T. Kunitake, Practical preparation of anionic mercapto ligand-stabilized gold nanoparticles and their immobilization, *Colloids Surf. A Physicochem. Eng. Aspects*, 149 (1999), 193–199.

24. M. A. Hayat (ed.), *Colloidal Gold: Principles, Methods and Applications*, Vols. 1 and 2 (1989) and Vol. 3 (1991), Academic Press, San Diego, CA.

25. M. C. Daniel and D. Astruc, Gold nanoparticles: assembly, supramolecular chemistry, quantum-size-related properties, and applications toward biology, catalysis, and nanotechnology, *Chem. Rev.*, 104 (2004), 293–346.

26. *Vogel's Textbook of Quantitative Inorganic Analysis*, ELBS edition. Longman, England (1991).
27. S. H. Wu and D. H. Chen, Synthesis and characterization of nickel nanoparticles by hydrazine reduction in ethylene glycol, *J. Colloid Interface Sci*, 259 (2003), 282–286.
28. B. A. Korgel, S. Fullam, S. Connolly, and D. Fitzmaurice, Assembly and self-organization of silver nanocrystal superlattices: ordered soft spheres, *J. Phys. Chem. B*, 102 (1998), 8379–8388.
29. G. B. Birrell, K. K. Hedberg, and O. H. Griffith, Pitfalls of immunogold labeling: analysis by light microscopy, transmission electron microscopy and photoelectron microscopy, *J. Histochem. Cytochem.*, 35 (1987), 843–853.
30. P. V. Kamat, M. Flumiani, and G. V. Hartland, Picosecond dynamics of silver nanoclusters: photoejection of electrons and fragmentation, *J. Phys. Chem. B*, 102 (1998), 3123–3128.
31. G. Frens, Controlled nucleation for the regulation of the particle size in monodisperse gold suspensions, *Nature*, 241 (1973), 20–22.
32. F. Bonet, S. Grugeon, R. H. Urbina, E. K. Tekaia, and J. M. Tarascon, In situ deposition of silver and palladium nanoparticles prepared by the polyol process, and their performance as catalytic converters of automobile exhaust gases, *Solid State Sci.*, 4 (2002), 665–670.
33. T. Teranishi, and M. Miyake, Size control of palladium nanoparticles and their crystal structures, *Chem. Mater.*, 10 (1998), 594–600.
34. A. Frattini, N. Pellegri, D. Nicastro, and O. de Sanctis, Effect of amine groups in the synthesis of Ag nanoparticles using aminosilanes, *Mater. Chem. Phys.*, 94 (2005), 148–152.
35. M. G. Dauge, J. S. Landers, H. L. Lewis, and J. L. Dye, Alkali-metal anions and trapped electrons formed by evaporating metal–ammonia solutions which contain cryptands, *Chem. Phys. Lett.*, 66 (1979), 169–182.
36. J. L. Dye, and K. L. Tsai, Small alloy particles formed by coreduction of soluble precursors with alkalides or electrides in aprotic solvents, *Faraday Discuss.*, 92 (1991), 45–55.
37. H. Bonnemann, R. A. Brand, W. Brijoux, H. W. Hofstadt, M. Frerichs, V. Kempter, W. Maus-Friedrichs, N. Matoussevitch, K. S. Nagabhushana, F. Voigts, and V. Caps, Air stable Fe and Fe–Co magnetic fluids: synthesis and characterization, *Appl. Organomet. Chem.*, 19 (2005), 790–796.
38. B. V. Enustun, and J. Turkevich, Coagulation of colloidal gold, *J. Am. Chem. Soc.*, 85 (1963), 3317–3328.
39. A. C. Templeton, W. P. Wuelfing, and R. W. Murray, Monolayer-protected cluster molecules, *Acc. Chem. Res.*, 33 (2000), 27–36.
40. K. V. Sarathy, G. U. Kulkarni, and C. N. R. Rao, A novel method of preparing thiol-derivatised nanoparticles of gold, platinum and silver forming superstructures, *Chem. Commun.* (1997), 537–538.
41. B. L. V. Prasad, S. I. Stoeva, C. M. Sorensen, and K. J. Klabunde, Digestive-ripening agents for gold nanoparticles: alternatives to thiols, *Chem. Mater.*, 15 (2003), 935–942.
42. A. Taleb, C. Petit, and M. P. Pileni, Optical properties of self-assembled 2D and 3D superlattices of silver nanoparticles, *J. Phys. Chem. B*, 102 (1998), 2214–2220.

43. A. C. Templeton, M. J. Hostetler, C. T. Kraft, and R. W. Murray, Reactivity of monolayer-protected gold cluster molecules: steric effects, *J. Am. Chem. Soc.*, 120 (1998), 1906–1911.

44. N. Sandhyarani, T. Pradeep, J. Chakrabarti, H. K. Sahu, and M. Yousuf, Distinct liquid phase in metal-cluster superlattice solids, *Phys. Rev. B*, 62 (2000), 739–742.

45. N. Sandhyarani, M. P. Antony, G. P. Selvam, and T. Pradeep, Melting of monolayer protected cluster superlattices, *J. Chem. Phys.*, 113 (2000), 9794–9803.

46. P. E. Laibinis, R. G. Nuzzo, and G. M. Whiteside, Structure of monolayers formed by coadsorption of two *n*-alkanethiols of different chain lengths on gold and its relation to wetting, *J. Phys. Chem.*, 96 (1992), 5097–5105.

47. A. Badia, S. Singh, L. Demers, L. Cuccia, G. R. Brown, and R. B. Lennox, Self-assembled monolayers on gold nanoparticles, *Chem. Eur. J.*, 2 (1996), 359–363.

48. A. C. Templeton, D. E. Cliffel, and R. W. Murray, Redox and fluorophore functionalization of water-soluble, tiopronin-protected gold clusters, *J. Am. Chem. Soc.*, 121 (1999), 7081–7089.

49. T. G. Schaaff, G. Knight, M. N. Shafigullin, R. F. Borkman, and R. L. Whetten, Isolation and selected properties of a 10. 4kDa gold: glutathione cluster compound, *J. Phys. Chem. B*, 102 (1998), 10643–10646.

50. S. H. Chen, and K. Kimura, Synthesis and characterization of carboxylate-modified gold nanoparticle powders dispersible in water, *Langmuir*, 15 (1999), 1075–1082.

51. K. Kimura, S. Sato, and H. Yao, Particle crystals of surface modified gold nanoparticles grown from water, *Chem. Lett*, 4 (2001), 372–373.

52. Y. Negishi, K. Nobusada, and T. Tsukuda, Glutathione-protected gold clusters revisited: bridging the gap between gold(I)-thiolate complexes and thiolate-protected gold nanocrystals, *J. Am. Chem. Soc.*, 127 (2005), 5261–5270.

53. M. T. Reetz and W. Helbig, Size-selective synthesis of nanostructured transition metal clusters, *J. Am. Chem. Soc.*, 116 (1994), 7401–7402.

54. L. R. Sánchez, M. C. Blanco, and M. A. L. Quintela, Electrochemical synthesis of silver nanoparticles, *J. Phys. Chem. B*, 104 (2000), 9683–9688.

55. M. B. Mohamed, Z. L. Wang, and M. A. El-Sayed, Temperature-dependent size-controlled nucleation and growth of gold nanoclusters, *J. Phys. Chem. A*, 103 (1999), 10255–10259.

56. J. Asenjo, R. Amigo, E. Krotenko, F. Torres, J. Tejada, E. Brillas, and G. Sardin, Electrochemical synthesis of nanoparticles of magnetic mixed oxides of Sr–Fe and Sr–Co–Fe, *J. Nanosci. Nanotechnol.*, 1 (2001), 441–449.

57. I. G. Draganic and Z. D. Draganic, *The Radiolysis of Water*, Academic Press, New York, 1971.

58. A. Henglein, Radiolytic preparation of ultrafine colloidal gold particles in aqueous solution: optical spectrum, controlled growth, and some chemical reactions, *Langmuir*, 15 (1999), 6738–6744.

59. S. Wang and H. Zin, Fractal and dendritic growth of metallic Ag aggregated from different kinds of γ-irradiated solutions, *J. Phys. Chem. B*, 104 (2000), 5681–5685.

60. A. Henglein, Formation and absorption spectrum of copper nanoparticles from the radiolytic reduction of $Cu(CN)_2-$, *J. Phys. Chem. B*, 104 (2000), 1206–1211.

61. B. G. Ershov, N. L. Sukhov, and E. Janata, Formation, absorption spectrum, and chemical reactions of nanosized colloidal cobalt in aqueous solution, *J. Phys. Chem. B*, 104 (2000), 6138.
62. J. H. Hodak, A. Henglein, and G. V. Hartland, Photophysics of nanometer sized metal particles: electron-phonon coupling and coherent excitation of breathing vibrational modes, *J. Phys. Chem. B*, 104 (2000), 9954–9965.
63. A. Henglein, Preparation and optical aborption spectra of $Au_{core}Pt_{shell}$ and $Pt_{core}Au_{shell}$ colloidal nanoparticles in aqueous solution, *J. Phys. Chem. B*, 104 (2000), 2201–2203.
64. P. Mulvaney, M. Giersig, and A. Henglein, Surface chemistry of colloidal gold: deposition of lead and accompanying optical effects, *J. Phys. Chem.*, 96 (1992), 10419–10424.
65. A. Henglein and D. Meisel, Radiolytic control of the size of colloidal gold nanoparticles, *Langmuir*, 14 (1998), 7392–7396.
66. Y. Kato, S. Sugimoto, K. Shinohara, N. Tezuka, T. Kagotani, and K. Inomata, Magnetic properties and microwave absorption properties of polymer-protected cobalt nanoparticles, *Mater. Trans.*, 43 (2002), 406–409.
67. T. W. Smith and D. Wychick, Colloidal iron dispersions prepared via the polymer-catalyzed decomposition of iron pentacarbonyl, *J. Phys. Chem.*, 84 (1980), 1621–1629.
68. S. W. Charles, S. Wells, and J. Villadsen, Formation and chemical stability of metallic glass particles prepared by thermolysis of $Fe(CO)_5$, *Hyperfine Interact.*, 27 (1986), 333–336.
69. J. van Wonterghem, S. Mørup, S. W. Charles, S. Wells, and J. Villadsen, Formation of a metallic glass by thermal decomposition of $Fe(CO)_5$, *Phys. Rev. Lett.*, 55 (1985), 410–413.
70. D. Dinega and M. G. Bawendi, A solution-phase chemical approach to a new crystal structure of cobalt, *Angew. Chem. Int. Ed. Engl.*, 38 (1999), 1788–1799.
71. V. F. Puntes, K. M. Krishnan, and A. P. Alivisatos, Colloidal nanocrystal shape and size control: the case of cobalt, *Science*, 291 (2001), 2115–2117.
72. V. F. Puntes, D. Zanchet, C. K. Erdonmez, and A. P. Alivisatos, Synthesis of hcp-Co nanodisks, *J. Am. Chem. Soc.*, 124 (2002), 12874–12880.
73. S. Sun, C. B. Murray, D. Weller, L. Folks, and A. Moser, Monodisperse FePt nanoparticles and ferromagnetic FePt nanocrystal superlattices, *Science*, 287 (2000), 1989–1992.
74. J.-I. Park, and J. Cheon, Synthesis of "solid solution" and "core-shell" type cobalt-platinum magnetic nanoparticles via transmetalation reactions, *J. Am. Chem. Soc.*, 123 (2001), 5743–5746.
75. C. Amiens, D. de Caro, B. Chaudret, J. S. Bradley, R. Mazel, and C. Roucau, Selective synthesis, characterization, and spectroscopic studies on a novel class of reduced platinum and palladium particles stabilized by carbonyl and phosphine ligands, *J. Am. Chem. Soc.*, 115 (1993), 11638–11639.
76. F. Dumestre, B. Chaudret, C. Amiens, M.-C. Fromen, M.-J. Casanove, P. Renaud, and P. Zurcher, Shape control of thermodynamically stable cobalt nanorods through organometallic chemistry, *Angew. Chem. Int. Ed.*, 41 (2002), 4286–4289.
77. K. J. Rao, B. Vaidhyanathan, M. Ganguli, and P. A. Ramakrishnan, Synthesis of inorganic solids using microwaves, *Chem. Mater.*, 11 (1999), 882–895.

78. I. S. Pastoriza and L. M. Liz-Marzán, Formation of PVP-protected metal nanoparticles in DMF, *Langmuir*, 18 (2002), 2888–2894.

79. F. Fiévet, J. P. Lagier, and M. Figlarz, Preparing monodisperse metal powders in micrometer and submicrometer sizes by the polyol process, *MRS Bull.*, 24 (1989), 29.

80. W. Yu, W. Tu, and H. Liu, Synthesis of nanoscale platinum colloids by microwave dielectric heating, *Langmuir*, 15 (1999), 6–9.

81. M. Tsuji, M. Hashimoto, and T. Tsuji, Fast preparation of nano-sized nickel particles under microwave irradiation without using catalyst for nucleation, *Chem. Lett.*, 31 (2002), 1232–1234.

82. W. Tu, and H. Liu, Continuous synthesis of colloidal metal nanoclusters by microwave irradiation, *Chem. Mater.*, 12 (2000), 564–567.

83. B. L. Cushing, V. L. Kolesnichenko, and C. J. O'Connor, Recent advances in the liquid-phase syntheses of inorganic nanoparticles, *Chem. Rev.*, 104 (2004), 3893–3946.

84. K. S. Suslick, S.-B. Choe, A. A. Cichowlas, and M. W. Grinstaff, Sonochemical synthesis of amorphous iron, *Nature*, 353 (1991), 414–417.

85. K. S. Suslick, M. Fang, and T. Hyeon, Sonochemical synthesis of iron colloids, *J. Am. Chem. Soc.*, 118 (1996), 11960–11961.

86. M. W. Grinstaff, M. B. Salmon, and K. S. Suslick, Magnetic properties of amorphous iron, *Phys. Rev. B*, 48 (1993), 269–273.

87. A. Gedanken, Using sonochemistry for the fabrication of nanomaterials, *Ultrason. Sonochem.*, 11 (2004), 47–55.

88. P. H. Borse, L. S. Kankate, F. Dassenoy, W. Vogel, J. Urban, and S. K. Kulkarni, Synthesis and investigations of rutile phase nanoparticles of TiO_2, *J. Mater. Sci. Mater. Electron.*, 13 (2002), 553–559.

89. Y. Gu, G. Z Li, G. Meng, and D. Peng, Sintering and electrical properties of coprecipitation prepared $Ce_{0.8}Y_{0.2}O_{1.9}$ ceramics, *Mater. Res. Bull.*, 35 (2000), 297–304.

90. P. C. Kuo and T. S. Tsai, New approaches to the synthesis of acicular alpha-FeOOH and cobalt modified iron-oxide nanoparticles, *J. Appl. Phys.*, 65 (1989), 4349–4356.

91. Z. X. Tang, C. M. Sorensen, K. J. Klabunde, and G. C. Hadjipanayis, Preparation of manganese ferrite fine particles from aqueous solution, *J. Colloid Interface Sci.*, 146 (1991), 38–54.

92. C. N. Chinnasamy, B. Jeyadevan, O. Perales-Perez, K. Shinoda, K. Tohji, and A. Kasuya, Growth dominant co-precipitation process to achieve high coercivity at room temperature in $CoFe_2O_4$ nanoparticles, *IEEE Trans. Magn.*, 38 (2002), 2640–2642.

93. T. C. Rojas and M. Ocana, Uniform nanoparticles of Pr(III)/ceria solid solutions prepared by homogeneous precipitation, *Scr. Mater.*, 46 (2002), 655–660.

94. J. Li, D. Dai, B. Zhao, Y. Lin, and C. Liu, Properties of ferrofluid nanoparticles prepared by coprecipitation and acid treatment, *J. Nanopart. Res.*, 4, (2002), 261–264.

95. K. T. Wu, P. C. Kuo, Y. D. Yao, and E. H. Tsai, Magnetic and optical properties of Fe_3O_4 nanoparticle ferrofluids prepared by coprecipitation technique, *IEEE Trans. Magn.*, 37 (2001), 2651–2653.

96. D. K. Kim, Y. Zhang, W. Voit, K. V. Rao, and M. Muhammed, Synthesis and characterization of surfactant-coated superparamagnetic iron oxide nanoparticles, *J. Mag. and Mag. Mater.*, 225 (2001), 30–36.

97. H. Chen, X. Qiu, W. Zhu, and P. Hagenmuller, Synthesis and high rate properties of nanoparticled lithium cobalt oxides as the cathode material for lithium-ion battery, *Electrochem. Commun.*, 4 (2002), 488–491.

98. P. Deb, T. Biswas, D. Sen, A. Basumallick, and S. Mazumder, Characteristics of Fe_2O_3 nanoparticles prepared by heat treatment of a nonaqueous powder precipitate, *J. Nanopart. Res.*, 4 (2002), 91–97.

99. S. O'Brien, L. Brus, and C. B. Murray, Synthesis of monodisperse nanoparticles of barium titanate: toward a generalized strategy of oxide nanoparticle synthesis, *J. Am. Chem. Soc.*, 123 (2001), 12085–12086.

100. D. Caruntu, Y. Remond, N. H. Chou, M.-J. Jun, G. Caruntu, J. He, G. Goloverda, C. O'Connor, and V. Kolesnichenko, Reactivity of 3d transition metal cations in diethylene glycol solutions: synthesis of transition metal ferrites with the structure of discrete nanoparticles complexed with long-chain carboxylate anions, *Inorg. Chem.*, 41 (2002), 6137–6146.

101. T. Trindade, P. O'Brien, and N. L. Pickett, Nanocrystalline semiconductors: synthesis, properties, and perspectives, *Chem. Mater.*, 13 (2001), 3843–3858.

102. C. D. Dushkin, S. Saita, K. Yoshie, and Y. Yamaguchi, The kinetics of growth of semiconductor nanocrystals in a hot amphiphile matrix, *Adv. Colloid Interface Sci.*, 88 (2000), 37–78.

103. C. Burda, X. B. Chen, R. Narayanan, and M. A. El-Sayed, Chemistry and properties of nanocrystals of different shapes, *Chem. Rev.*, 105 (2005), 1025–1102.

104. M. L. Steigerwald and C. R. Sprinkle, Application of phosphine tellurides to the preparation of group II–VI (2–16) semiconductor materials, *Organometallics*, 7 (1988), 245–246.

105. S. M. Stuczynski, J. G. Brennan, and M. L. Steigerwald, Formation of metal–chalcogen bonds by the reaction of metal-alkyls with silyl chalcogenides, *Inorg. Chem.*, 28 (1989), 4431–4432.

106. M. L. Steigerwald and L. E. Brus, Synthesis, stabilization and electronic structure of quantum semiconductor nanoclusters, *Ann. Rev. Mater. Sci.*, 19 (1989) 471–495.

107. J. G. Brennan, T. Siegrist, P. J. Carroll, S. M. Stuczynski, L. E. Brus, and M. L. Steigerwald, The preparation of large semiconductor clusters via the pyrolysis of a molecular precursor, *J. Am. Chem. Soc.*, 111 (1989), 4141–4143.

108. J. G. Brennan, T. Siegrist, P. J. Carroll, S. M. Stuczynski, P. Reynders, L. E. Brus, and M. L. Steigerwald, Bulk and nanostructure group II–VI compounds from molecular organometallic precursors, *Chem. Mater.*, 2 (1990), 403–409.

109. S. M. Stuczynski, R. L. Opila, P. Marsh, J. G. Brennan, and M. L. Steigerwald, Formation of indium phosphide from trimethylindium ($In(CH_3)_3$) and tris(trimethylsilyl) phosphine ($P(Si(CH_3)_3)_3$), *Chem. Mater.*, 3 (1991), 379–381.

110. C. B. Murray, D. J. Norris, and M. G. Bawendi, Synthesis and characterization of nearly monodisperse CdE (E = sulfur, selenium, tellurium) semiconductor nanocrystallites, *J. Am. Chem. Soc.*, 115 (1993), 8706–8715.

111. L. R. Becerra, C. B. Murray, R. G. Griffin, and M. G. Bawendi, Investigation of the surface-morphology of capped CdSe nanocrystallites by P-31 nuclear-magnetic resonance, *J. Chem. Phys.*, 100 (1994), 3297–3399.

112. C. B. Murray, C. R. Kagan, and M. G. Bawendi, Self-organisation of CdSe nanocrystallites into 3-dimensional quantum-dot superlattices, *Science*, 270 (1995), 1335–1338.

113. M. Danek, K. F. Jensen, C. B. Murray, and M. G. Bawendi, Synthesis of luminescent thin-film CdSe/ZnSe quantum dot composites using CdSe quantum dots passivated with an overlayer of ZnSe, *Chem. Mater.*, 8 (1996), 173–180.

114. A. A. Guzelian, U. Banin, A. V. Kadavanich, X. Peng, and A. P. Alivisatos, Colloidal chemical synthesis and characterization of InAs nanocrystal quantum dots, *Appl. Phys. Lett.*, 69 (1996), 1432–1434.

115. A. A. Guzelian, J. E. B. Katari, A. V. Kadavanich, U. Banin, K. Hamad, E. Juban, A. P. Alivisatos, R. H. Wolters, C. C. Arnold, and J. R. Heath, Synthesis of size-selected, Surface-Passivated InP nanocrystals, *J. Phys. Chem.*, 100 (1996), 7212–7219.

116. X. Peng, J. Wickham, and A. P. Alivisatos, Kinetics of II–VI and III–V colloidal semiconductor nanocrystal growth: "focusing" of size distributions, *J. Am. Chem. Soc.*, 120 (1998), 5343–5344.

117. X. Peng, L. Manna, W. Yang, J. Wickham, E. Scher, and A. Kadavanich, Shape control of CdSe nanocrystals, *Nature*, 404 (2000), 59–61.

118. L. Manna, E. C. Scher, and A. P. Alivisatos, Synthesis of soluble and processable rod-, arrow-, teardrop-, and tetrapod-shaped CdSe nanocrystals, *J. Am. Chem. Soc.*, 122 (2000), 12700–12706.

119. Z. A. Peng and X. Peng, Mechanisms of the shape evolution of CdSe nanocrystals, *J. Am. Chem. Soc.*, 123 (2001), 1389–1395.

120. Z. A. Peng and X. Peng, Formation of high-quality CdTe, CdSe, and CdS nanocrystals using CdO as precursor, *J. Am. Chem. Soc.*, 123 (2001), 183–184.

121. L. Qu, Z. A. Peng, and X. Peng, Alternative routes toward high quality CdSe nanocrystals, *Nano Lett.*, 1 (2001), 333–337.

122. O. Palchik, S. Avivi, D. Pinkert, and A. Gedanken, Preparation and characterization of Ni/NiO composite using microwave irradiation and sonication, *Nanostruct. Mater.*, 11 (1999), 415–420.

123. J. Zhu, O. Palchik, S. Chen, and A. Gedanken, Microwave assisted preparation of CdSe, PbSe, and Cu_2-xSe nanoparticles, *J. Phys. Chem. B*, 104 (2000), 7344–7347.

124. O. Palchik, J. Zhu, and A. Gedanken, Microwave assisted preparation of binary oxide nanoparticles, *J. Mater. Chem.*, 10 (2000), 1251–1254.

125. O. Palchik, R. Kerner, A. Gedanken, A. M. Weiss, M. A. Slifkin, and V. Palchik, Microwave-assisted polyol method for the preparation of CdSe "nanoballs," *J. Mater. Chem.*, 11 (2001), 874–878.

126. H. Grisaru, O. Palchik, A. Gedanken, V. Palchik, M. A. Slifkin, and A. M. Weiss, Preparation of the $Cd_1{-x}Zn_x$ Sealloys in the nanophase form using microwave irradiation, *J. Mater. Chem.*, 12 (2002), 339–344.

127. R. Kerner, O. Palchik, and A. Gedanken, Sonochemical and microwave-assisted preparations of PbTe and PbSe: a comparative study, *Chem. Mater.*, 13 (2001), 1413–1419.

128. X. Zheng, Y. Xie, L. Zhu, X. Jiang, and A. Yan, Formation of vesicle-templated CdSe hollow spheres in an ultrasound-induced anionic surfactant solution, *Ultrason. Sonochem.*, 9 (2002), 311–316.

129. R. E. Bailey, A. M. Smith, and S. Nie, Quantum dots in biology and medicine, *Physica E*, 25 (2004), 1–12.

130. K. Holmberg, Surfactant-templated nanomaterials synthesis, *J. Colloid Interface Sci.*, 274 (2004), 355–364.

131. C. J. O'Connor, J. A. Sims, A. Kumbhar, V. L. Kolesnichenko, W. L. Zhou, and J. A. Wiemann, Magnetic properties of $FePt_x$/Au and $CoPt_x$/Au core–shell nanoparticles, *J. Magn. Magn. Mater.*, 226–230 (2001), 1915–1917.

132. J. Lin, W. Zhou, A. Kumbhar, J. Wiemann, J. Fang, E. E. Carpenter, and C. J. O'Connor, Gold-coated iron (Fe–Au) nanoparticles: synthesis, characterization, and magnetic field-induced self-assembly, *J. Solid State Chem.*, 159 (2001), 26–31.

133. B. Ravel, E. E. Carpenter, and V. G. Harris, Oxidation of iron in iron/gold core/shell nanoparticles, *J. Appl. Phys.*, 91 (2002), 8195–8197.

134. E. E. Carpenter, A. Kumbhar, J. A. Wiemann, H. Srikanth, J. Wiggins, W. Zhou, and C. J. O'Connor, Synthesis and magnetic properties of gold–iron–gold nanocomposites, *Mater. Sci. Eng.* A, 286 (2000), 81–86.

135. J. Lin, W. Zhou, A. Kumbhar, J. Wiemann, J. Fang, E. E. Carpenter, and C. J. O'Connor, Gold-coated iron (Fe–Au) nanoparticles: synthesis, characterization, and magnetic field-induced self-assembly, *J. Solid State Chem.*, 159 (2001) 26–31.

136. C. J. O'Connor, C. T. Seip, E. E. Carpenter, S. Li, and V. T. John, Synthesis and reactivity of nanophase ferrites in reverse micellar solutions, *Nanostruct. Mater.*, 12 (1999), 65–70.

137. S. Santra, R. Tapec, N. Theodoropoulou, J. Dobson, A. Hebard, and W. Tan, Synthesis and characterization of silica-coated iron oxide nanoparticles in microemulsion: the effect of nonionic surfactants, *Langmuir*, 17 (2001), 2900–2906.

138. L. M. Liz-Marzán, M. Giersig, and P. Mulvaney, Synthesis of nanosized gold–silica core–shell particles, *Langmuir*, 12 (1996), 4329–4335.

139. W. Stöber, A. Fink, and E. Bohn, Controlled growth of monodisperse silica spheres in the micron size range, *J. Colloid Interface Sci.*, 26 (1968), 62–69.

140. S. I. Pastoriza, D. S. Koktysh, A. A. Mamedov, M. Giersig, N. A. Kotov, and L. M. Liz-Marzán, One-pot synthesis of Ag–TiO_2 core–shell nanoparticles and their layer-by-layer assembly, *Langmuir*, 16 (2000), 2731–273.

141. R. T. Tom, A. S. Nair, N. Singh, M. Aslam, C. L. Nagendra, R. Philip, K. Vijayamohanan, and T. Pradeep, Freely dispersible Au–TiO_2, Au–ZrO_2, Ag–TiO_2, and Ag–ZrO_2 core–shell nanoparticles: one-step synthesis, characterization, spectroscopy, and optical limiting properties, *Langmuir*, 19 (2003), 3439–3445.

142. A. S. Nair, T. Pradeep, and I. MacLaren, An investigation of the structure of stearate monolayers on Au–ZrO_2 and Ag–ZrO_2 core–shell nanoparticles, *J. Mater. Chem*, 14 (2004), 857–862.

143. L. M. Liz-Marzán, Tailoring surface plasmons through the morphology and assembly of metal nanoparticles, *Langmuir*, 22 (2006), 32–41.

144. S.-H. Yu, Hydrothermal/solvothermal processing of advanced ceramic materials, *J. Ceram. Soc. Jpn.*, 109 (2001), S65.

145. F. Cansell, B. Chevalier, A. Demourgues, J. Etourneau, C. Even, Y. Garrabos, V. Pessey, S. Petit, A. Tressaud, and F. Weill, Supercritical fluid processing: a new route for materials synthesis, *J. Mater. Chem.*, 9 (1999), 67–75.

146. M. Rajamathi and R. Seshadri, Oxide and chalcogenide nanoparticles from hydrothermal/solvothermal reactions, *Curr. Opin. Solid State Mater. Sci.*, 6 (2002), 337–345.

147. U. K. Gautam, M. Ghosh, M. Rajamathi, and R. Seshadri, Solvothermal routes to capped oxide and chalcogenide nanoparticles, *Pure Appl. Chem.*, 74 (2002), 1643–1649.

148. G. Demazeau, Solvothermal processes: a route to the stabilization of new materials, *J. Mater. Chem.*, 9 (1999), 15–18.

149. J. Li, Z. Chen, R.-J. Wang, and D. M. Proserpio, Low temperature route towards new materials: solvothermal synthesis of metal chalcogenides in ethylenediamine, *Coord. Chem. Rev.*, 190–192, (1999), 707–735.

150. M. Niederberger, G. Garnweitner, N. Pinna, and G. Neri, Non-aqueous routes to crystalline metal oxide nanoparticles: formation mechanisms and applications, *Prog. Solid State Chem.*, 33 (2005), 59–70.

151. C. N. R. Rao, V. V. Agrawal, K. Biswas, U. K. Gautam, M. Ghosh, A. Govindaraj, G. U. Kulkarni, K. R. Kalyanikutty, K. Sardar, and S. R. C. Vivekchandi, Soft chemical approaches to inorganic nanostructures, *Pure Appl. Chem.*, 78 (2006), 1619–1650.

152. B. L. Newalkar, S. Komarneni, and H. Katsuki, Microwave-hydrothermal synthesis and characterization of barium titanate powders, *Mater. Res. Bull.*, 36 (2001), 2347–2355.

153. T. Adschiri, Y. Hakuta, and K. Arai, Hydrothermal synthesis of metal oxide fine particles at supercritical conditions, *Ind. Eng. Chem. Res.*, 39 (2000), 4901–4907.

154. Y. Oguri, R. E. Riman, and H. K. Bowen, Processing of anatase prepared from hydrothermally treated alkoxy-derived hydrous titania, *J. Mater. Sci.*, 23 (1988), 2897–2904.

155. M. Kondo, K. Shinozaki, R. Ooki, and N. Mizutani, Crystallisation behaviour and microstructre of hydrothermally treated monodispersed titanium-dioxide particles, *J. Ceram. Soc. Jpn.*, 102 (1994), 742–746.

156. H. Cheng, J. Ma, Z. Zhao, and L. Qi, Hydrothermal preparation of uniform nanosize rutile and anatase particles, *Chem. Mater.*, 7 (1995), 663–671.

157. Z. Yanquing, S. Erwei, C. Zhizhan, L. Wenjun, and H. Xingfang, Influence of solution concentration on the hydrothermal preparation of titania crystallites, *J. Mater. Chem.*, 11 (2001), 1547–1551.

158. H. Yin, Y. Wada, T. Kitamura, T. Sumida, Y. Hasegawa, and S. Yanagida, Novel synthesis of phase-pure nano-particulate anatase and rutile TiO_2 using $TiCl_4$ aqueous solutions, *J. Mater. Chem.*, 12 (2002), 378–383.

159. M. Wu, J. Long, A. Huang, Y. Luo, S. Feng, and R. Xu, Microemulsion-mediated hydrothermal synthesis and characterization of nanosize rutile and anatase particles, *Langmuir*, 15 (1999), 8822–8825.

160. T. Masui, H. Hirai, R. Hamada, N. Imanaka, G. Adachi, T. Sakata, and H. Mori, Synthesis and characterization of cerium oxide nanoparticles coated with turbostratic boron nitride, *J. Mater. Chem.*, 13 (2003), 622–627.

161. M. Inoue, M. Kimura, and T. Inui, Transparent colloidal solution of 2nm ceria particles, *Chem. Commun.* (1999), 957–958.

162. S. Thimmaiah, M. Rajamathi, N. Singh, P. Bera, F. Meldrum, N. Chandrasekhar, and R. Seshadri, A solvothermal route to capped nanoparticles of α-Fe_2O_3 and $CoFe_2O_4$, *J. Mater. Chem.*, 11 (2001), 3215–3221.

163. S. Komarneni, M. C. D'Arrigo, C. Leionelli, G. C. Pellacani, and H. Katsuki, Microwave–hydrothermal synthesis of nanophase ferrites, *J. Am. Ceram. Soc.*, 81 (1998), 3041–3043.
164. H. Katsuki and S. Komarneni, Microwave–hydrothermal synthesis of monodispersed nanophase α-Fe_2O_3, *J. Am. Ceram. Soc.*, 84 (2001), 2313–2317.
165. M. D. Kadgaonkar, S. C. Laha, R. K. Pandey, P. Kumar, S. P. Mirajkar, and R. Kumar, Cesium containing MCM-41 materials as selective acylation and alkylation catalysts, *Catalysis Today*, 97 (2004) 225–231.
166. F. Schuth and W. Schmidt, Microporous and mesoporous materials, *Adv. Mater.*, 14 (2002), 629–638.
167. X. He and D. Antonelli, Recent advances in synthesis and applications of transition metal containing mesoporous molecular sieves, *Angew. Chem. Int. Ed.*, 41 (2002), 214–229.
168. L. M. Bronstein, Nanoparticles made in mesoporous solids, *Top. Curr. Chem.*, 226 (2003), 55–89.
169. W. S. Sheldrick and M. Wachold, Solventothermal synthesis of solid-state chalcogenidometalates, *Angew. Chem. Int. Ed.*, 36 (1997), 207–224.
170. Q. Peng, Y. Dong, Z. Deng, X. Sun, and Y. Li, Low-temperature elemental-direct-reaction route to II–VI semiconductor nanocrystalline ZnSe and CdSe, *Inorg. Chem.*, 40 (2001), 3840–3841.
171. U. K. Gautam, M. Rajamathi, F. Meldrum, P. Morgan, and R. Seshadri, A solvothermal route to capped CdSe nanoparticles, *Chem. Commun.* (2001), 629–630.
172. X. F. Qian, X. M. Zhang, C. Wang, W. Z. Wang, Y. Xie, and Y. T. Qian, Solvent-thermal preparation of nanocrystalline tin chalcogenide, *J. Phys. Chem. Solids*, 60 (1999), 415–417.
173. X. H. Chen and R. Fan, Low-temperature hydrothermal synthesis of transition metal dichalcogenides, *Chem. Mater.*, 13 (2001), 802–805.
174. G. O. Mallory, and J. B. Hajdu, *Electroless Plating: Fundamentals and Applications*, American Electroplaters and Surface Finishers Society, Orlando, FL, 1990.
175. M. P. Pileni, Fabrication and properties of nanosized material made by using colloidal assemblies as templates, *Cryst. Res. Technol.*, 33 (1998), 1155–1186.
176. N. Pradhan, N. R. Jana, K. Mallick, and T. J. Pal, *Surf. Sci. Technol.*, (2000), 188.
177. A. S. Nair, T. T. Renjis, V. Suryanarayanan, and T. Pradeep, ZrO_2 bubbles from core–shell nanoparticles, *J. Mater. Chem.*, 13 (2003), 297–300.
178. Z. Liang, A. Susha, and F. Caruso, Gold nanoparticle-based core-shell and hollow spheres and ordered assemblies thereof, *Chem. Mater.*, 15 (2003), 3176–3183.
179. R. W. J. Scott, O. M. Wilson, and R. M. Crooks, Synthesis, characterization, and applications of dendrimer-encapsulated nanoparticles, *J. Phys. Chem. B*, 109 (2005), 692–704.
180. K. R. Gopidas, J. K. Whitesell, and M. A. Fox, Nanoparticle-cored dendrimers: synthesis and characterization, *J. Am. Chem. Soc.*, 125 (2003), 6491–6502.
181. J. C. Hulteen, and C. R. Martin, in *Nanoparticles and Nanostructured Films: Preparation, Characterization and Applications*, J. H., Fendler, Ed., Wiley, New York, 1998.
182. C. A. Foss, Jr., in *Metal Nanoparticles: Synthesis, Characterization, and Applications*, D. L. Feldheim, and C. A. Foss, Jr., Eds., Marcel Dekker, New York, 2001.

183. M. Ichikawa, in *Metal Clusters in Chemistry*, P. Brunstein, L. A. Oro, and P. R. Raithby (eds.), Vol. 3, Wiley-VCH, Weinheim (1999), p. 1273.

184. J. P. Kuczynski, B. H. Milosavljevic, and J. K. Thomas, Photophysical properties of cadmium sulfide in Nafion film, *J. Phys. Chem.*, 88 (1984), 980–984.

185. J. C. Hoh and I. I. Yaacob, Polymer matrix templated synthesis: cobalt ferrite nanoparticles preparation, *J. Mater. Res.*, 17 (2002), 3105–3109.

186. G. L. Hornyak, C. J. Patrissi, and C. R. Martin, Fabrication, characterization, and optical properties of gold nanoparticle/porous alumina composites: the nonscattering Maxwell–Garnett limit, *J. Phys. Chem. B*, 101 (1997), 1548–1555.

187. C. Schonenberger, B. M. I. van der Zande, L. G. J. Fokkink, M. Henny, C. Schmid, M. Kruger, A. Bachtold, R. Huber, H. Birk, and U. Staufer, Template synthesis of nanowires in porous polycarbonate membranes: electrochemistry and morphology, *Phys. Chem. B*, 101 (1997), 5497–5505.

188. P. M. Ajayan, O. Stephan, P. Redlich, and C. Colliex, Carbon nanotubes as removable templates for metal-oxide nanocomposites and nanostructures, *Nature*, 375 (1995), 564–567.

189. B. C. Satishkumar, A. Govindaraj, E. M. Vogl, L. Basumallick, and C. N. R. Rao, Oxide nanotubes prepared using carbon nanotubes as templates, *J. Mater. Res.*, 12 (1997), 604–606.

190. C. N. R. Rao, F. L. Deepak, G. Gundiah, and A. Govindaraj, Inorganic nanowires, *Prog. Solid State Chem.*, 31 (2003), 5–147.

191. K. Simkiss and K. M. Wilbur, *Biomineralization*, Academic Press, New York, 1989.

192. R. P. Blakemore, D. Maratea, and R. S. Wolfe, Isolation and pure culture of a freshwater magnetic spirillum in chemically defined medium, *J. Bacteriol.*, 140(2) (1979), 720–729.

193. P. Mukherjee A. Ahmad, D. Mandal, S. Senapati, S. R. Sainkar, M. I. Khan, R. Ramani, R. Parischa, P. V. Ajaykumar, M. Alam, M. Sastry, and R. Kumar, Bioreduction of $AuCl_4-$ ions by the fungus Verticillium sp. and surface trapping of the gold nanoparticles formed, *Angew. Chem. Int. Ed.*, 40 (2001), 3585–3588.

194. B. Nair and T. Pradeep, Coalescence of nanoclusters and the formation of sub-micron crystallites assisted by Lactobacillus strains, *Cryst. Growth Des.*, 2 (2002), 293–298.

195. Anshup, J. S. Venkataraman, C. Subramaniam, R. R. Kumar, S. Priya, T. R. S. Kumar, R. V. Omkumar, A. John, and T. Pradeep, Growth of gold nanoparticles in human cells, *Langmuir*, 21 (2005), 11562–11567.

196. D. Pum and U. B. Sleytr, The application of bacterial S-layers in molecular nanotechnology, *Trends Biotechnol.*, 17 (1999) 8–12.

197. U. B. Sleytr, P. Messner, D. Pum, and M. Sara, Crystalline bacterial cell surface layers (S layers): from supramolecular cell structure to biomimetics and nanotechnology, *Angew. Chem. Int. Ed.*, 38 (1999), 1034–1054.

198. C. A. Mirkin, R. L. Letsinger, R. C. Mucic, and J. J. Storhoff, A DNA-based method for rationally assembling nanoparticles into macroscopic materials, *Nature*, 382 (1996), 607–608.

199. R. Bashir, DNA-mediated artificial nanobiostructures: state of the art and future directions, *Superlattices Microstruct.*, 29 (2001), 1–16.

200. J. J. Storhoff and C. A. Mirkin, Programmed materials synthesis with DNA, *Chem. Rev.*, 99 (1999), 1849–1862.

201. T. Douglas, in *Biomimetic Materials Chemistry*, S. Mann, Ed., Wiley, New York, 1996.

202. M. Han, C. R. Vestal, and Z. J. Zhang, Quantum couplings and magnetic properties of $CoCr_xFe_{2-x}O_4$ ($0 < x < 1$) spinel ferrite nanoparticles synthesized with reverse micelle method, *J. Phys. Chem. B*, 108 (2004), 583–587.

203. T. Hyeon, Chemical synthesis of magnetic nanoparticles, *Chem. Commun.* (2003), 927–934.

204. M. Green, Organometallic based strategies for metal nanocrystal synthesis, *Chem. Commun.* (2005), 3002–3011.

205. J. Zhang, M. Post, T. Veres, Z. J. Jakubek, J. Guan, D. Wang, F. Normandin, Y. Deslandes, and B. Simard, Laser-assisted synthesis of superparamagnetic Fe–Au core–shell nanoparticles, *J. Phys. Chem. B*, 110 (2006), 7122–7128.

206. S. H. Sun, Recent advances in chemical synthesis, self-assembly, and applications of FePt nanoparticles, *Adv. Mater.*, 18 (2006), 393–403.

207. M. Turker, Effect of production parameters on the structure and morphology of Ag nanopowders produced by inert gas condensation, *Mater. Sci. Eng. A*, 367 (2004), 74–81.

208. M. W. Swihart, Vapor-phase synthesis of nanoparticles, *Curr. Opinion Colloid Interface Sci.*, 8 (2003), 127–133.

209. F. E. Kruis, H. Fissan, and A. Peled, Synthesis of nanoparticles in the gas phase for electronic, optical and magnetic applications: a review, *J. Aerosol Sci.*, 29 (1998), 511–535.

210. H. Akoh, Y. Tsukasaki, S. Yatsuya, and A. Tasaki, Ferromagnetic ultrafine particles prepared by vacuum evaporation on running oil substrate, *J. Cryst. Growth*, 45 (1978), 495.

211. J. A. Eastman, S. U. S. Choi, S. Li, W. Yu, and L. J. Thompson, Anomalously increased effective thermal conductivities of ethylene glycol–based nanofluids containing copper nanoparticles, *Appl. Phys. Lett.*, 78 (2001), 718–720.

212. R. Prasher, P. Bhattacharya, and P. E. Phelan, Thermal conductivity of nanoscale colloidal solutions (nanofluids), *Phys. Rev. Lett.*, 94 (2005), art. 025901.

213. A. Ceylan, K. Jastrzembski, and S. I. Shah, Enhanced solubility Ag–Cu nanoparticles and their thermal transport properties, *Metall. Mater. Trans. A*, 37A (2006), 2033–2038.

214. F. Kim, J. H. Song, and P. Yang, Photochemical synthesis of gold nanorods, *J. Am. Chem. Soc.*, 124 (2002), 14316–14317.

215. C. J. Murphy, T. K. Sau, A. M. Gole, C. J. Orendorff, J. Gao, L. Gou, S. E. Hunyadi, and T. Li, Anisotropic metal nanoparticles: synthesis, assembly, and optical applications, *J. Phys. Chem. B*, 109 (2005), 13857–13870.

216. K. Aslan, Z. Leonenko, J. R. Lakowicz, and C. D. Geddes, Fast and slow deposition of silver nanorods on planar surfaces: application to metal-enhanced fluorescence, *J. Phys. Chem. B*, 109 (2005), 3157–3162.

217. H. Jia, W. Xu, J. An, D. Li, and B. Zhao, A simple method to synthesize triangular silver nanoparticles by light irradiation, *Spectrochim. Acta A*, 64 (2006), 956–960.

218. J. P. Juste, I. P. Santos, L. M. Liz-Marzán and P. Mulvaney, Gold nanorods: synthesis, characterization and applications, *Coord. Chem. Rev.*, 249 (2005) 1870–1901.

219. R. Jin, Y. W. Cao, C. A. Mirkin, K. L. Kelly, G. C. Schatz, and J. G. Zheng, Photoinduced conversion of silver nanospheres to nanoprisms, *Science*, 294 (2001), 1901–1903.

220. J. E. Millstone, S. Park, K. L. Shuford, L. Qin, G. C. Schatz, and C. A. Mirkin, Observation of a quadrupole plasmon mode for a colloidal solution of gold nanoprisms, *J. Am. Chem. Soc.*, 127 (2005), 5312–5313.

221. S. S. Sankar, A. Rai, B. Ankamwar, A. Singh, A. Ahmad, and M. Sastry, Biological synthesis of triangular gold nanoprisms, *Nature Mater.*, 3 (2004), 482–488.

222. H. W. Kroto, J. R. Heath, S. C. O'Brien, R. F. Curl, and R. E. Smalley, C_{60}: buckminsterfullerene, *Nature*, 318 (1985), 162–163.

223. W. Kratschmer, L. D. Lamb, K. Fostiropoulos, and D. R. Huffman, Solid C60: a new form of carbon, *Nature*, 347, (1990), 354–358.

224. C. J. Welch and W. H. Pirckle, Progress in the design of selectors for buckminsterfullerene, *J. Chromatogr.*, 609, (1992) 89–101.

225. S. Iijima, Helical microtubules of graphitic carbon, *Nature*, 354 (1991), 56–58.

226. T. W. Ebbesen and P. M. Ajayan, Large-scale synthesis of carbon nanotubes, *Nature*, 358 (1992), 220–222.

227. M. Takizawa, S. Bandow, M. Yudasaka, Y. Ando, H. Shimoyama, and S. Iijima, Change of tube diameter distribution of single-wall carbon nanotubes induced by changing the bimetallic ratio of Ni and Y catalysts, *Chem. Phys. Lett.*, 326 (2000), 351–357.

228. M. Yudasaka, Formation of single-wall carbon nanotubes catalyzed by Ni separating from Y in laser ablation or in arc discharge using a C target containing a NiY catalyst, *Chem. Phys. Lett.*, 312 (1999), 155–160.

229. A. G. Rinzler, J. H. Hafner, P. Nikolaev, L. Lou, S. G. Kim, D. Tomanek, P. Nordlander, D. T. Colbert, and R. E. Smalley, Unraveling nanotubes-field emission from an atomic wire, *Science*, 269 (1995), 1550–1553.

230. J. Kong, A. M. Cassel, and H. Dai, Chemical vapor deposition of methane for single-walled carbon nanotubes, *Chem. Phys. Lett.*, 292 (1998), 567–574.

231. H. Dai, Carbon nanotubes: synthesis, integration, and properties, *Acc. Chem. Res.*, 35 (2002), 1035–1044.

232. P. Nikolaev, M. J. Bronikowski, R. K. Bradley, F. Rohmund, D. T. Colbert, K. A. Smith, and R. E. Smalley, Gas-phase catalytic growth of single-walled carbon nanotubes from carbon monoxide, *Chem. Phys. Lett.*, 313 (1999), 91–97.

233. M. J. Bronikowski, P. A. Willis, D. T. Colbert, K. A. Smith, and R. E. Smalley, Gas-phase production of carbon single-walled nanotubes from carbon monoxide via the HiPco process: a parametric study, *J. Vac. Sci. Technol. A*, 19(4) (2001), 1800–1805.

234. I. W. Chiang, B. E. Brinson, A. Y. Huang, P. A. Willis, M. J. Bronikowski, J. L. Margrave, R. E. Smalley and R. H. Hauge, Purification and characterization of single-wall carbon nanotubes (SWNTs) obtained from the gas-phase decomposition of CO (HiPco process), *J. Phys. Chem. B*, 105, (2001), 8297–8301.

235. C. N. R. Rao, R. Sen, B. C. Satishkumar, and A. Govindaraj, Large aligned-nanotube bundles from ferrocene pyrolysis, *Chem. Commun.* (1998), 1525–1526.

236. V. C. Moore, M. S. Strano, E. H. Haroz, R. H. Hauge, and R. E. Smalley, Individually suspended single-walled carbon nanotubes in various surfactants, *Nano Lett.*, 3 (2003), 1379–1382.

3 Conduction Heat Transfer in Nanofluids

The property that created the most interest in nanofluids during the past decade was its thermal conductivity. Before discussing thermal conduction in nanofluids it will not be out of context to discuss some aspects of conduction heat transfer. The fundamentals of heat transfer can be found in standard textbooks, such as those by Incropera and DeWitt (1998) or Holman (1997), so no effort is made to repeat those elements in similar detail in this book. In this chapter, fundamental equations and useful correlations of conduction heat transfer are presented, which will be useful in subsequent sections. The derivation of these correlations is not included here; readers can turn to standard textbooks for them. Instead, the basic concepts behind these correlations, their physical significance, and their range of applicability are described. This approach will be relevant especially for interdisciplinary readers.

3.1. CONDUCTION HEAT TRANSFER

Usually, we talk about three modes of heat transfer: (1) conduction, (2) convection, and (3) radiation. Boiling and condensation are a combination of these modes. It has to be kept in mind that in reality, there is seldom a process with just one pure mode of heat transfer. However, one mode of heat transfer may be dominant enough that the other modes can be neglected.

In conduction, heat is transferred due to molecular vibration. Hence, it takes place even if the medium is at rest. The heat transfer mechanism for various types of media is different in conduction (Table 3.1).

The French scientist Joseph Fourier was the first person to develop a comprehensive theory for heat conduction, in his well-known book, *Théorie analytique de la chaleur* (1822). According to Fourier, the heat flux in the direction of heat flow is proportional to the temperature gradient in that direction. Let us consider heat flowing through a solid slab whose faces are maintained at different temperatures. *Fourier's law* for this slab can be written as

---

Nanofluids: Science and Technology, By Sarit K. Das, Stephen U. S. Choi, Wenhua Yu, and T. Pradeep

$$q_x = \frac{Q_x}{A} = -k\frac{dT}{dx} \tag{3.1}$$

q_x is the heat flux in the x direction, as shown in Fig. 3.1, Q_x is the heat flowing in the x direction, A is the area perpendicular to x through which it flows, and k is the constant of proportionality, known as the *thermal conductivity* of the medium. Since the thermal conductivity is a measure of the capability of a medium to conduct heat, it is always positive. Hence it is necessary to introduce a negative sign because the quantity dT/dx is negative (since the temperature decreases in the direction of heat flow). This ensures that the heat flux q_x is positive, thus indicating that heat is flowing from left to right (Fig. 3.1).

Conduction heat transfer is the most thoroughly understood phenomenon among the three modes of heat transfer, due to the fact that the medium remains at rest and the proportionality constant k of equation (3.1) is a property of the material independent of factors such as geometry. The unit of thermal conductivity is W/m·K in the SI system. Thermal conductivity depends on the material and the temperature. Figure 3.2 shows the range of thermal conductivity of different materials, and Fig. 3.3 shows the effect of temperature. There are two important observations relevant to nanofluids to be made in these two figures.

Table 3.1 Mechanism of Conduction in various Media

Medium	Carrier of Heat
Metallic solids and liquids	Free electrons
Nonmetallic solids and liquids	Electrons and phonons
Gases	Atoms and molecules

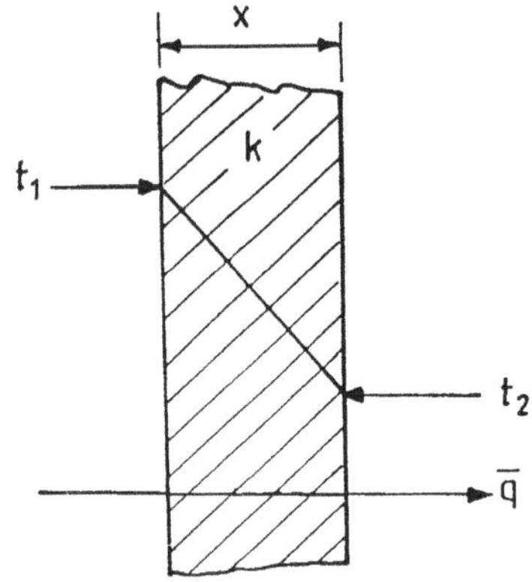

Fig. 3.1 Heat flow through a solid slab by conduction.

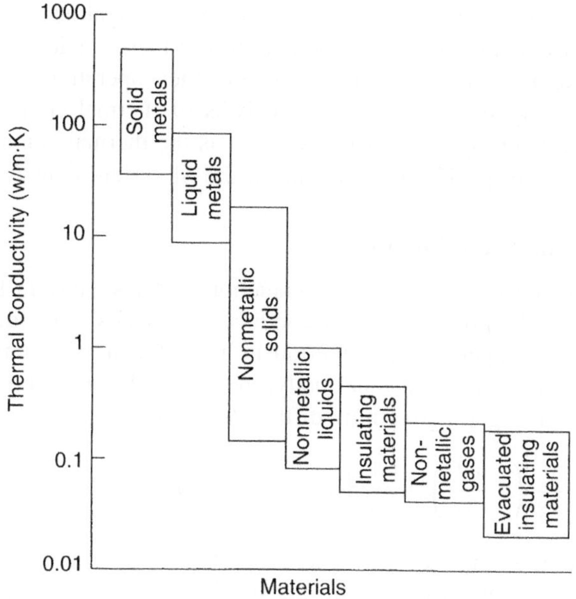

Fig. 3.2 Ranges of values of thermal conductivity of various materials.

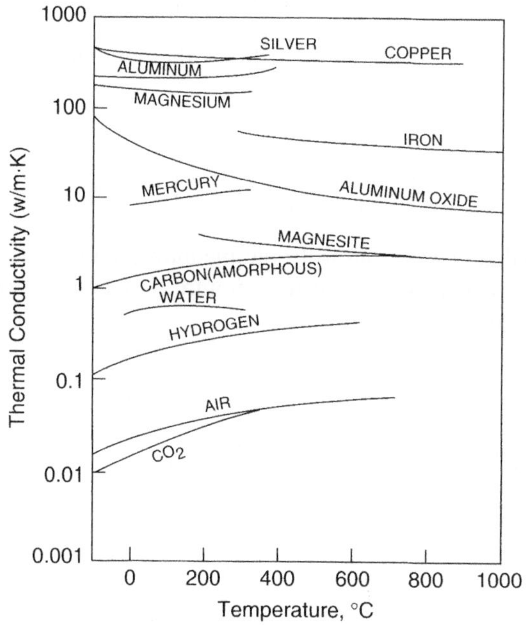

Fig. 3.3 Effect of temperature on thermal conductivity.

First, it is clear that the thermal conductivities of nonmetallic liquids (which are generally used as coolants) are one to three orders of magnitude lower than those of the solids. Second, the thermal conductivities of these liquids are very weakly dependent on temperature in the usual range of their operations.

Another quantity often used in the analysis of thermal conduction is *thermal diffusivity*, given by $\alpha = k/\rho C_p$ where k is the thermal conductivity, ρ the density, and C_p the specific heat. Its unit in the SI system is m^2/s.

3.1.1. Heat Conduction Equations

The quantity of heat transferred by conduction is measurable in terms of temperature. Hence, the primary task of conduction analysis is to frame equations that describe the temperature distribution in a conducting medium. In the most general case of a three-dimensional body, the energy balance and Fourier's law can be combined to get a governing differential equation in the form

$$\rho C_p \frac{\partial T}{\partial t} = \frac{\partial}{\partial x}\left(k_x \frac{\partial T}{\partial x}\right) + \frac{\partial}{\partial y}\left(k_y \frac{\partial T}{\partial y}\right) + \frac{\partial}{\partial z}\left(k_z \frac{\partial T}{\partial z}\right) + q'''(x, y, z) \quad (3.2)$$

where q''' is the volumetric heat generation. This is an extremely general case applicable to three-dimensional transient conduction in an anisotropic medium with internal heat generation. Anisotropic conductions are observed in fibrous materials such as wood or fiber-reinforced polymer (FRP). This can be of relevance to nanofluids with anisotropic material, such as carbon nanotubes (CNTs). However, some special cases are of practical importance:

1. *Isotropic medium*. If the medium has a thermal conductivity independent of direction, $k_x = k_y = k_z = k$
This results in

$$\rho C_p \frac{\partial T}{\partial t} = \frac{\partial}{\partial x}\left(k \frac{\partial T}{\partial x}\right) + \frac{\partial}{\partial y}\left(k \frac{\partial T}{\partial y}\right) + \frac{\partial}{\partial z}\left(k \frac{\partial T}{\partial z}\right) + q''' \quad (3.3)$$

It may be noted that even for an isotropic medium the thermal conductivity will not be constant—it may vary with temperature. For constant conductivity (the variation of k is small) we get

$$\frac{1}{\alpha}\frac{\partial T}{\partial t} = \left(\frac{\partial^2 T}{\partial x^2} + \frac{\partial^2 T}{\partial y^2} + \frac{\partial^2 T}{\partial z^2}\right) + \frac{q'''}{k} \quad (3.4)$$

2. *Steady state*. Under steady-state conditions the temporal variation disappears, giving

$$\frac{\partial^2 T}{\partial x^2} + \frac{\partial^2 T}{\partial y^2} + \frac{\partial^2 T}{\partial z^2} + \frac{q'''}{k} = 0 \quad (3.5)$$

3. *Simpler cases*. There are a number of simpler cases that are important from practical view point:

a. Two-dimensional,

$$\frac{1}{\alpha}\frac{\partial T}{\partial t} = \left(\frac{\partial^2 T}{\partial x^2} + \frac{\partial^2 T}{\partial y^2}\right) + \frac{q'''}{k} \tag{3.6}$$

b. Two-dimensional without heat generation,

$$\frac{1}{\alpha}\frac{\partial T}{\partial t} = \frac{\partial^2 T}{\partial x^2} + \frac{\partial^2 T}{\partial y^2} \tag{3.7}$$

c. Two-dimensional steady-state without heat generation,

$$\frac{\partial^2 T}{\partial x^2} + \frac{\partial^2 T}{\partial y^2} = 0 \tag{3.8}$$

d. One-dimensional,

$$\frac{1}{\alpha}\frac{\partial T}{\partial t} = \frac{\partial^2 T}{\partial x^2} + \frac{q'''}{k} \tag{3.9}$$

e. One-dimensional without heat generation,

$$\frac{1}{\alpha}\frac{\partial T}{\partial t} = \frac{\partial^2 T}{\partial x^2} \tag{3.10}$$

f. One-dimensional steady-state without heat generation,

$$\frac{\partial^2 T}{\partial x^2} = 0 \tag{3.11}$$

In the polar coordinate system, three-dimensional conduction with constant thermal conductivity is described by

$$\frac{1}{r}\frac{\partial}{\partial r}\left(r\frac{\partial T}{\partial r}\right) + \frac{1}{r^2}\left(\frac{\partial^2 T}{\partial \phi^2}\right) + \frac{\partial^2 T}{\partial z^2} + \frac{q'''}{k} = \frac{1}{\alpha}\frac{\partial T}{\partial t} \tag{3.12}$$

Here the axial, radial, and azimuthal coordinates are z, r, and ϕ, respectively. In two and one dimensions this equation reduces, respectively, to

$$\frac{1}{r}\frac{\partial}{\partial r}\left(r\frac{\partial T}{\partial r}\right) + \frac{q'''}{k} + \frac{1}{r^2}\frac{\partial^2 T}{\partial \phi^2} = \frac{1}{\alpha}\frac{\partial T}{\partial t} \tag{3.13}$$

$$\frac{1}{r}\frac{\partial}{\partial r}\left(r\frac{\partial T}{\partial r}\right) + \frac{q'''}{k} = \frac{1}{\alpha}\frac{\partial T}{\partial t} \tag{3.14}$$

It is important to note that even the one-dimensional conduction equation is important for heat transfer applications, particularly for measurement of thermal conductivity of liquids by the transient hot-wire method.

$$\frac{\partial}{\partial r}\left(r\frac{\partial T}{\partial r}\right) = 0 \tag{3.15}$$

3.1.2. Boundary Conditions

The different differential equations described above require boundary conditions for their solution. The boundary conditions used depend on the application and the physical features. Some commonly used boundary conditions are the following:

1. *Dirichlet boundary conditions.* When the boundary of the conducting medium is held at constant temperature, we put $T = T_0$ at $x = x_0$, where x_0 and T_0 are known values of the position of the boundary and its temperature, respectively.

2. *Neumann boundary conditions.* At the boundary of a medium, when the heat flux remains constant we have Neumann boundary conditions. Here

$$q = q_0 \qquad \text{at } x = x_0$$

or

$$-k\frac{\partial T}{\partial x} = q_0 \qquad \text{at } x = x_0 \tag{3.16}$$

where q_0 is the prescribed heat flux at the boundary.

3. *Convective boundary conditions.* In the more general convective boundary conditions that occur in most applications, a convective heat loss is taken by the fluid adjacent to the wall. Here

$$h(T - T_f) = -k\frac{\partial T}{\partial x} \qquad \text{at } x = x_0 \tag{3.17}$$

where T_f is the temperature of the surrounding fluid and h is the convective heat transfer coefficient to the fluid.

4. *Adiabatic boundary conditions.* These are special boundary conditions where the heat flux at the boundary (q_0) is zero. This gives

$$\frac{\partial T}{\partial x} = 0 \qquad \text{at } x = x_0 \tag{3.18}$$

This boundary condition occurs if the boundary of a medium is perfectly insulated, to eliminate heat loss.

3.1.3. Steady Conduction

Steady conduction is of importance, and some cases of practical relevance are given below. The solutions stated here are obtained by solving the appropriate form of differential equation subject to the relevant boundary conditions described in Section 3.1.2.

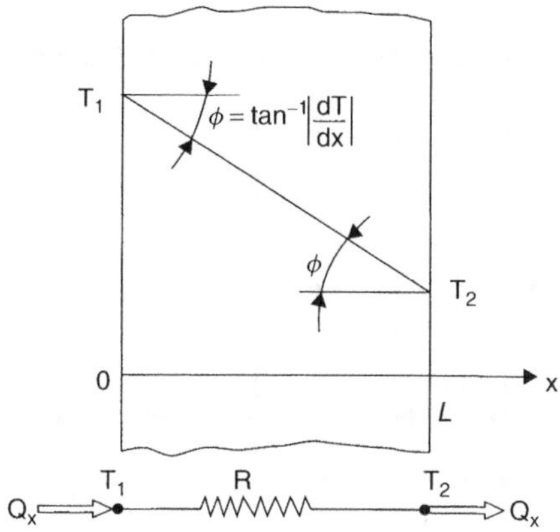

Fig. 3.4 Slab with one-dimensional conduction.

Slab of Uniform Thickness A slab of uniform thickness is shown in Fig. 3.4. When the two ends are held at constant temperatures, with end temperatures T_1 and T_2 (Fig. 3.4), the temperature profile is given by

$$\frac{T - T_1}{T_2 - T_1} = \frac{x}{L} \tag{3.19}$$

This essentially means that the temperature profile is linear inside the slab. One important observation from this result is that the temperature gradient, and hence the heat flux is constant at each cross section because the cross-sectional area of the slab remains unchanged. The heat transfer rate is given by

$$Q = -kA\frac{\partial T}{\partial x} = -kA\frac{T_2 - T_1}{L} = \frac{T_1 - T_2}{L/kA} \tag{3.20}$$

Thus, this equation brings out the analogy between heat flow and electric current where the heat flow is analogous to electric current, temperature difference, potential difference, and *L*/*kA* (the resistance). Thus, *L*/*kA* can be called the thermal resistance.

Composite Slab Composite slabs are often encountered in practice. The most realistic boundary conditions in such a case are the convective boundary conditions. Each slab may have a different material and different thickness, as shown in Fig. 3.5.

The temperature sketch indicates a different temperature gradient in each layer due to different values of thermal conductivity. The analysis of this problem is quite complex as the solution of a differential equation, since each layer is to be treated separately and the interfacial conditions, are not known. However, the problem is simplified using a thermal resistance approach. The equivalent

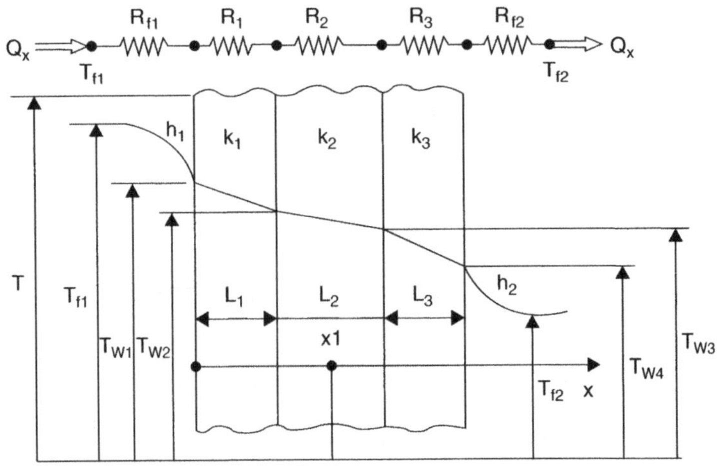

Fig. 3.5 Composite wall with convective boundary conditions on both sides.

electrical circuit is drawn in the figure. Since the same heat flux passes through all the layers, we can write

$$Q_x = \frac{T_{fi} - T_{w1}}{1/h_1 A} = \frac{T_{w1} - T_{w2}}{L_1/k_1 A} = \frac{T_{w2} - T_{w3}}{L_2/k_2 A} = \frac{T_{w3} - T_{w4}}{L_3/k_3 A} = \frac{T_{w4} - T_{f2}}{1/h_2 A} \tag{3.21}$$

Here the convective film resistance on each side is taken as $1/hA$, which we will describe later. Now, since intermediate temperatures are not known, the heat transfer can also be calculated by combining the thermal resistances in series and putting it in the form

$$Q_x = \frac{T_{f1} - T_{f2}}{R} \qquad \text{where} \quad R = \frac{1}{h_1 A} + \frac{L_1}{k_1 A} + \frac{L_2}{k_2 A} + \frac{L_3}{k_3 A} + \frac{1}{h_2 A} \tag{3.22}$$

Any intermediate temperature can also be evaluated by taking the resistance up to that point.

Hollow Cylinder For a hollow cylinder maintained at two constant temperatures at two end surfaces (Fig. 3.6), the solution is to be obtained from a cylindrical form of the conduction equation. It should be noted that the heat transfer area is not constant here and increases from the inner surface to the outer surface. As a consequence, the temperature distribution is also nonlinear, given by

$$\frac{T - T_1}{T_2 - T_1} = \frac{\ln (r/r_i)}{\ln (r_o/r_i)} \tag{3.23}$$

Hollow cylindrical walls are common in heat exchanger tubes, cylindrical reactors, and storage tanks. In this case, the thermal resistance can be calculated as

$$Q_r = \frac{T_1 - T_2}{R} \qquad \text{where} \quad R = \frac{\ln(r_o/r_i)}{2\pi k L} \tag{3.24}$$

L being the length of the cylinder.

Composite Cylinder Similar to the composite slab, composite hollow cylinders are also of immense importance. They can also be analyzed using the concept of thermal resistance. For the three-layer composite cylinder shown in Fig. 3.7, heat transfer rate may be given by

$$Q_r = \frac{T_{fi} - T_{fo}}{\frac{1}{2\pi L}\left[\frac{1}{h_i r_i} + \frac{\ln(r_1/r_i)}{k_1} + \frac{\ln(r_2/r_1)}{k_2} + \frac{\ln(r_o/r_2)}{k_3} + \frac{1}{h_o r_o}\right]} \tag{3.25}$$

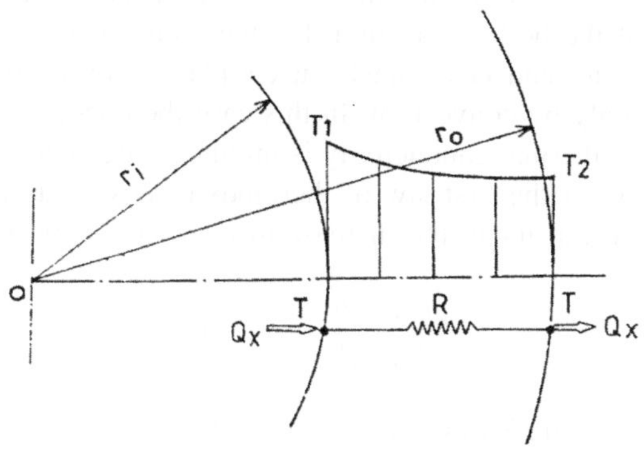

Fig. 3.6 Temperature distribution in a hollow cylinder.

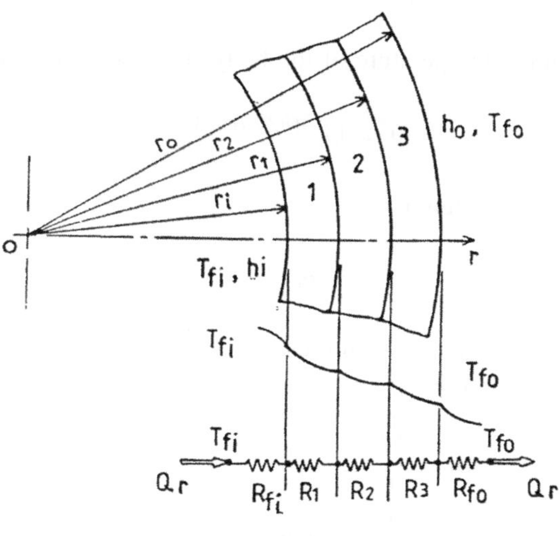

Fig. 3.7 Temperature distribution in a composite cylinder.

3.1.4. Transient Conduction

The cases discussed in Section 3.1.3 concern heat flow under steady conditions where temperatures vary along space coordinates. In many practical problems, the temperatures of a medium vary along both the space and time coordinates. These situations are designated as transient or unsteady conduction. Such cases are of immense practical importance. The majority of thermal conductivity measurements in nanofluids are carried out with a transient hotwire method, where the mode of conduction is transient in nature. It goes without saying that transient conduction is more complex than steady conduction. In the following section some simple cases of transient conduction are presented.

Lumped-Parameter Method The simplest type of transient analysis is lumped-parameter analysis, whose main assumption is that the temperature of a solid body is a function of time alone and not space. This essentially means that the conductivity of the body is so high that the entire body is at uniform temperature at any instant and heat transfer takes place between the body and the surrounding fluid only by convection. In this case the conduction equation is of no help because the thermal conductivity is infinitely large here. The law of conservation of energy (or the first law of thermodynamics) is used here to equate the convective heat transfer to the increase in internal energy of the body. This gives

$$\frac{\rho C_p V}{A_s h}\frac{d\theta}{dt} + \theta = 0 \tag{3.26}$$

where V/A_s is the volume/surface area ratio of the body. The condition required for the solution of this equation is the initial condition

$$\theta = \theta_i = T_i - T_\infty \qquad \text{at} \quad t = 0 \tag{3.27}$$

where T_i is the initial temperature of the body. This gives a solution

$$\theta = \theta_i e^{-(hA_s t/\rho C_p V)} \tag{3.28}$$

This solution can be written as

$$\theta = \theta_i \, e^{-Bi\cdot \text{Fo}} \tag{3.29}$$

where the two dimensionless quantities, the Biot number (Bi) and the Fourier number (Fo), are defined as

$$\text{Bi} = \frac{hL}{k}$$

$$\text{Fo} = \frac{\alpha t}{L^2} \qquad \text{where} \quad L = V/A_s$$

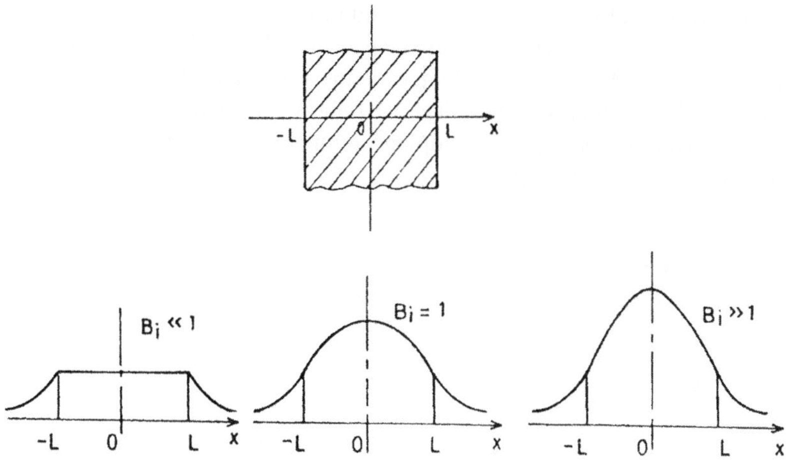

Fig. 3.8 Temperature distribution in a slap at several Biot numbers.

The Fourier number is dimensionless time and the Biot number is the ratio of conductive to convective resistance. The lower the Biot number, the lower the conductive resistance and the closer it is to the lumped-parameter assumption. Usually, a body can be treated as a lumped parameter if the Biot number is less than 0.1, which limits the error due to the lumped assumption within 5%. Figure 3.8 shows the temperature distribution inside a slab being cooled by fluid on both sides, showing the effect of the Biot number on temperature distribution.

One-Dimensional Transient Conduction In most applications the lumped-parameter assumptions are not valid, so analysis has to be carried out with temperature as a function of both space and time. The simple case of one-dimensional transient conduction can be used on geometries such as slab, cylinder, or sphere, which are important in practice. For a slab the one-dimensional transient equation is

$$\frac{1}{\alpha}\frac{\partial T}{\partial t} = \frac{\partial^2 T}{\partial x^2} \tag{3.30}$$

The solution of this equation with appropriate boundary conditions yields Fig. 3.8 Temperature distribution in a slab at several Biot numbers.

$$\frac{T - T_\infty}{T_i - T_\infty} = \sum_{x=1}^{\infty} \frac{2 \sin \lambda_x L \cos \lambda_x (x/L)}{\lambda_x L + \sin \lambda_x L \cos \lambda_x L} e^{-\lambda_x^2 L^2 Fo} \qquad \text{where} \qquad \cot \lambda_x L = \frac{\lambda_x L}{\text{Bi}} \tag{3.31}$$

Similar expressions can be obtained for cylinders and spheres, which instead of sinusoidal functions involve Bessel functions.

It is very difficult to use these complex equations for practical purposes. Hence, standard charts known as *Heisler charts* are available for these temperature distributions. Figure 3.9 shows the change of the midplane temperature of a slab with time and Biot number, and Fig. 3.10 shows the temperature distribution at various locations for a given midplane temperature. Similar charts are also available for cylinders and spheres.

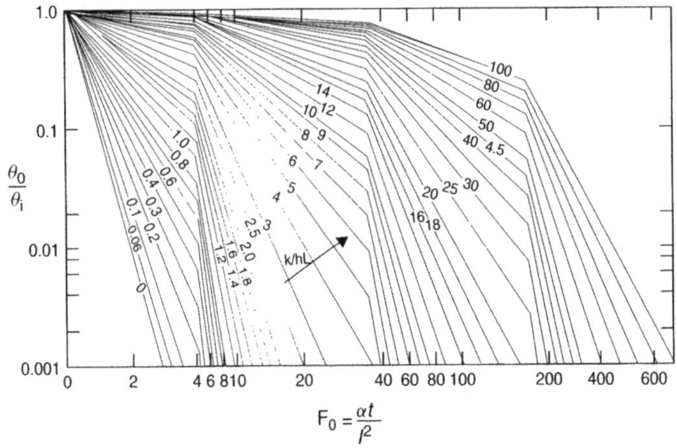

Fig. 3.9 Midplane temperature of a uniform slab.

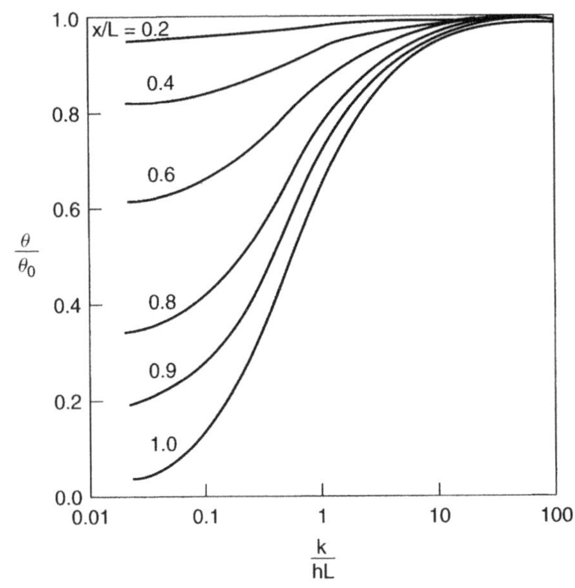

Fig. 3.10 Temperature distribution in a slab.

3.2. MEASUREMENT OF THERMAL CONDUCTIVITY OF LIQUIDS

In this section the basic principle of measurement of the thermal conductivity of liquids, design of apparatus and its fabrication, experimental procedure, and data reduction are discussed. There are two types of methods of measuring thermal conductivity of liquids: steady-state methods and transient methods. The disadvantages of steady-state methods are that heat lost cannot be quantified and may give considerable inaccuracy, and natural convection may set in, which gives higher apparent values of conductivity.

Therefore, to measure thermal conductivity accurately it is best to use transient methods. Some of the transient methods that have been described in the literature are discussed below.

3.2.1. Transient Hot-Wire Method

Principle of Measurement A method used widely to measure thermal conductivity is the transient hot-wire method. In this method, a thin metallic wire is used as both a line heat source and a temperature sensor. The wire is surrounded by the liquid whose thermal conductivity is to be measured. The wire is then heated by sending current through it. Now, the higher the thermal conductivity of the surrounding liquid, the lower will be the temperature rise of the wire. This principle is used to measure the thermal conductivity. The experiment lasts for a maximum of 2 to 8 seconds, hence is very fast, and in such a brief duration, natural convection cannot set in. Hence, in conjunction with advanced electronic data acquisition equipment, the method gives very accurate values of thermal conductivity. The mathematical model for this method is described below.

An infinitely long line heat source is suspended vertically in liquid whose thermal conductivity is to be measured. The method is called *transient* because heat is supplied suddenly, so that eventually the wire gets heated. The working equation is based on a specific solution of Fourier's law for radial (one-dimensional) transient heat conduction with a line heat source at the axis of the cylindrical domain.

The differential equation of conduction of heat in Cartesian coordinates is

$$\frac{\partial^2 T}{\partial x^2} + \frac{\partial^2 T}{\partial y^2} + \frac{\partial^2 T}{\partial z^2} = \frac{1}{\alpha}\frac{\partial T}{\partial t} \tag{3.32}$$

From the solution presented by Carslaw and Jaeger (1967), we can get the temperature distribution equation for a line heat source by integrating it over the entire length (z direction) of the line (i.e., from $-\infty$ to $+\infty$ in cylindrical coordinates):

$$T = \frac{Q'}{4\pi\alpha} \ln \frac{4\alpha t}{r^2} - \frac{\gamma Q'}{4\pi\alpha} \tag{3.33}$$

Equation (3.33) can be written as

$$T = \frac{q}{4\pi k} \ln \frac{4kt}{r^2 \rho c_p} - \frac{\gamma q}{4\pi k} \tag{3.34}$$

where, q is the heat liberated per unit time per unit length of the line source in W/m and k is the conductivity of the liquid in W/m·K. If temperatures of the heat source at time t_1 and t_2 are T_1 and T_2, respectively, then putting these conditions in equation (3.34) gives the conductivity of the liquid as

$$\frac{q}{4\pi(T_2 - T_1)} \ln \frac{t_2}{t_1} \tag{3.35}$$

Experimental Setup and Procedure A typical experimental setup for measuring thermal conductivity of nanofluids by the transient hot-wire method is shown in Fig. 3.11. A wire is placed along the axis of the cell, which will be surrounded by the liquid whose thermal conductivity is to be measured. As the wire is to be used both for heating and for temperature sensing, the wire material generally chosen is platinum. Platinum has high electrical resistivity [i.e., $1.06 \times 10^{-7} \Omega \cdot m$ (at 20°C)] an order of magnitude higher than that of other metals. Also, it has a temperature coefficient of resistance of 0.0003925°C^{-1} (for pure platinum), which is much higher than that of other metals. It is an ideal metal for temperature sensing because of its strictly linear variation of resistance over a large temperature range. The wire is to be used as a line heat source, so the wire diameter is usually kept within 100 μm. The length of the wire is kept to just a few centimeters, which compared to the wire diameter represents an infinitely long line heat source, assuring one directional (radial) heat transfer.

It will be helpful to describe the apparatus with a specific example used by Patel et al. (2003), shown in Fig. 3.11. The platinum wire is 15 cm in length. The surrounding fluid acts as a semi-infinite medium. For this, a cylindrical cavity

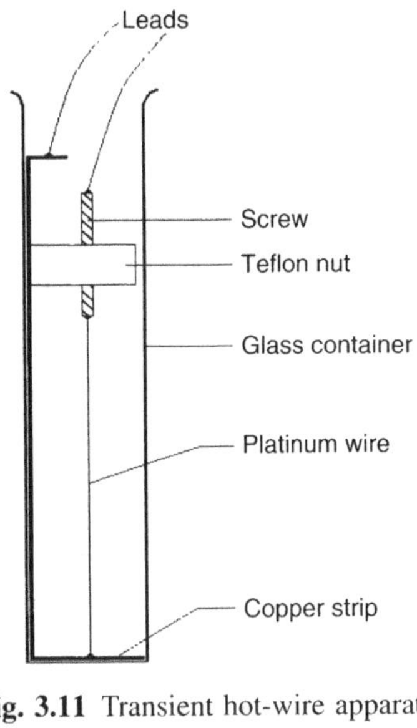

Fig. 3.11 Transient hot-wire apparatus.

2 cm in diameter and 15 cm in length is provided, which will be filled with fluid whose thermal conductivity is to be measured. Because the experiment lasts for just 4 to 5 seconds and total quantity of heat going into liquid per second is very small (maximum of 0.75 J), the diameter of the container provided is sufficient to assume the liquid to be a thermally semi-infinite medium. The wire is soldered to a copper screw at one end and a copper strip at the other end, as shown in the figure. Both the screw and the strip are thick enough to provide the least electrical resistance. As the wire has to work simultaneously as line-heating source and temperature sensor, it is made an arm of a Wheatstone bridge circuit (Fig. 3.12). The Wheatstone bridge is balanced before starting the experiment.

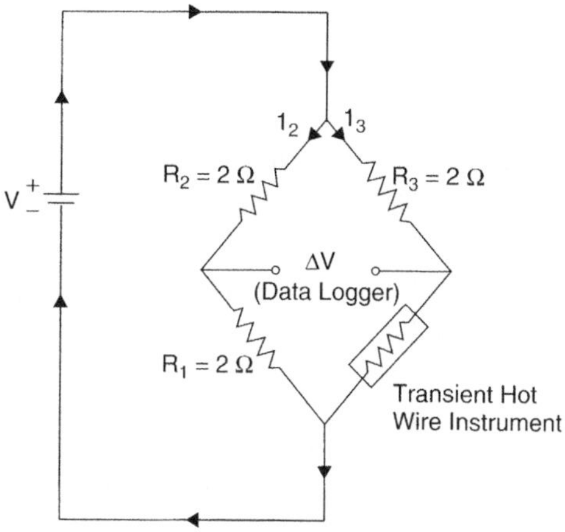

Fig. 3.12 Electrical circuit for the transient hot-wire apparatus.

Fig. 3.13 Transient hot-wire setup.

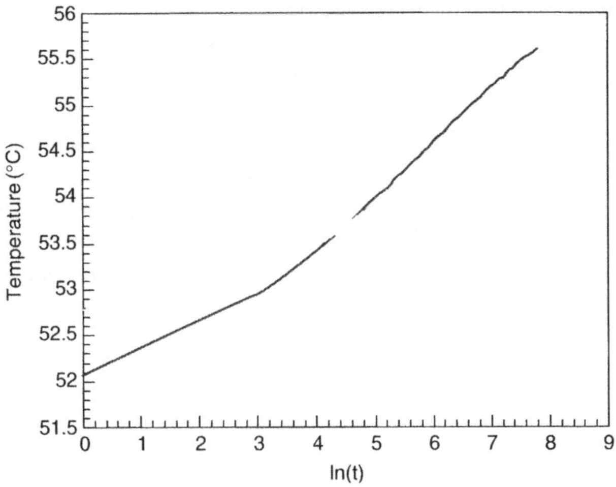

Fig. 3.14 Temperature measurement by the transient hot-wire method The steeper curve is for pure water, and the flatter curve is for water–alumina nanofluids.

A variable but stable voltage source is used to supply current. The electrical circuit is connected to the data acquisition system, which is connected to a computer. Figure 3.13 shows the transient hot-wire setup.

The electrical circuit is shown in Fig. 3.12. The constant-voltage source is connected to the grid lines. All the resistances are measured accurately. The platinum wire is heated by sending current through the Wheatstone bridge for 5 to 10 seconds. Data are acquired for voltage supplied by the voltage source as well as the voltage difference across the bridge at a small time interval (less than 40 ms). The measurement cell can be immersed in a constant-temperature bath to measure the thermal conductivity at high temperature. The natural log of the time (t) is plotted against the temperature of the wire, which is calculated from the data acquired. The initial portion of the graph, which is a straight line, is chosen and its slope $[\ln(t_2/t_1)/(T_2 - T_1)]$ is measured to calculate the thermal conductivity from the formula derived earlier [equation (3.35)]. An example of this slope variation for pure water and nanofluids is shown in Fig. 3.14. Typical accuracy of the transient hot-wire measurement for pure water is shown in Fig. 3.15.

3.2.2. Temperature Oscillation Method

Principle of Measurement The principle of measurement of thermal conductivity in this method is based on the propagation of a temperature oscillation inside a cylindrical liquid volume. The measurement of thermal diffusivity and thermal conductivity is based on the energy equation for conduction:

$$\frac{1}{\alpha}\frac{\partial T}{\partial t} = \nabla^2 T \tag{3.36}$$

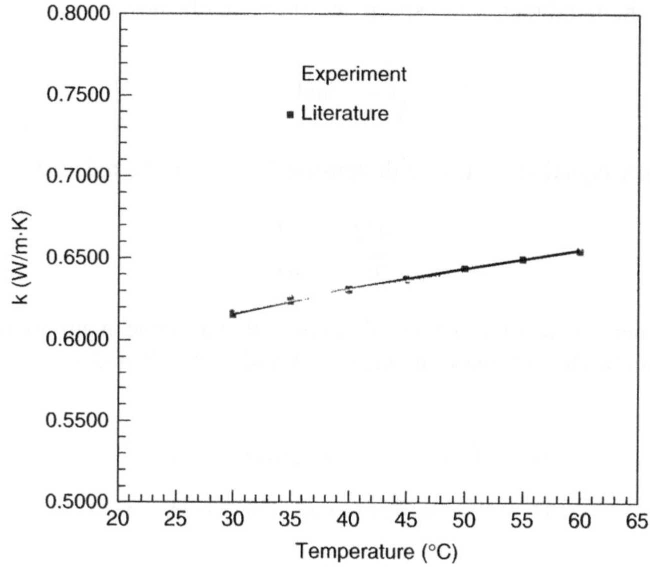

Fig. 3.15 Typical accuracy of transient hot-wire measurement.

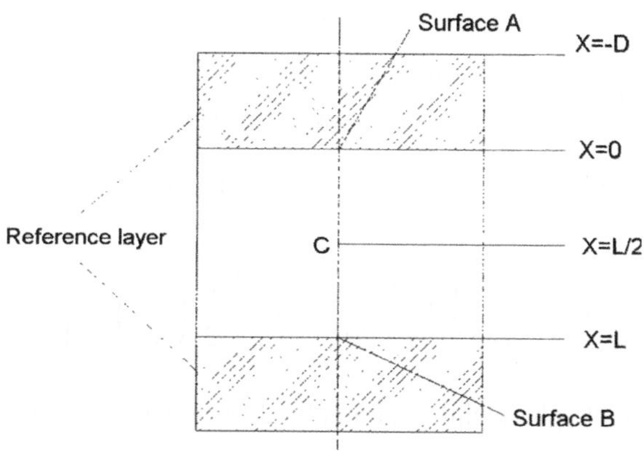

Fig. 3.16 Fluid volume for analysis.

In the present case this equation is applied with the assumption that the test fluid is isotropic and the thermophysical properties are uniform and constant with time throughout the entire specimen volume. The cylindrical fluid volume considered for analysis with its boundaries is shown in Fig. 3.16. At surfaces *A* and *B*, periodic temperature oscillations are generated with an angular velocity given by

$$\omega = \frac{2\pi}{t_p} \tag{3.37}$$

Thus, with the nondimensional space and time coordinates

$$\xi = x\sqrt{\frac{\omega}{\alpha}} \quad \text{and} \quad \tau = \omega t \tag{3.38}$$

the governing equation in its one-dimensional form can be reduced to

$$\frac{\partial^2 T}{\partial \xi^2} = \frac{\partial T}{\partial \tau} \tag{3.39}$$

For the general case of input oscillations with the same main frequency but different amplitude and phase at surfaces A and B, the boundary conditions are given by

$$T(\xi_0 = 0, \tau) = T_m + u_0 \cos(\tau + G_0) \tag{3.40}$$

$$T(\xi_L = L\sqrt{\omega/a}, \tau) = T_m + u_L \cos(\tau + G_L) \tag{3.41}$$

Under steady periodic conditions the solution of the differential equation (3.39) with boundary conditions given by equations (3.40) and (3.41) can be obtained by using the Laplace transform method. The solution can be written in complex form as

$$T(\xi, \tau) = T_m + \frac{u_L e^{iG_L} \sinh(\xi\sqrt{i}) - u_0 e^{iG_0} \sinh[\sqrt{i}\ (\xi_{-}\xi_L)]}{\sinh(\xi_L \sqrt{i})} e^{i\tau} \tag{3.42}$$

The complex amplitude ratio between the midpoint of the specimen and the surface can be given by

$$B^* = \frac{2u_L e^{iG_L}}{u_L e^{iG_L} + u_0 e^{iG_0}} \cosh\left[\frac{L}{2}\left(\frac{i\omega}{\alpha}\right)^{1/2}\right] \tag{3.43}$$

The real measurable phase shift and amplitude ratio can be expressed as

$$\Delta G = \arctan \frac{\mathrm{Im}(B^*)}{\mathrm{Re}(B^*)} \tag{3.44}$$

$$\frac{u_L}{u_{L/2}} = \sqrt{\mathrm{Re}(B_R^*)^2 + \mathrm{Im}(B_R^*)^2} \tag{3.45}$$

By measuring phase and amplitude of temperature oscillation at the two surfaces as well as at the center (point C), the thermal diffusivity can be determined from either equation (3.42) or (3.43).

To measure the thermal conductivity directly from experiment, we must consider the temperature oscillation in the reference layer at the two boundaries of the test fluid. The frequency of temperature oscillation in this layer is also the same as the frequency generated in the Peltier element (described later) and that in the test fluid. The one-dimensional heat conduction in the reference layer is given by

$$\frac{\partial^2 T_R}{\partial \zeta^2} = \frac{\partial T_R}{\partial \tau} \qquad \text{where} \quad \zeta = x\sqrt{\frac{\omega}{\alpha_R}} \tag{3.46}$$

The boundary conditions for the reference layer are

$$T_R(\zeta = 0, \tau) = T(\xi = 0, \tau) \tag{3.47}$$

The interface temperature balance

$$\lambda_R\sqrt{\frac{\omega}{\alpha_R}}\frac{\partial T_R}{\partial \zeta}|_{\zeta=0} = \lambda\sqrt{\frac{\omega}{\alpha}}\frac{\partial T}{\partial \xi}|_{\xi=0} \tag{3.48}$$

The solution of equation (3.46) along with boundary conditions (3.47) and (3.48) is given by

$$T_R^*(\zeta, \tau, \xi_L) = T_m + u_0 e^{i(\tau+G_0)}\cosh(\zeta\sqrt{i}) + C[u_R e^{i(\tau+G_R)} - u_0 e^{i(\tau+G_0)}\cosh(\xi_L\sqrt{i})]\frac{\sinh(\zeta\sqrt{i})}{\sinh(\xi_L\sqrt{i})} \qquad \text{where} \quad C = \frac{\lambda}{\lambda_R}\sqrt{\frac{\alpha_R}{\alpha}} \tag{3.49}$$

In this case the complex amplitude ratio between $x = -D$ (D being the thickness of reference layer) and $x = 0$ is given by

$$B_R^* = \cosh(\zeta_D\sqrt{i}) - C\sinh(\zeta_D\sqrt{i})\frac{(u_L/u_0)e^{i(G_L-G_0)} - \cosh(\xi_L\sqrt{i})}{\sinh(\xi_L\sqrt{i})} \tag{3.50}$$

The real phase shift and amplitude attenuation are given by

$$\Delta G_R = \arctan\frac{\text{Im}(B_R^*)}{\text{Re}(B_R^*)} \tag{3.51}$$

$$\frac{u_D}{u_0} = \sqrt{\text{Re}(B_R^*)^2 + \text{Im}(B_R^*)^2} \tag{3.52}$$

The thermal diffusivity of the test fluid has already been measured as described earlier and that of the reference layer being known, the thermal conductivity of the specimen can be evaluated as described above.

The temperature oscillation generated by the Peltier element must be strictly periodic. The shape of the oscillation is immaterial because any periodic oscillation can be expanded by a Fourier series in the form

$$T(\tau) = \frac{a_0}{2} + \sum_{k=1}^{\infty} A_k \sin(k\tau + \theta_k) \tag{3.53}$$

where

$$a_k = \frac{1}{\pi} \int_0^{2\pi} T(\tau)\cos(k\tau)\, d\tau, \qquad k = 0, 1, 2, 3, \ldots \tag{3.54}$$

$$b_k = \frac{1}{\pi} \int_0^{2\pi} T(\tau)\sin(k\tau)\, d\tau, \qquad k = 0, 1, 2, 3, \ldots \tag{3.55}$$

and

$$A_k = \sqrt{a_k^2 + a_k^2} \qquad \tan G_k = \frac{a_k}{b_k} \tag{3.56}$$

In the solution presented earlier for fluid as well as the reference layer, the fundamental oscillation is considered and the coefficients a_k and b_k are evaluated by numerical integration at the appropriate location to yield the corresponding amplitude attenuation and phase shift.

Experimental Setup and Procedure The experimental setup is shown schematically in Fig. 3.17. A temperature oscillation technique which is a modification of that used by Czarnetzky and Roetzel (1995) has been described here. This technique requires a specially fabricated test cell (1) which is cooled by cooling water (2) on both of the ends coming from a thermostatic bath (3). An electrical

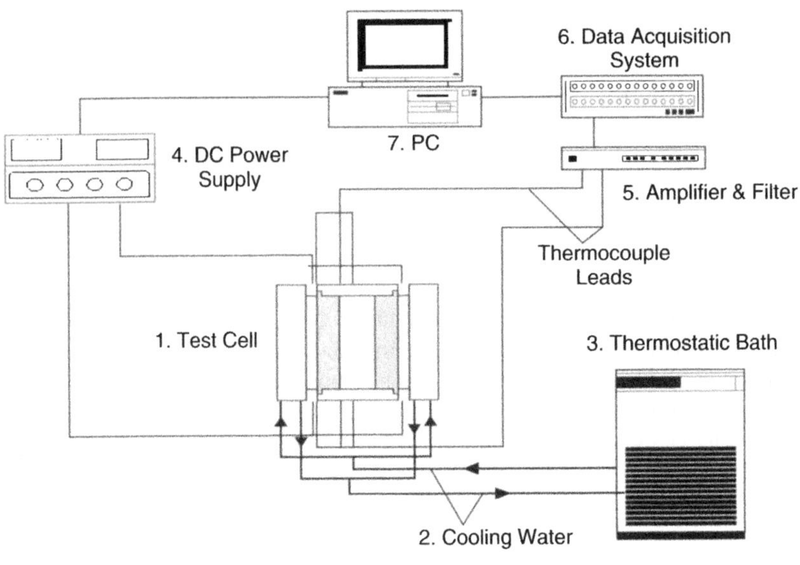

Fig. 3.17 Experimental setup.

connection provides dc power obtained through a converter (4) to the Peltier element. The temperatures are measured in the test section (discussed later) through a number of thermocouples, and the responses are amplified using an amplifier (5) followed by a filter, which is finally fed to a data acquisition system (6) comprising a card for logging the data measured. The data logger is, in turn, connected to a computer (7). Fluid temperature control is effected by proper adjustment of the cooling water from the thermostatic bath. However, for higher temperatures it is sometimes necessary to increase the input voltage to attain the required temperature level, which is then fine-tuned to the required temperature by controlling the cooling water temperature.

The test section is a flat cylindrical cell, shown in Fig. 3.18. The cell is mounted with its axis horizontal. The frame of the cell is made of poly (oxymethylene), which acts as the first layer of insulation. The frame consists of the main part with a 40-mm hole that acts as a cavity to hold the test fluid, and two end plates, which sandwich the water cooler and the Peltier element. The hole in the main frame is closed from both sides with disk type reference material 40 mm in diameter and 15 mm thick. The space formed for the test fluid is 40 mm in diameter and 8 mm thick. The fluid is filled through a small hole in the body of the cell. Temperatures are measured at three locations: at the interface of the Peltier element and the reference layer, at the interface of the reference layer and the test fluid, and at the central axial plane of the test fluid. For this purpose, Ni–CrNi thermocouples 0.1 mm in diameter were used at the interfaces and those 0.5 mm in diameter were used at the central plane, for stability considerations. The thermocouples at the interfaces are put in a small groove and welded at the tip, whereas the thermocouple at the center is hung from the wall.

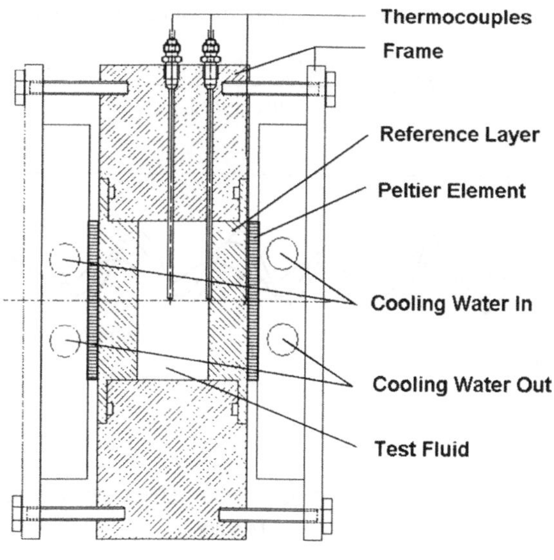

Fig. 3.18 Test cell construction.

Before placing the end reference plates, the central position of the thermocouple is ensured through a precision measurement and the entire cell is insulated further. The temperature of the reference material is oscillated periodically by two Peltier elements (40 mm x 40 mm square) from both ends. The temperature oscillations in this element are controlled to obtain two objectives:

1. The oscillation amplitude is adjusted to keep it small (on the order of 1.5 K) within the test fluid to retain constant fluid properties and both to avoid natural convection and not allow the amplitude to be decreased so much that the accuracy of the measurement would be affected. The Grashof number was calculated to be 850, which is below the asymptotic limit for the onset of natural convection. Measurement with pure water of known conductivity reconfirmed that no natural convection was present.
2. The smaller amplitude and accurate adjustment of the mean temperature of oscillation ensures that for the conducting fluid, the test is made at the temperature selected.

For example, a typical temperature oscillation recorded at the locations after steady oscillation are reached is shown in Fig. 3.19. It can be seen that the amplitude of the temperature oscillation produced by the Peltier element gets attenuated and its phase gets shifted as it crosses the reference material. It is further attenuated and shifted as it reaches the center of the test fluid. The theoretical principle presented earlier reveals that it is possible to evaluate the thermal diffusivity of the fluid very accurately by considering amplitude attenuating of thermal oscillation from the boundary (fluid reference material interface) to the center of the fluid. However, for direct measurement of thermal conductivity, one has to consider the attenuation at the reference material as well.

Also, the density can be measured and the specific heat calculated from a handbook:

$$C_{p,nf} = \frac{m_s C_{p,s} + m_w C_{p,w}}{m_s + m_w} \tag{3.57}$$

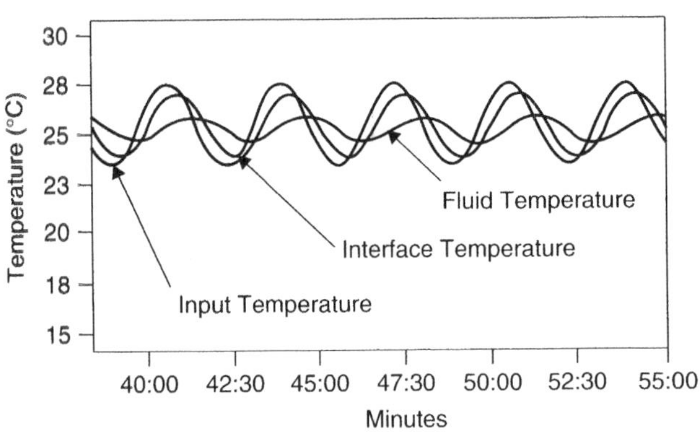

Fig. 3.19 Temperature oscillations recorded by Das et al. (2003).

Finally, the thermal conductivity can be calculated:

$$k_{nf} = \alpha_{nf} \rho_{nf} C_{p,nf} \tag{3.58}$$

The maximum error in the conductivity measurement was limited to 7% up to 50°C.

3.3. THERMAL CONDUCTIVITY OF OXIDE NANOFLUIDS

The fact that the thermal conductivity of the suspensions is higher than that of the base fluid is nothing novel. In the nineteenth century, Maxwell (1881) proposed a model for thermal conductivity of suspensions which clearly indicated the higher value of thermal conductivity. This is primarily due to the fact that the solids have orders-of-magnitude higher thermal conductivity than that of liquids. With the exception of liquid metals (which cannot be used in most cooling applications due to their high melting points), the liquids have poor thermal conductivity (e.g., water has a thermal conductivity of 0.6 W/m·K which is lower than that of solid oxides, which have thermal conductivities of order 20 to 50 W/m·K). However, pure metals have much higher thermal conductivities (e.g., Cu ≈ 400 W/m·K). However, although these higher conductivities are attractive from the viewpoint of cooling capabilities, other problems associated with suspensions, such as sedimentation, clogging, fouling, erosion, and excessive pressure drop, make them unsuitable for cooling applications (discussed in Chapter 1).

When nanofluids were invented by Choi and his group at the Argonne National Laboratory, they first tried to use oxide particles of nanometer size to suspend in the common coolants (e.g., water, ethylene glycol). Oxides were tried mainly for ease of manufacture and stabilization compared to pure metallic particles, which are difficult to suspend without agglomeration. Subsequently, many investigators carried out experiments with oxide particles, predominantly Al_2O_3 particles, as well as CuO, TiO_2, and stable compounds such as SiC. In this section we discuss the results of these experiments and their general trends. Although the invention at Argonne was disseminated at a number of conferences (Choi, 1995; Eastman et al., 1997), the major archival publication came in 1999 (Lee et al., 1999), in which Choi and his co-workers described using the transient hot-wire method to measure the thermal conductivity of Al_2O_3 and CuO nanoparticles suspended in water and ethylene glycol. First, they found that the enhancement of thermal conductivity is linear, as shown in Fig. 3.20. They chose the modified Maxwell theory of Hamilton and Crosser (1962) as the basis for comparing the experimental results. This theory gives the enhancement of thermal conductivity of suspension in the form

$$k_{\text{eff}} = k_0 \frac{k_p + (n-1)\, k_0 - (n-1)\, \varepsilon\, (k_0 - k_p)}{k_p + (n-1)\, k_0 + \varepsilon\, (k_0 - k_p)} \tag{3.59}$$

where k_{eff} is the effective conductivity of the nanofluid, k_p the particle conductivity, k_0 the base fluid conductivity, ε the particle volume fraction, and n the particle shape factor:

$$n = \frac{3}{\psi} \tag{3.60}$$

where ψ is the sphericity of the particles. Based on this model, they compared their experiments, which show good agreement with the model for Al_2O_3 nanofluids in both water and ethylene glycol (Figs. 3.21 and 3.22).

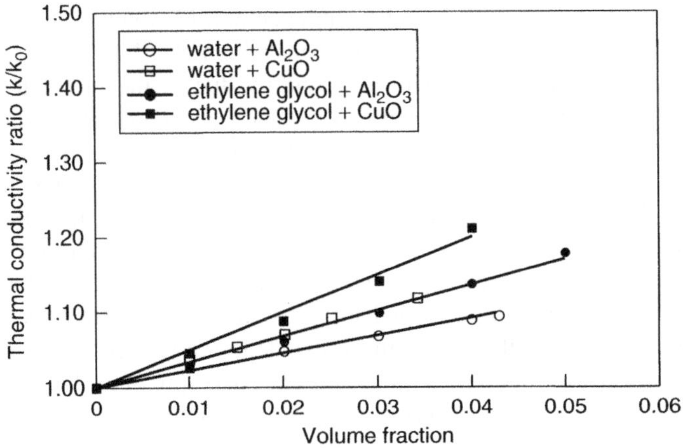

Fig. 3.20 Enhanced thermal conductivity of oxide nanofluids (Lee et al., 1999).

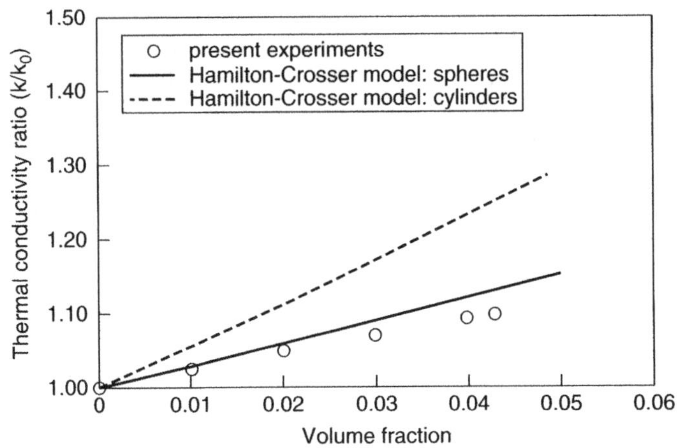

Fig. 3.21 Comparison of Al_2O_3–water nanofluid conductivity using Hamilton–Crosser theory.

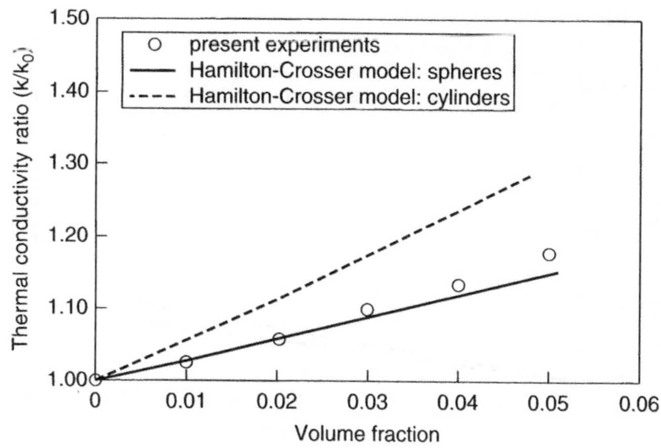

Fig. 3.22 Comparison of Al_2O_3–ethylene glycol nanofluid conductivity using Hamilton–Crosser theory.

When they conducted the same experiments with CuO-particle-based nanofluids, the results were astonishing. The thermal conductivity of both water and ethylene glycol nanofluids were much higher than predicated by Hamilton—Crosser theory (Figs. 3.23 and 3.24). This result is surprising because of its sharp contrast with the Al_2O_3 nanofluid results. One has to keep in mind that although the thermal conducivity of Al_2O_3 and CuO do not differ widely, the average particle size used for Al_2O_3 was 38 nm whereas that for CuO was only 24 nm, which introduces the possibility of a nanoparticle size effect. This apprehension deepens when comparing their Al_2O_3 nanofluid results with those of Masuda

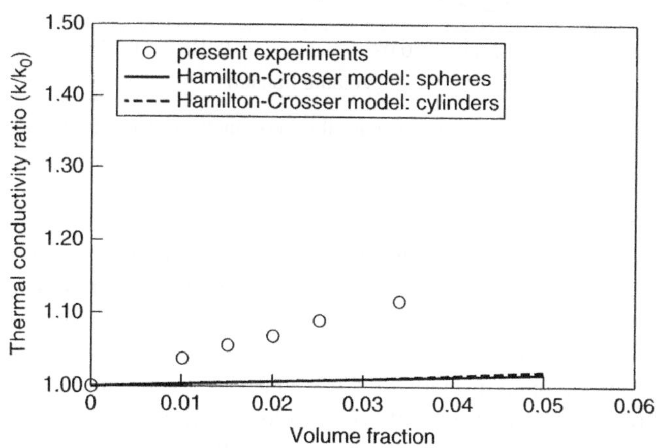

Fig. 3.23 Comparison of CuO–water nanofluid conductivity using Hamilton–Crosser theory.

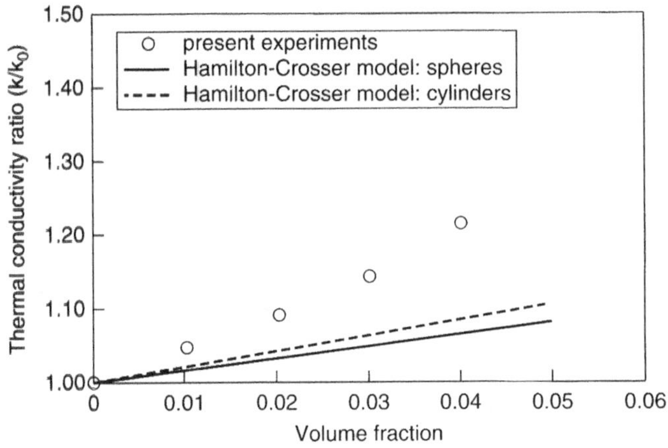

Fig. 3.24 Comparison of CuO–ethylene glycol nanofluid conductivity using Hamilton–Crosser theory.

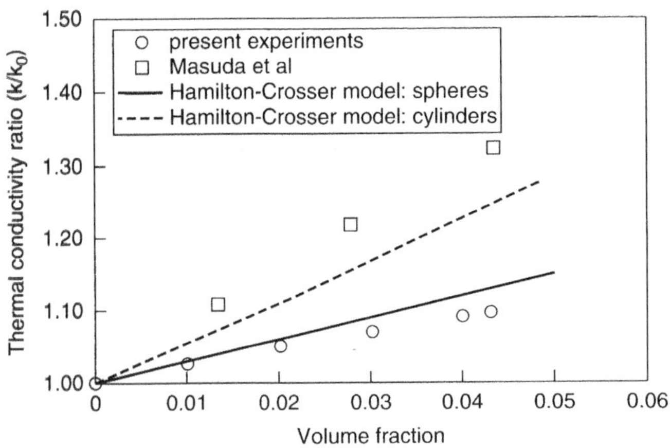

Fig. 3.25 Comparison of Al_2O_3–water nanofluid conductivity from Lee et al. (1999) and Masuda et al. (1993).

et al. (1993), in which Al_2O_3 nanofluid with 13-nm particles showed a much higher thermal conductivity than when using the Hamilton–Crosser model, (Fig. 3.25). These results seem to indicate that the nanoparticle size effect may be the reason for the abnormal increase in the thermal conductivity of nanofluids.

After the initial breakthrough of Choi and his group, summed up in their work described above, a stream of work followed, much of which used Al_2O_3 nanoparticles, due both to its availability and to the urge to verify and confirm the data of Lee et al. (1999). These publications also dealt with other aspects of thermal conductivity enhancement. Work by Xie et al. (2002b) is interesting in this respect. They used α and γ alumina of various sizes (between 12 and 30.2 nm) and focused on two other factors: the pH value of the suspension and the specific surface area (SSA). The pH value is important because of the isoelectric point (IEP). At the pH value of the isoelectric point, the repulsive forces between the particles are reduced to zero, which increases the possibility of agglomeration. On the other hand, hydration forces among the particles increase with increasing difference in pH from the value at the isolelectric point (pH_{IEP}). This gives greater mobility to the nanoparticles and increases the thermal transport capability. With the decreasing size of the particles, the surface area of the particles per unit volume increases:

$$\mathrm{SSA} = \frac{\text{particle surface area}}{\text{particle volume}} \tag{3.61}$$

$$\frac{\pi\, d_p^2}{\pi/6\, d_p^3} = \frac{6}{d_p} \tag{3.62}$$

This clearly indicates that a decrease in particle diameter (d_p) causes the SSA to increase. Table 3.2 gives the variation in SSA for the particles used by Xie et al. ((2002)b).

Thus, with a decrease in particle size, the surface area of nanoparticles increases, giving more heat transfer area between the phases, which may be helpful in increasing the thermal transport. The results presented by Xie et al. (2002b) for thermal conductivity enhancement against pH value are shown in Fig. 3.26. The figure clearly indicates that the enhancement increases with particle volume fraction (ε) and decreases with increasing pH. For these particles, the pH_{IEP} is 9.2, and hence the suspension in most unstable at this value of pH, which can be the primarily reason for decreased conductivity enhancement at higher pH values. With respect to SSA, they found the nonmonotonic behavior of enhancement shown in Fig. 3.27. Here we can observe that enhancement of thermal conductivity of nanofluids first increases with SSA and then decreases. The reason for this seems to be that at lower SSA values (i.e., large particle diameters), as the particle diameter is reduced, the heat transfer surface area increases, and due to this dominating surface area effect, the enhancement of thermal conductivity increases. However, below a particle size of 35 nm, which is the phonon mean free path for polycrystalline Al_2O_3, the size effect begins to

Table 3.2 Properties of Nanosized Al_2O_3 Particles

Sample Symbol	αA-5	αA-25	αA-58	γA-58	αA-101	γA-122	γA-124
Specific area (m^2/g)	5	25	58	58	101	122	124
Particle size (nm)	302	60.4	26.0	26.0	15.0	12.4	12.2
Crystalline phase	α	α	α	γ	γ	γ	γ

Source: Xie et al. (2002b), with permission from the American Institute of Physics.

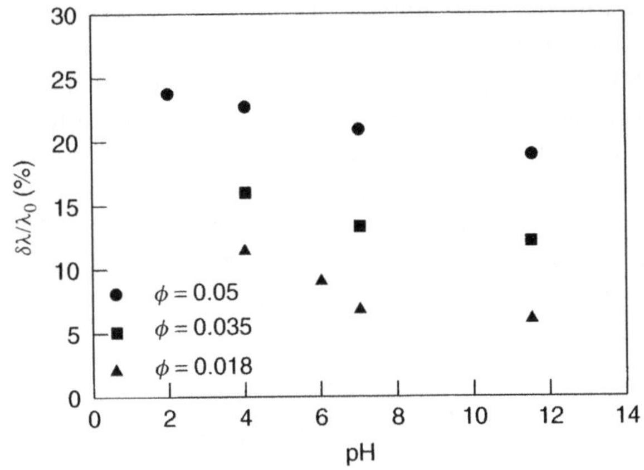

Fig. 3.26 Thermal conductivity enhancement of a 25-nm Al_2O_3 suspension against the pH value. [From Xie et al. (2002b), with permission from the American Institute of Physics.]

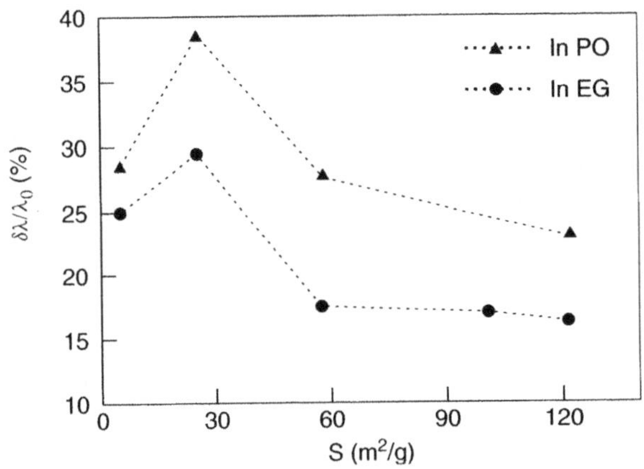

Fig. 3.27 Thermal conductivity enhancement with SSA ($\phi = 0.05$). [FromXie et al. (2002b), with permission from the American Institute of Physics.]

dominate and the conductivity of nanoparticles is reduced substantially. This is due to the scattering of phonons at the particle boundary. Thus, the enhancement is reduced due to the nanoscale effects at low particle sizes. They also observed that the crystalline structure (α or γ) has very little effect on the enhancement of thermal conductivity, which was found to be greater for low-conductivity liquids, which is consistent with Lee et al. (1999). They also compared their data with the theoretical model of Davis (1986), in the form

$$\frac{k_{\text{eff}}}{k_f} = 1 + \frac{3(a-1)}{a+2-(a-1)\,\phi}[\phi + f(a)\phi^2 + O(\phi^3)] \tag{3.63}$$

where a is the thermal conductivity ratio of the solid particle and f is a special function. Their measurements showed much higher thermal conductivity than that found using the model described above.

Murshed et al. (2005) carried out experiments with spherical and rod-shaped T_iO_2 nanoparticles. The spherical particles were 15 nm in diameter and the rod-shaped paricles were 10 nm in diameter and 40 nm in length. The base fluid was deionized water. The measurement method was transient hot wire. It should be mentioned here that they used oleic acid and cetyltrimethylammonium bromide (CTAB) surfactants (0.01 to 0.02 vol %). They maintained a nearly neutral (pH 6.2 to 6.8) suspension. The results from their measurements are shown in Fig. 3.28.

For the first time, a nonlinear correlation between the volume fraction and conductivity enhancement was observed here at lower concentrations. This is interesting with respect to the temperature effect and pure metallic particles discussed in Sections 3.3 and 3.5. They found that the conductivity enhancement was higher for rod-shaped particles than for spherical particles. Enhancement up to 29.7% was found with 5% spherical particles and up to 32.8% with rod-shaped particles. They attributed this to the higher shape factor ($n = 6$) of the rods than of the spheres ($n = 3$) in the Hamilton–Crosser (1962) model. This is somewhat confusing because subsequent comparison of such particles with theoretical models by Bruggeman (1935), Wasp (1977), and Hamilton–Crosser (1962) showed clearly that these models are inadequate to predict the data, as shown in Figs. 3.29 and 3.30.

An interesting study with nonoxide, nonmetallic particles was carried out by Xie et al. (2002a) using silicon carbide (SiC) particles of 26 and 600 nm (0.6 μm) in water and ethylene glycol. The work reported linear variation of

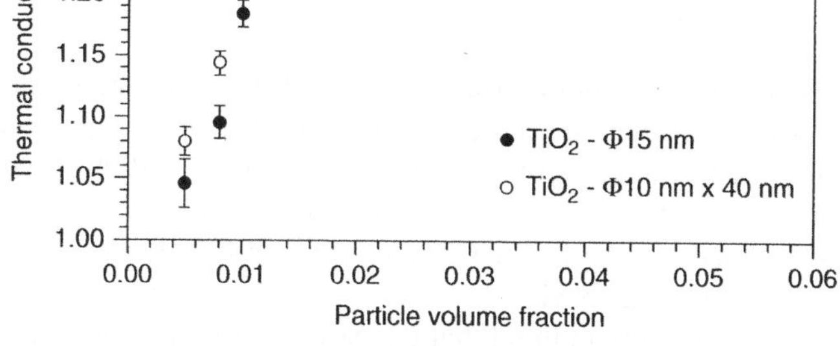

Fig. 3.28 Enhancement of T_iO_2-water nanofluid conductivity (with CTAB surfactant). [From Murshed et al. (2005), with permission from Elsevier.]

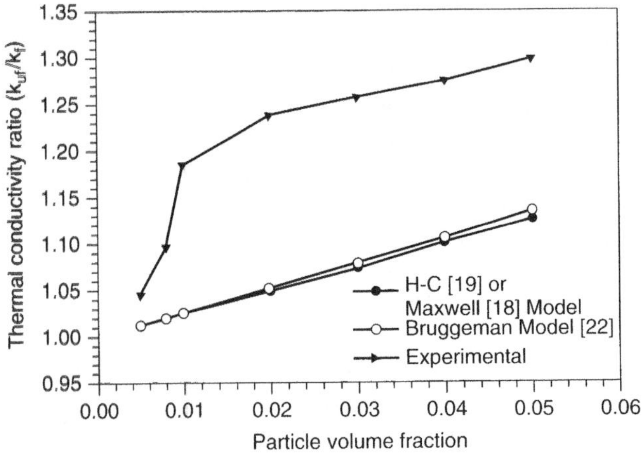

Fig. 3.29 Comparison of spherical T_iO_2–water nanofluid conductivity enhancement with theoretical models. [Form Murshed et al. (2005), with permission from Elsevier.]

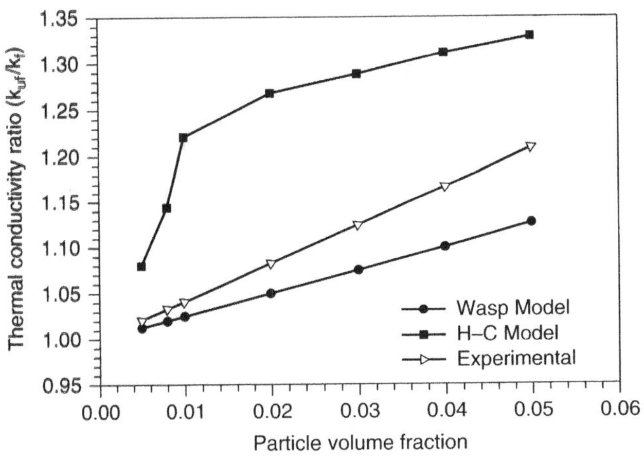

Fig. 3.30 Comparison of rod-shaped T_iO_2–water nanofluid conductivity enhancement with theoretical models. [From Murshed et al. (2005), with permission from Elsevier.]

thermal conductivity with volume fraction for both particle sizes, and more surprisingly, almost identical enhancement using water versus ethylene glycol as the base fluid. This is in direct contradiction with Lee et al. (1999), who reported higher enhancement with ethylene glycol. The other surprising result of Xie et al. (2002a) was the fact that the larger particles showed higher enhancement at the same volume fraction (Fig. 3.31).

However, it must be kept in mind that the particle morphologies were different. The 600-nm particles were rod shaped and the 26 nm particles were spherical. They also attributed this to the morphology and found that the results conform

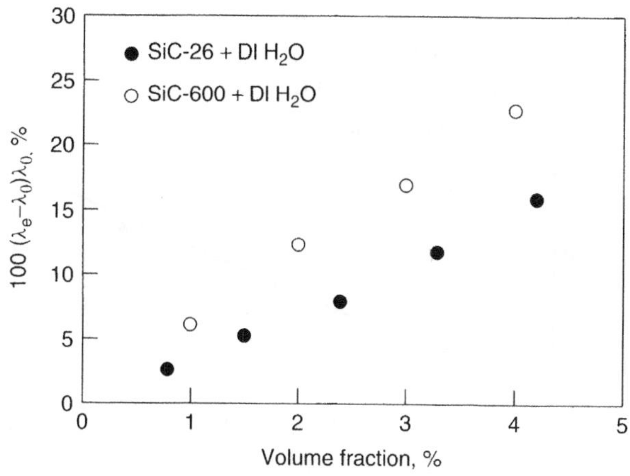

Fig. 3.31 Comparison of conductivities of SiC–water nanofluid for two different particles. [From Xie et al. (2002a), with permission from Springer.]

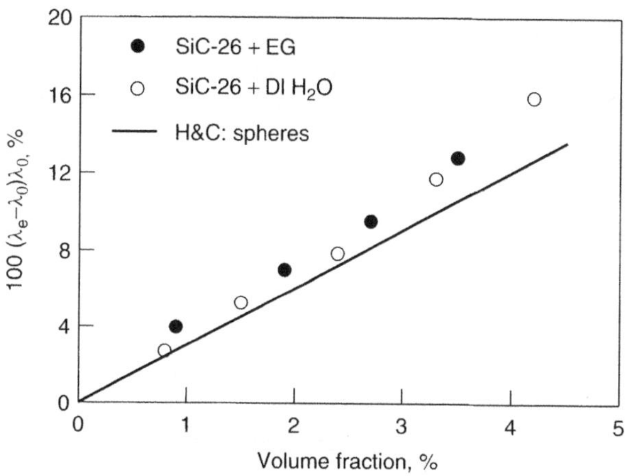

Fig. 3.32 Measured and calculated thermal conductivities of SiC 26-nm suspension. [From Xie et al. (2002a), with permission from Springer.]

to those of the Hamilton–Crosser (1962) model for 600 nm particles, while the model underpredicts the spherical particle results by 20%, as shown in Figs. 3.32 and 3.33. This is expected because the 600-nm particles are too large to be called nanoparticles and hence they conform to the results of microslurry theory, whereas the suspension with 26-nm particles are within the nanofluid range and show nanofluid behavior.

However, the majority of studies confirmed that at lower particle sizes, there is a significant increase in the effective conductivity of nanofluids. Chon and

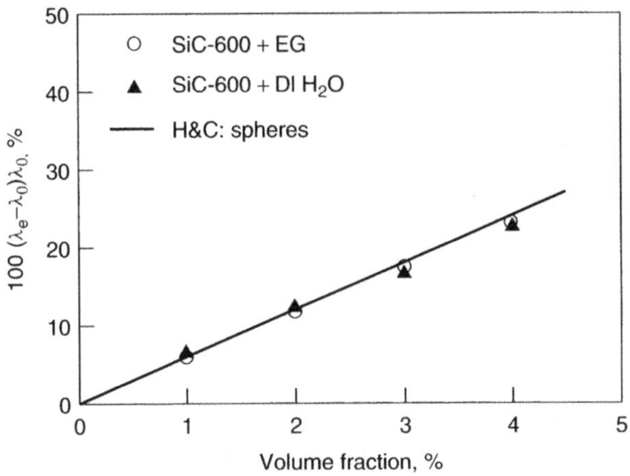

Fig. 3.33 Measured and calculated thermal conductivity of SiC 600-nm suspension. [From Xie et al. (2002a), with permission from Springer.]

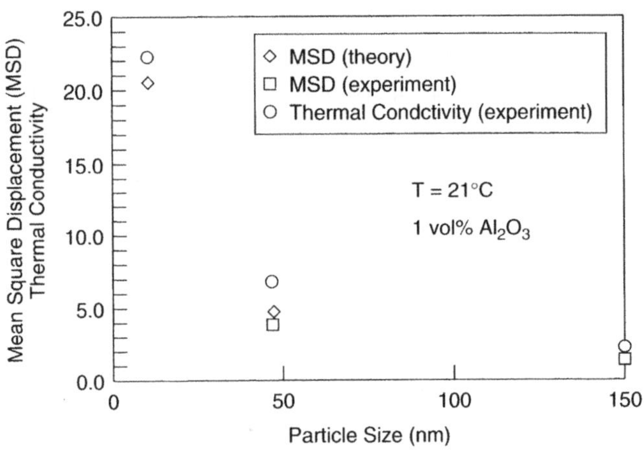

Fig. 3.34 Increase in thermal conductivity with decreasing size. [From Chon and Kihm (2005), with permission from ASME Publishing.]

Kihm (2005) showed clearly that enhancement with particle size was smaller for Al_2O_3 particles of 11, 47, and 150 nm (Fig. 3.34).

3.4. TEMPERATURE DEPENDENCE OF THERMAL CONDUCTIVITY ENHANCEMENT

Although the majority of measurements, including those discussed in Section 3.3 have been carried out at room temperature, comments by many investigators regarding the possibility that particle movement plays a role in thermal conductivity enhancement remained merely a guess. Work by Das et al. (2003) opened up a new direction in this area by showing that there exists an extremely strong temperature effect on the thermal conductivity enhancement of nanofluids. The starting point of their work was an apparent anomaly in the results of Lee et al (1999). They found that while Al_2O_3 particles in water showed agreement with the Hamilton–Crosser (1962) model, CuO particles showed greater enhancement. The only difference between the two particles was the size (38.4 nm for Al_2O_3 and 23.6 nm for CuO). This induced Das et al. (2003) to think that there may be a temperature corresponding to each particle size below which the particle movement induces them to show nanobehavior. The only way to confirm this hypothesis was to carry out measurements at different temperatures.

They used the transient temperature oscillation technique to carry out experiments with the same particles as those used by Lee et al. (1999). The agglomerated and dispersed (by ultrasonication) particles are shown in Figs. 3.35 and 3.36. A typical particle-size distribution for their sample is shown in Fig. 3.37. They

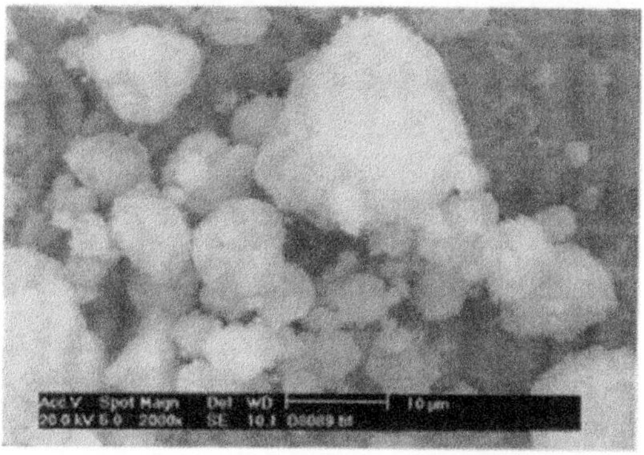

(a)

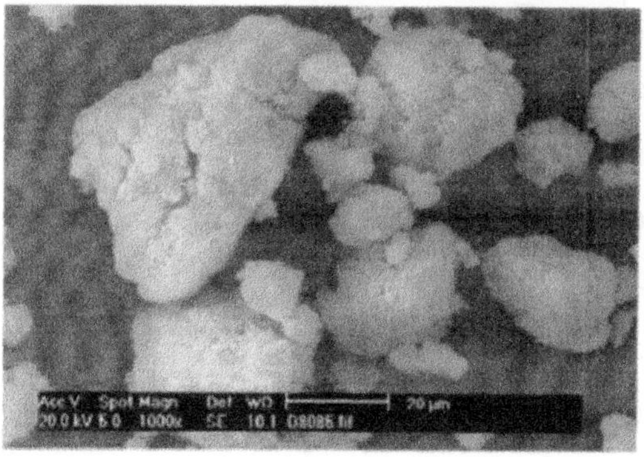

(b)

Fig. 3.35 TE micrographs of agglomerated (a) Al_2O_3 and (b) CuO particles.

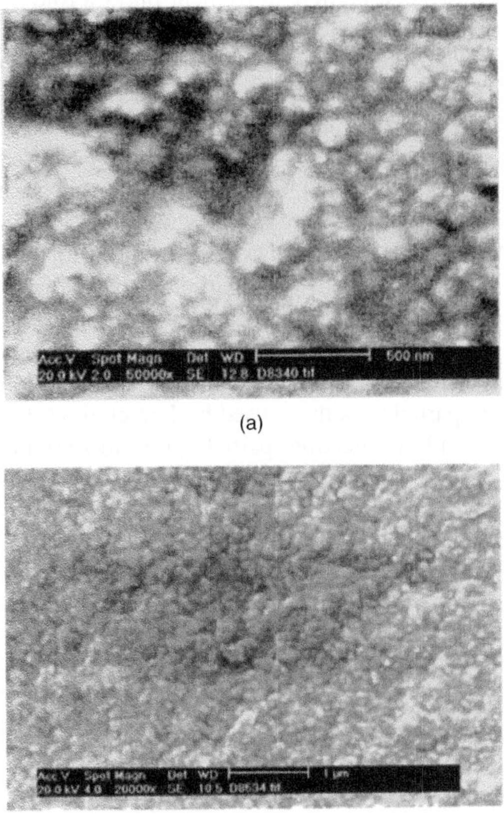

(a)

(b)

Fig. 3.36 TE micrographs of dispersed (a) Al_2O_3 and (b) CuO particles.

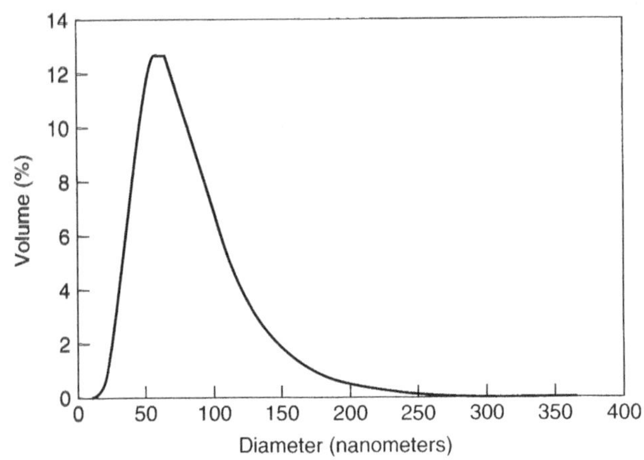

Fig. 3.37 Particle-size distribution of the Al_2O_3 sample.

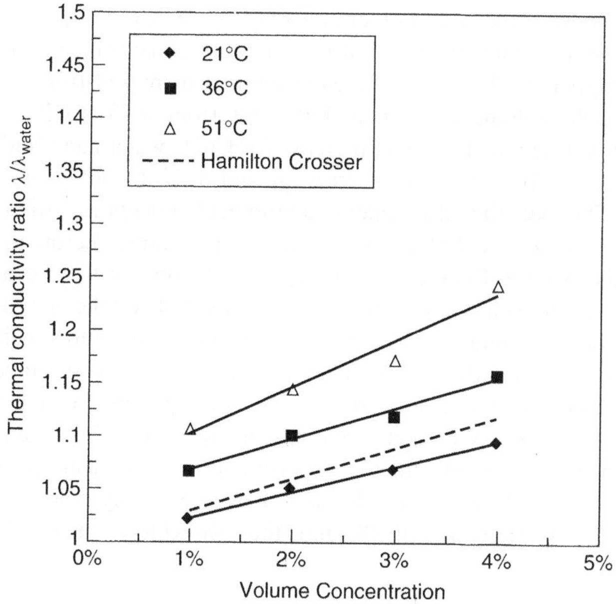

Fig. 3.38 Temperature effects of the thermal conductivity of water–Al_2O_3 nanofluids.

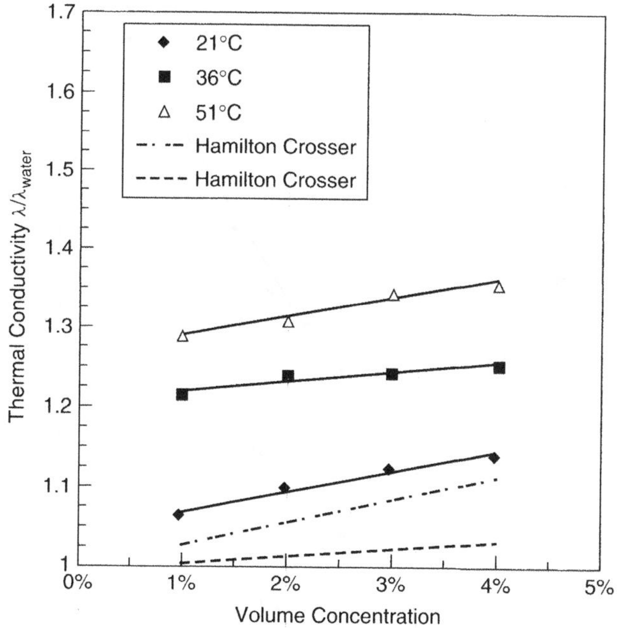

Fig. 3.39 Temperature effects of the thermal conductivity of water–CuO nanofluids.

measured thermal conductivities at temperatures between 21 and 55°C, and the results were nothing less than miraculous (Figs. 3.38 and 3.39). Over this small 34°C rise in temperature, the thermal conductivity enhancement was more than three times higher. With Al_2O_3, the enhancement increased from 2% to 10.8% at a 1% particle volume fraction and it went from 9.4% to 24.3% at a 4% particle-volume fraction. The same increase for CuO–water nanofluids was 6.5% to 29% for a 1% particle-volume fraction and 14% to 36% for a 4% particle fraction. This puts the entire phenomenological concept regarding nanofluids completely in perspective. In fact, all the theories proclaimed before this work was published (discussed in Chapter 4), crumpled at this observation because none of them could predict such a strong temperature effect. The other important observation from the preceding figures is that at elevated temperatures, neither Al_2O_3- nor CuO-based nanofluids comply with the Hamilton–Crosser model. This is because the model is completely insensitive to temperature variations between 21 and 55°C. This clearly indicates that agreement of the Al_2O_3 nanofluids with the Hamilton–Crosser model in the experiments of Lee et al. (1999) at room temperature was purely accidental because it was probably below the threshold temperature at a particle size of 38.4 nm, thus showing no nanobehavior. They

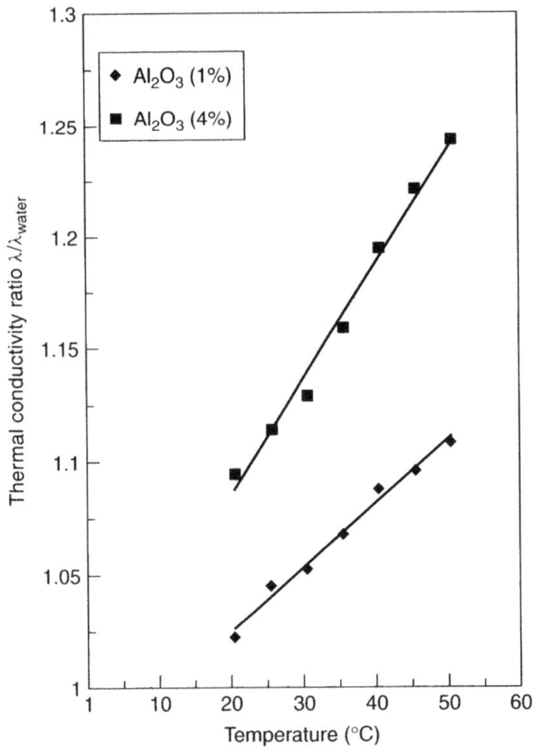

Fig. 3.40 Temperature dependence of Al_2O_3–water nanofluids (Das et al., 2003).

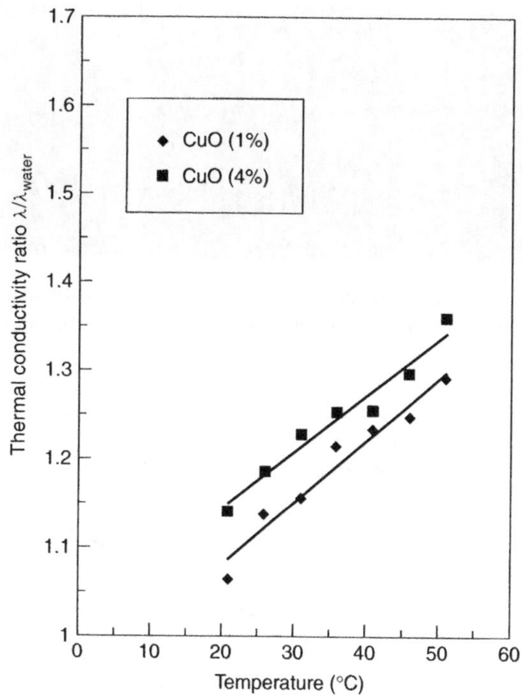

Fig. 3.41 Temperature dependence of CuO–water nanofluids (Das et al., 2003).

also plotted the same data against temperature for 1 and 4% particle-volume fractions (Figs. 3.40 and 3.41). They commented that some Brownian-like motion may be responsible for the behavior but did not expend on it due to lack of further evidence. However, a recent photographic study by Chon and Kihm (2005) has confirmed this motion. They presented the optical microscopy image of the Brownian motion shown in Fig. 3.42. They also measured the thermal conductivity at temperatures between 20 and 70°C and found thermal conductivity enhancement similar to that found by Das et al. (2003) as shown in (Fig. 3.43).

The work by Das et al. (2003) and Chon and Kihm (2005) both showed one more feature. Although their enhancement data apparently showed linear variation with volume fraction, at higher temperatures, extrapolating the straight lines backward does not yield 0% enhancement at zero percent volume fraction, which is a clear discrepancy because if there are no particles there should be no enhancement. This clearly indicates that there should be nonlinear enhancement with volume fraction at the lower values of enhancement, as shown by Murshed et al. (2005).

The temperature effect was confirmed further by Li and Peterson (2006). They used a steady state method called the *cut bar method* for the measurement of thermal conductivity. They found a huge effect of temperature on the thermal conductivity of Al_2O_3- and CuO-based nanofluids with water as the

Fig. 3.42 Optical microscopic images (× 630) of nanoparticle Brownian motion in water. [From Chon and Kihm (2005), with permission from ASME Publishing.]

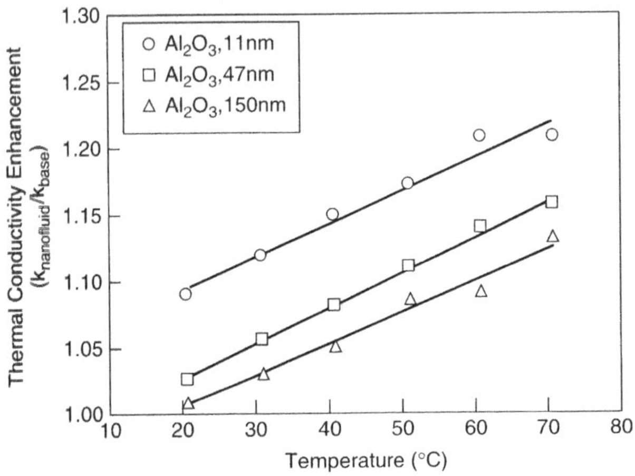

Fig. 3.43 Temperature dependence of thermal conductivity enhancement of Al_2O_3–water nano–fluids at varoius particle sizes. [From Chon and Kihm (2005), with permission from ASME Publishing.]

base fluid. Their results are shown in Figs. 3.44 and 3.45, respectively. Based on their observations, they suggested the following equations for thermal conductivity enhancement, obtained by linear regression analysis for Al_2O_3–water nanofluids:

$$\frac{k_{\text{eff}} - k_f}{k_f} = 0.764481\phi + 0.018688867t - 0.462147175 \tag{3.64}$$

Here ϕ is the particle-volume fraction and t is the temperature in °C. The r^2 value for this equation was 0.9171. For CuO–water nanofluids, the equation was

$$\frac{k_{\text{eff}} - k_f}{k_f} = 3.76108\phi + 0.017924t - 0.30734 \tag{3.65}$$

This equation has an r^2 value of 0.9078.

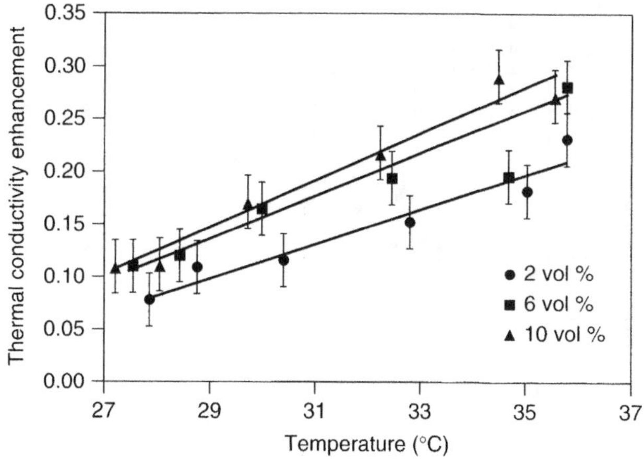

Fig. 3.44 Thermal conductivity enhancement of Al_2O_3–water nanofluids against temperature. [From Li and Peterson (2006), with permission from the American Institute of Physics.]

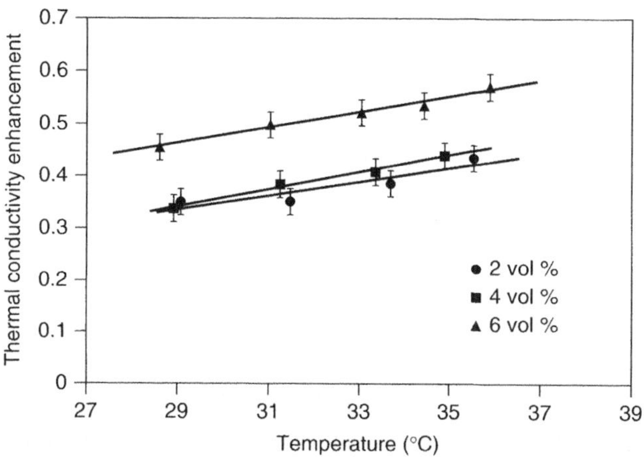

Fig. 3.45 Thermal conductivity enhancement of CuO–water nanofluids against temperature. [From Li and Peterson (2006), with permission from the American Institute of Physics.]

Comparing their data with those of Das et al. (2003), Li and Peterson (2006) found that the temperature effect of their data is stronger (Figs. 3.46 and 3.47). They found a threefold increase in conductivity over just 10°C for Al_2O_3. Thus, it is clear that unlike parameters such as particle size and particle concentration, the temperature effect has been observed by all the investigators without exception, and they lie in a similar range. Also, there in no controversy about the nonapplicability of traditional models such as those of Maxwell (1882) and

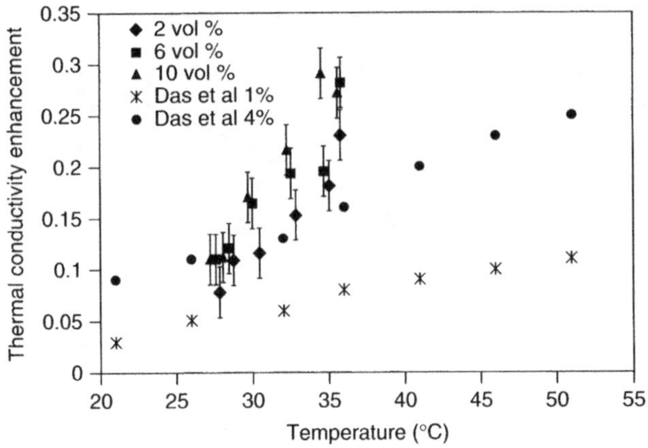

Fig. 3.46 Experimental data for thermal conductivity enhancement of Al_2O_3–water nanofluids with temperature. [From Li and Peterson (2006), with permission from the American Institute of Physics.]

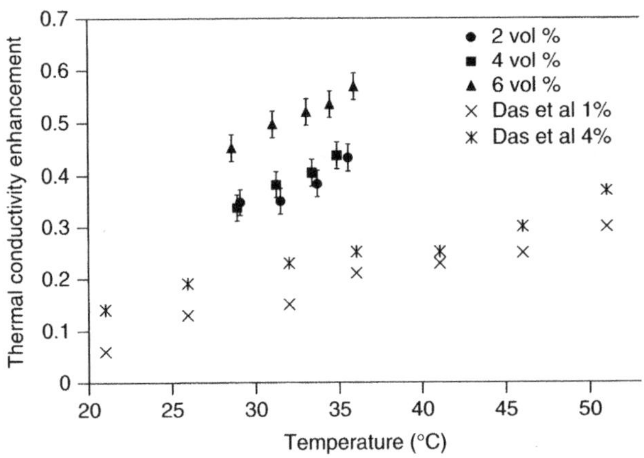

Fig. 3.47 Experimental data for thermal conductivity enhancement of CuO–water nanofluids with temperature. [From Li and Peterson (2006), with permission from the American Institute of Physics.]

Hamilton–Crosser (1962) at elevated temperatures. The foergoing observations will play an important role in the future for modeling of nanofluid conductivity, as discussed in Chapter 4.

3.5. METALLIC NANOFLUIDS

The investigations in nanofluids began with oxide nanoparticles, due to their stability, ease of synthesis, ease of dispersion in fluids, and other advantages. The resulting thermal conductivities were higher than the usual suspensions, but the increase in thermal conductivity was not so drastic that it could have attracted the attention of the heat transfer community as much as it has done. It is the metallic nanofluids that showed tremendous prospects in this respect. Work by Choi's Argonne group led to the astonishing result (Eastman et al., 2001) that just 0.3% of Cu nanoparticles showed an increase in thermal conductivity on the order of 40%. They used the one-step method to make nanosuspensions of copper particles of average diameter less than 10 mm. For stabilization, they added less than 1% thioglycolic acid. Three samples were taken for measurement. The suspension without thioglycolic acid, stored for two months, was labeled "old" and the suspension prepared within the last two days was labeled "fresh". The results of a conductivity measurement using the transient hot-wire method as the base fluid with ethylene glycol are shown in Fig. 3.48.

The results clearly indicate that whole "fresh" samples provide higher conductivity than "old" samples. Samples with stabilizer show an astonishing 40% increase in thermal conductivity with just a 0.3% volume fraction of particles. They also confirmed that thioglycolic acid alone (without particles) does not enhance the thermal conductivity. This results in two conclusions: that metallic nanofluids are promising and that in these nanofluids, the particle dispersion and stability critically affect the thermal conduction. They also compared oxide and metallic nanofluids, as shown in Fig. 3.49. It clearly demonstrates that metallic nanofluids have an order-of-magnitude-higher enhancement of conductivity than that of oxide nanofluids. Although the oxide nanoparticles used here were much larger (≈ 35 nm) than the metallic nanoparticles (< 10 nm), the variation in thermal conductivity cannot by explained by the size effect alone. In their

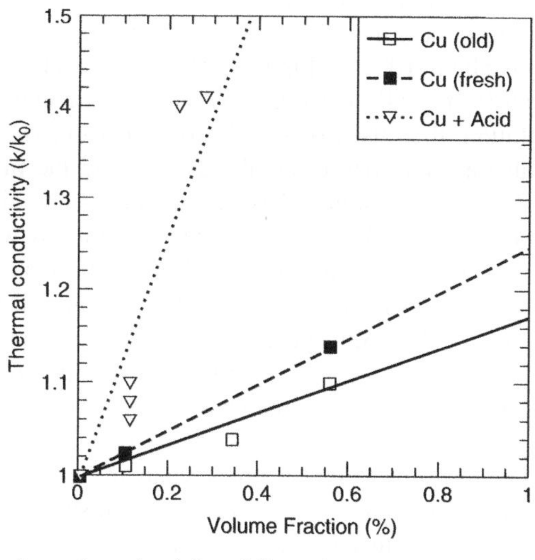

Fig. 3.48 Effective thermal conductivity of Cu–ethylene glycol nanofluid. [From Eastman et al. (2001).]

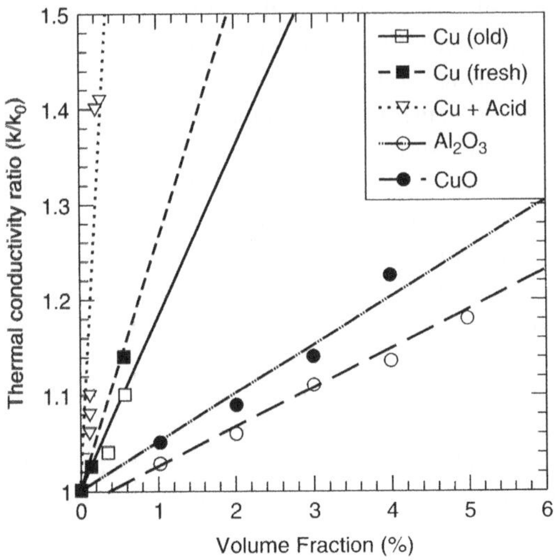

Fig. 3.49 Comparison of oxide (Al_2O_3 and CuO) and metallic (Cu) nanofluids with ethylene glycol. [From Eastman et al. (2001).]

discussion they clearly indicated that the Hamilton–Crosser model is inadequate for nanofluids since they only take care of volume fraction and shape, not of particle size. Also, the dependence of the effective conductivity on particle conductivity is very weak for the model, whereas the results above clearly indicate a higher dependence on particle conductivity.

However, the Argonne group was not the only one to develop metallic nanoparticle-based nanofluids. Xuan and Li (2000) experimented with transformer oil–Cu and water–Cu nanofluids. The volume fraction was 2 to 5% and the particle size was about 100 nm. Figure 3.50 shows a TE micrograph of Cu nanoparticles in transformer oil, and Fig. 3.51 shows Cu nanoparticles in water. To stabilize the solution, they used oleic acid for transformer-oil-based nanofluids and laureate salt for water-based nanofluids, keeping the pH value near the natural region. A better dispersion behavior in transformer oil was attributed to the higher viscosity of oil. They carried out experiments with the transient hot-wire technique and compared their data with the preliminary experiments of Eastman et al. (1997). The results for water- and transformer-oil-based nanofluids are shown in Figs. 3.52 and 3.53, respectively. The results clearly indicated very high enhancement (on the order of 40 to 50%). However, their results showed the attainment of such enhancement at a somewhat higher particle concentration than that shown by Eastman et al. (1997). This is clearly due to the fact that Eastman et al. (1997) used 18-nm Cu particles, whereas Xuan and Li (2000) used particles of about 100 nm. It is encouraging to see that even with such large particle sizes, the enhancement was remarkably high, indicating the strong influence of particle conductivity. The practical significance of this work lies in the inexpensive method of nanoparticle preparation, which is commercially exploitable.

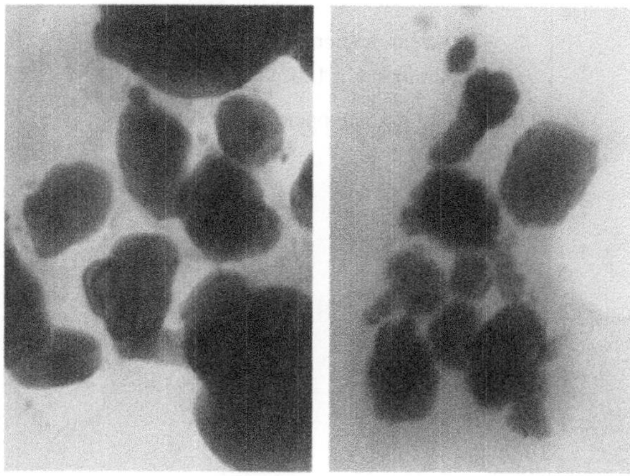

(a) (b)

Fig. 3.50 Micrograph of Cu–transformer oil nanofluid at pH 6.3: (a) 2 vol % (× 100,000); (b) 5 vol % (× 100,000). [From Xuan and Li (2000), with permission from Elsevier.]

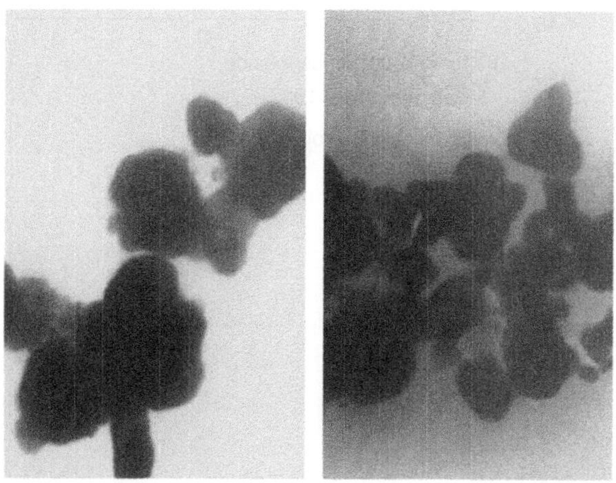

(a) (b)

Fig. 3.51 Micrograph of Cu–water nanofluid at pH 6.8: (a) 5 vol % (× 50,000); (b) 7.5 vol % (× 30,000). [From Xuan and Li (2000), with permission from Elsevier.]

In a very interesting study by Patel et al. (2003), they used naked and polymer monolayer-protected gold and silver particles of 10 to 20 nm suspended in water and toluene. Monolayer-protected particles were produced by the Brust method, whereas naked particles were produced by the citrate reduction method. Due to the limitations of the chemical synthesis technique used, the volume concentration

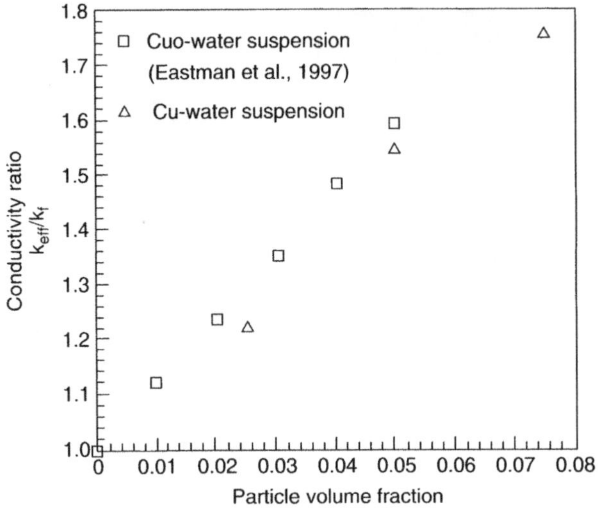

Fig. 3.52 Thermal conductivity enhancement of Cu–water nanofluids. [From Xuan and Li (2000), with permission from Elsevier.]

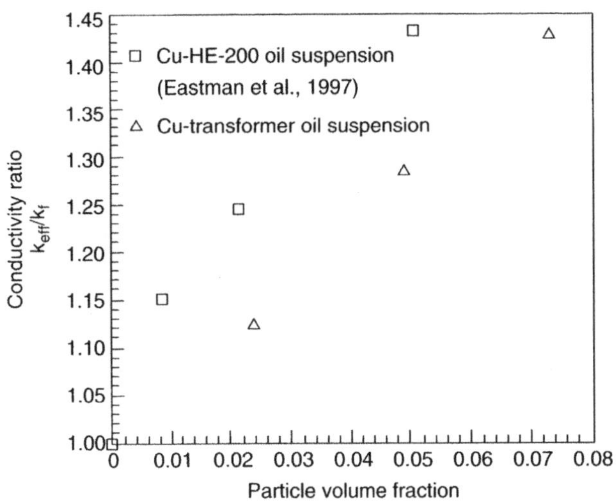

Fig. 3.53 Thermal conductivity enhancement of Cu–oil suspension. [From Xuan and Li (2000), with permission from Elsevier.]

was extremely low: a maximum of 0.011% for gold and 0.001% for silver. Even at such vanishing concentrations, they demonstrated a 3 to 10% increase in thermal conductivity, which in surprising because at their concentration, no enhancement at all is expected. They also showed considerable temperature effects. Figures. 3.54 and 3.55 show the thermal conduction enhancement achieved by

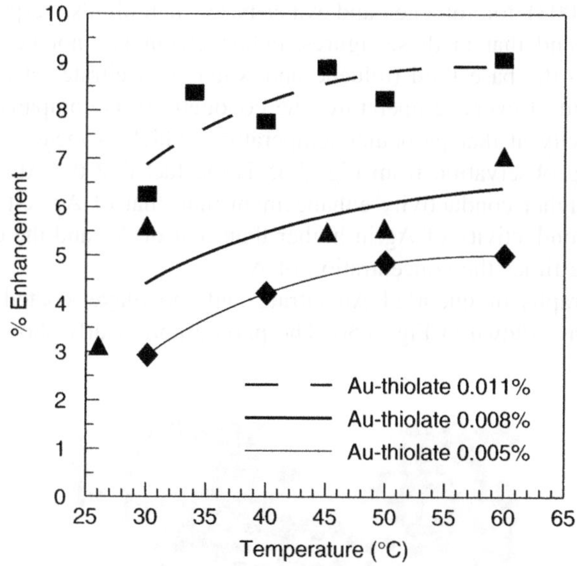

Fig. 3.54 Enhancement of thermal conductivity of Au thiolate particles in toluene with respect to toluene at respective temperatures.

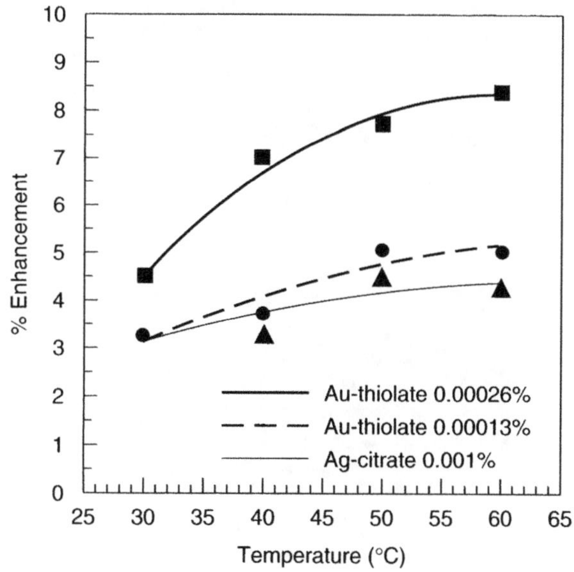

Fig. 3.55 Enhancement of thermal conductivity of Au and Ag citrate particles in water (with 5 mM sodium citrate) at respective temperatures.

Patel et al. (2003) for toluene- and water-based nanofluids, respectively. One has to understand that in these figures, enhancement has not been considered with respect to the base fluid (toluene and water with citrate) at base temperature (30°C), but at every temperature, the conductivity is compared to the base fluid conductivity at that particular temperature, which is measured separately. One interesting observation from Fig. 3.55 is the fact that the Au–citrate particles showed higher conductivity enhancement than that of Ag–citrate particles, although the conductivity of Ag in higher than that of Au and the concentration of Ag was five times the concentration of Au.

TE micrographs of uncorted Au citrate and monolayer-coated Au–thiolate nanoparticles are shown in Fig. 3.56. The photograph clearly shows particles of

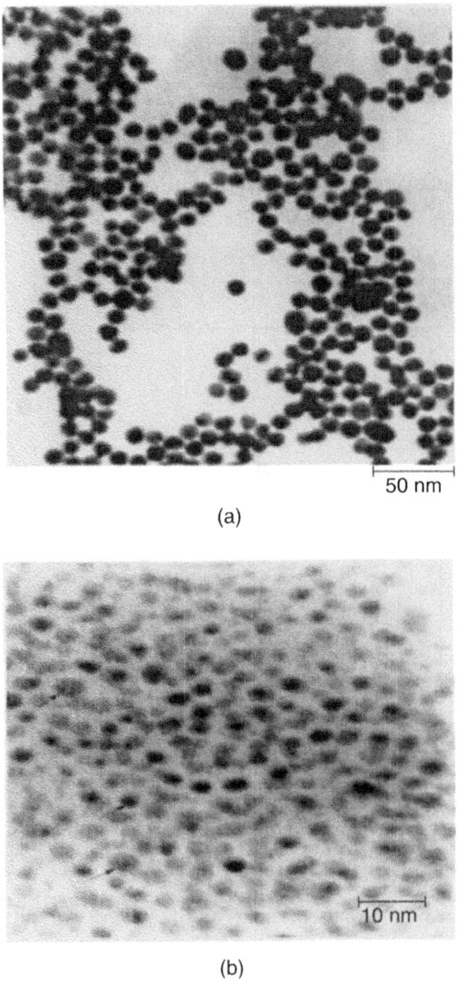

Fig. 3.56 TE micrograph of (a) Au–citrate and (b) Au–thiolate nanoparticles.

uniform size, well dispersed in the fluid. The only explanation for this can be the particle size. Here the Au–citrate paricles were 10 to 20 nm in diameter, while the Ag–citrate particles were 80 to 90 nm in average diameter, which might be at the reason for the higher enhancement that they showed.

Fe nanofluids were used in a recent study by Hong et al. (2005). The particles were around 10 nm in size, dispersed in ethylene glycol using ultrasonic vibration. Using the transient hot-wire method, they found that the enhancement of thermal conductivity depends on the sonication time (Fig. 3.57). With the increase in sonication time, the enhancement increases and finally reaches a saturation value at around 50 minutes of sonication time. The enhancement was found to be best with 0.55 vol % of particles.

Against volume fraction, their results showed few interesting features. First, the volume concentration used by them was low (0.2 to 0.55%), and at these concentrations they observed nonlinear behavior of enhancement against concentration, which is in keeping with the discussion in Section 3.4. Second, the enhancement they achieved was higher than that obtained by Eastman et al. (2001) with Cu nanoparticles (Fig. 3.58).

The explanation for the fact that Fe particles, which have a lower bulk conductivity than Cu, showed greater enhancement was investigated from different points of view by Hong et al. (2005). First, they indicated that intrinsic properties such as bulk conductivity may not be an indicator of nanosuspension behavior because at nanosize, optical, thermal, and magnetic properties are known to vary. They also indicated that their nanofluids have some clusters, as evident from the TE micrographs, which may play a positive role in terms of providing paths of less resistance for heat transport. In liquids, the conductivity is much smaller because of the absence of long-range vibrational mode. Clusters may improve

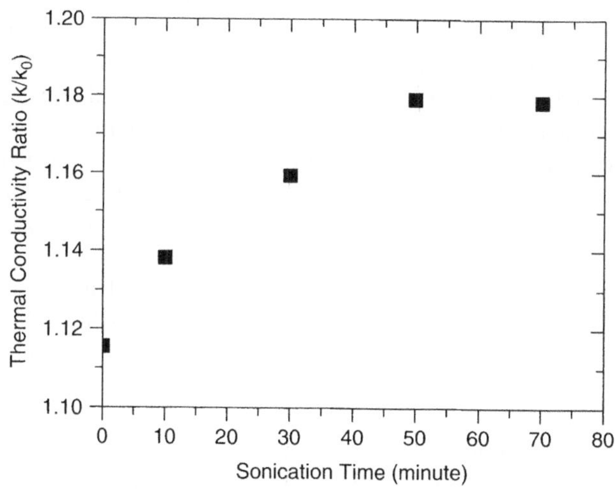

Fig. 3.57 Thermal conductivity of 0.55% Fe nanofluid as a function of sonication time. [From Hong et al. (2005), with permission from the American Institute of Physics.]

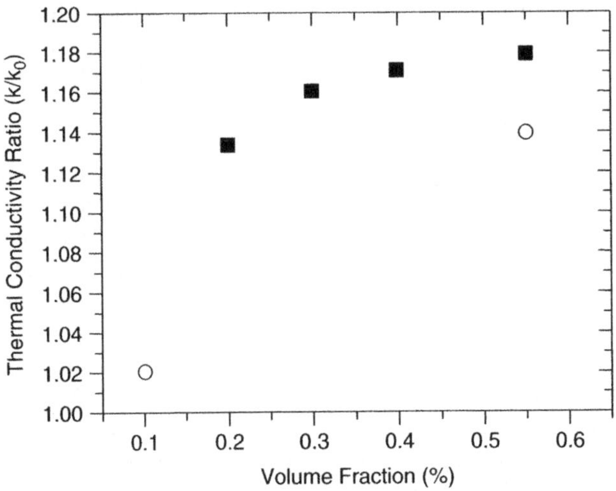

Fig. 3.58 Thermal conductivity of Fe–ethylene glycol nanofluid compared with Cu–ethylene glycol nanofluid. [From Eastman et al. (2001), with permission from the American Institute of Physics.]

this by providing additive paths. They also observed that the presence of a stabilizing agent does not improve the clustered nanofluid conductivity (presumably due to less mobility) in contrast to well-dispersed nanofluids of Eastman et al. (2001), where they played an important role in enhancement.

A nanofluid of Cu nanoparticles in water was produced by Liu et al. (2006) by the chemical reduction method. They got a variety of sizes, from 50 to 250 nm (also some particles of needle and square shape). They also showed some dependence of enhancement of conductivity on measurement time, which must be a problem of thermal equilibrium. For the ranges in their experiment, the enhancement in conductivity they obtained is given in Table 3.3. It is interesting to see that with just 0.05 to 0.2 vol %, they obtained enhanced thermal conductivity of between 3.6 and 23.8%, which in amazing. They compared their data with those of Patel et al. (2003) and Eastman et al. (2001), as shown in Fig. 3.59, which shows almost a linear relationship among them (although the particles were different). The one novelty of their nanofluid was that they did not use a stabilizer.

In very encouraging work by Chopkar et al. (2006), they developed mechanically alloyed particles of $Al_{70}Cu_{30}$ and used them in ethylene glycol for dispersion. They obtained the particles by mechanical alloying using ball milling. The particle sizes were in the range 20 to 40 nm. The particles dispersed excellently, as shown in Fig. 3.60. They used a comparator developed by them to measure the thermal conductivity of fluids and suspensions. Their results were very encouraging, as shown in Fig. 3.61, which shows more than 100% enhancement with just 2.5 vol % particle, which is even higher than Xuan and Li's (2000) value. However, one must consider the smaller particle size that they used. The enhancement shows nonlinear behavior with maximum enhancement between

Table 3.3 Grain Size Effect on Thermal Conductivity Enhancement of Cu–Water Nanofluids

Specimen	Volume Fraction (vol %)	Thermal Conductivity Increased Ratio (%)	Grain Size and Shape from SEM
1	0.05	11.6	100–200 nm, spherical and square
2	0.05	3.6	Not available
3	0.05	8.5	130–200 nm, spherical and square
4	0.1	23.8	75–100 nm, spherical and square
5	0.1	23.8	50–100 nm, spherical and square
6	0.1	11.0	100–300 nm, spherical square and needle
7	0.2	9.7	130–300 nm, spherical
8	0.2	13.2	200 nm × 500 nm needle
9	0.2	3.6	250 nm, spherical, square, and needle

Source: Liu et al. (2006), with permission from Elsevier.

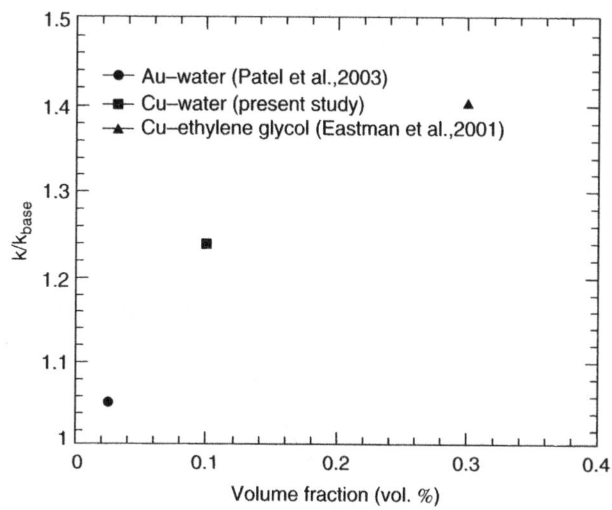

Fig. 3.59 Thermal conductivity enhancement for Au and Cu nanofluids. [From Liu et al. (2006), with permission from Elsevier.]

0.75 and 15%. Using this nanofluid, they carried out a quenching experiment on a copper block which showed significantly enhanced quenching performance of the fluid (Fig. 3.62). Finally, they presented the effect of particle size. Figure 3.63 shows an excellent inverse relationship with particle size, which confirms the previous observations and theories from many investigators. Thus, it can be concluded that studies on metallic nanofluids have opened a new horizon with high enhancement of thermal conductivity at low particle-volume fractions.

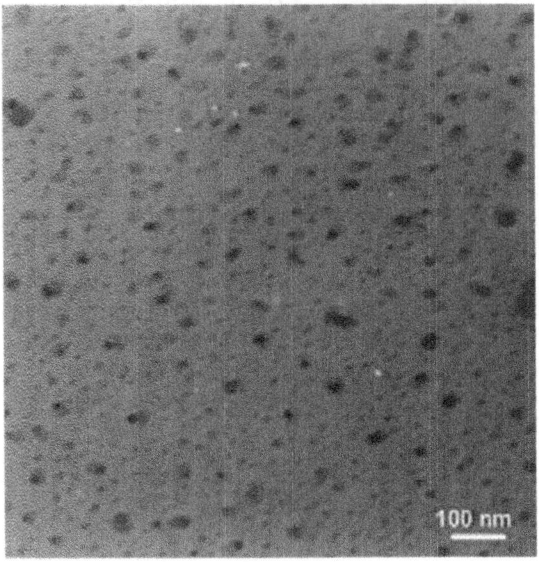

Fig. 3.60 TE micrograph of $Al_{70}Cu_{30}$ particles in ethylene glycol. [From Chopkar et al. (2006), with permission from Elsevier.]

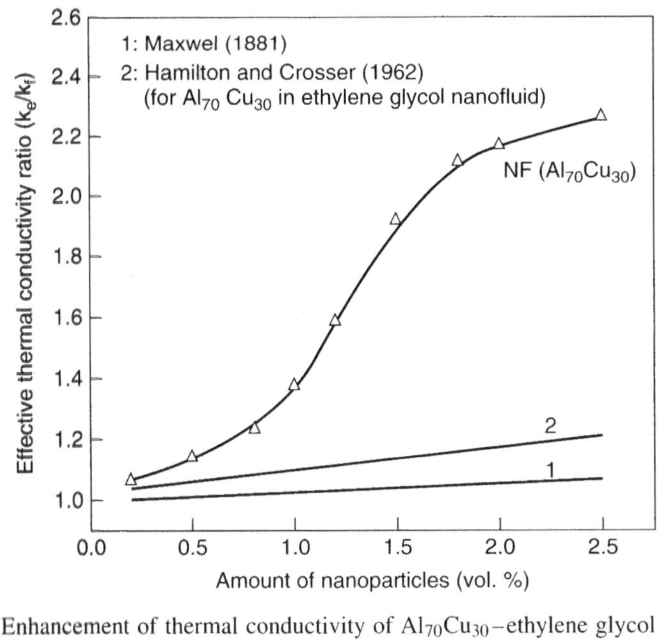

Fig. 3.61 Enhancement of thermal conductivity of $Al_{70}Cu_{30}$–ethylene glycol nanofluids. [From Chopkar et al. (2006), with permission from Elsevier.]

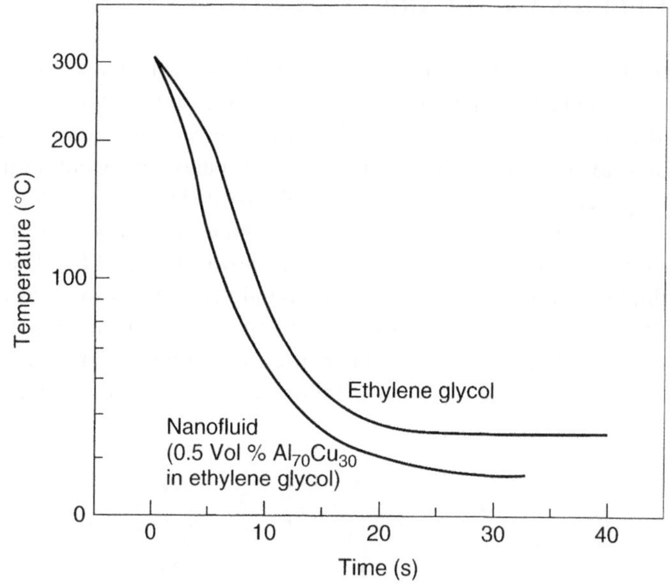

Fig. 3.62 Comparison between the quenching efficiency of pure ethylene glycol and a nanofluid with $Al_{70}Cu_{30}$–ethylene glycol. [From Chopkar et al. (2006), with permission from Elsevier.]

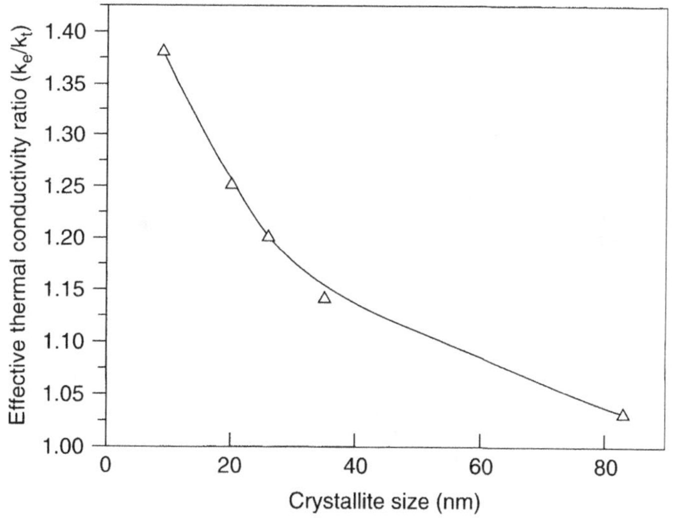

Fig. 3.63 Effect of particle size on the thermal conductivity ratio of $Al_{70}Cu_{30}$ nanofluids. [From Chopkar et al. (2006), with permission from Elsevier.]

3.6. NANOFLUIDS WITH CARBON NANOTUBES

The next breakthrough in the innovation of nanofluids, which also came from the group at the Argonne National Laboratory, was with carbon nanotubes. They used multiwalled carbon nanotubes (MWCNTs) of diameter 25 nm and length 50 μm, giving a large average aspect ratio of 2000. The nanotubes were straight (Fig. 3.64) and they were dispersed is synthetic α-olefin oil. They got stable suspension with 1 vol % nanotube loading. The thermal conductivity was measured by the transient hot-wire method, and the results are shown in Fig. 3.65. There results took nanofluid researchers by storm. It was found that with as small as a 1 vol % fraction, 159% enhancement of thermal conductivity could be obtained, which is unheard-of using either oxide or metallic particles. A comparison with all the predictive models proves to be a complete failure since none of the models predicts

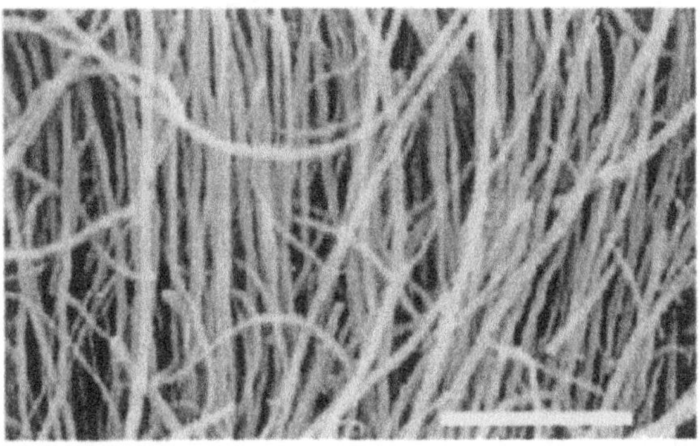

Fig. 3.64 SE micrograph of MWCNTs used by Choi et al. (2001).

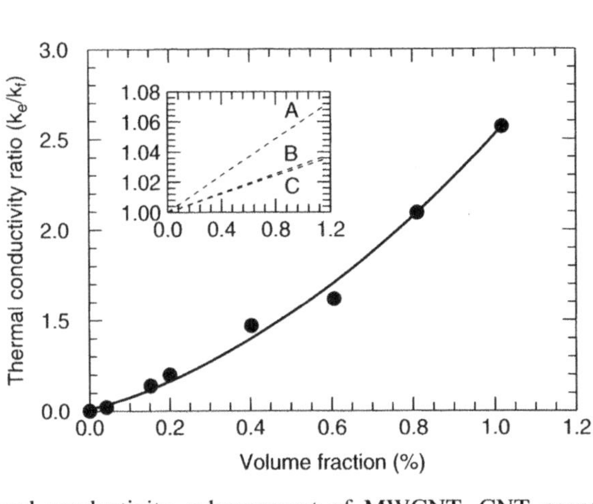

Fig. 3.65 Thermal conductivity enhancement of MWCNT–CNT nanotube nanofluids. The inset shows conductivity prediction by the (A) Hamilton–Crosser (1962), (B) Bonnecaze–Brady (1990), and (C) Maxwell (1881) models.

an enhancement of even 10% (MWCNT conductivity was taken as 2000 W/m·K, which in an acceptable mean value). The other important feature of this enhancement was the increasing slope of the enhancement curve. Due to this, although the conductivity at a 0.3% volume fraction is close to the value obtained for Cu nanoparticles containing nanofluids (Eastman et al., 2001), the value at a 1% volume fraction is much higher than that can be expected for Cu nanofluids.

They attributed this sudden rise in conductivity to the ballistic transport of heat in nanotubes due to the large phonon mean free path in them rather than to diffusive transport in a bulk medium. The fact that the acoustic mismatch of impedance at the solid–liquid interface can seriously limit ballistic transport was explained in the following way. They postulated that the liquid layer at the interface behaves as an organized layer like the solid, which makes up this mismatch. About the nonlinearity of enhancement with the volume fraction, they argued that nonlinearity usually arises out of particle interaction. Usually, at such a low volume fraction, such interactions are not expected for spherical particles, but since the MWCNTs have a large aspect ratio (~2000), some amount of interaction and consequent percolation of heat energy are inevitable: thus, the CNT nanofluids not only show higher enhancement but indicate new physical phenomena. In fact, percolation as a probable cause of increased thermal conductivity is also evidenced by the thermal conductivity measurement of epoxy composite single-walled CNTs as shown by Biercuk et al. (2002), where about 120% enhancement was reported with 1% SWCNTs.

The use of multiwalled carbon nanotubes for forming nanofluids was also demonstrated by Xie et al. (2003), who found that as-received pristine CNTs (PCNTs) were aggregated and entangled. To get rid of these to form stable nanofluids, they treated these nanotubes using concentrated nitric acid and subsequently introduced oxygen containing a functional group on the CNT surface to make them hydrophilic. The MWCNTs were dispersed in water and ethylene glycol without a surfactant, whereas in nonpolar fluids such as decene it was dispersed with oleylamine as a surfactant. This treated CNT (TNCT) was used to form the nanofluids, which were found to be quite different in XRD pattern and Raman spectra from the PCNTS. It was found that nanofluids were formed by PCNTS sediment within 5 minutes, whereas TCNTs are stable in both water and ethylene glycol for more than two months. The addition of oxygen containing groups alter the isoelectric point, which is evident from a plot of the zeta potential (Fig. 3.66). A zero zeta potential gives pH_{IEP}, which is at 7.3 (almost neutral) for PCNTs whereas the zeta potential of the TCNTs, and as a consequence, the pH_{IEP} for TCNTs, are much lower, which gives good stability in a nearly neutral solution.

With respect to the measurement of thermal conductivity, Xie et al. found that the thermal conductivity enhancement is higher for lower-conductivity fluids such as decene (DE) and ethylene glycol (EG) than for higher-conductivity fluids such as deionized water (DW) (Fig. 3.67). This conclusion is identical to that of nanofluids with oxide or metal particles. Further, they found that the enhancement is higher than that of the Hamilton–Crosser (1962) and Davis (1986) model but

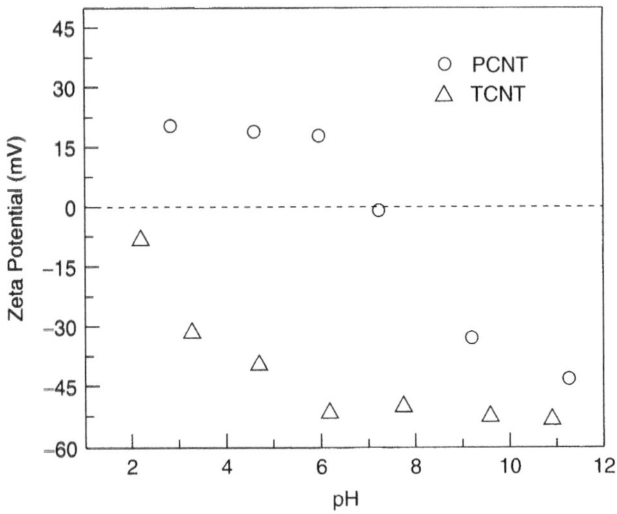

Fig. 3.66 Zeta potentials of PCNTs and TCNTs as a fraction of pH values. [From Xie et al. (2003), with permission from the American Institute of Physics.]

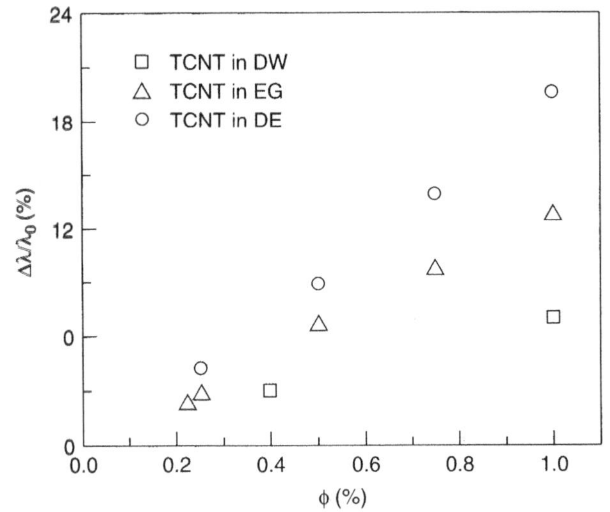

Fig. 3.67 Thermal conductivity enhancement of TCNT suspensions in various fluids. [From Xie et al. (2003), with permission from the American Institute of Physics.]

only on the order of 20 to 30% (Fig. 3.68) compared to the 159% obtained by Choi et al. (2001).

The large difference in enhancement between the two results is attributed to the method of preparation of Xie et al. (2003). However, the interesting point is the similarity of the nature of the nonlinear enhancement curve for both of them.

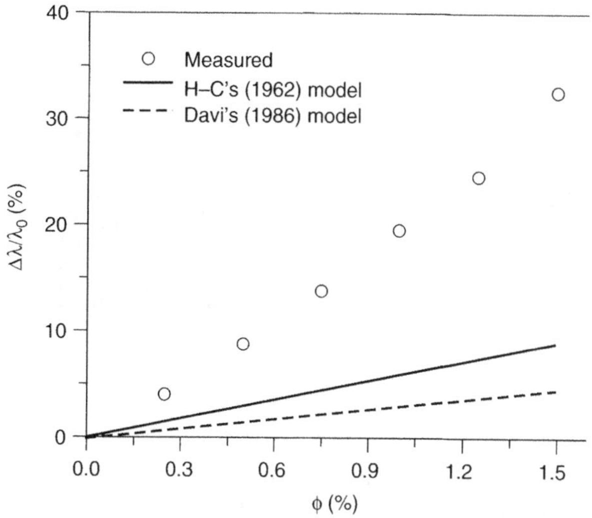

Fig. 3.68 Thermal conductivity enhancement of TCNT nanotubes in decene. [From Xie et al. (2003), with permission from the American Institute of Physics.]

Study on multiwalled CNTs (MWCNTs) and double-walled CNTs (DWCNTs) with detailed characterization of nanotubes by TEM and SEM was carried out by Assael et al. (2005). The two dispersants they used were hexdecyltrimethylammonium bromide (CTAB) and Nanosperse AQ. With these, three combinations of nanofluids were formed.

1. 0.6 vol % MWCNTs with 6 mass % CTAB (group A nanofluid)
2. 0.6 vol % MWCNTs with 0.7 mass % Nanosperse AQ (group B nanofluid)
3. 0.75 vol % DWCNTs with 1 mass % CTAB (group C nanofluid)

Thermal conductivity was measured for the MWCNT-contained nanofluid (groups A and B) as a fraction of the homogenization time for each sample (Fig. 3.69). The results clearly indicate enhancement on the order of 20 to 34% and the effect of homogenization time. Similar results for DWCNTs containing nanofluids are shown in Fig. 3.70, where the conductivity increases with homogenization time.

From this study Assael et al concluded that the enhancement of thermal conductivity increases with the aspect (length / diameter) ratio of the nanotubes, and hence more homogenization by sonication should be avoided since it may shorten the nanotubes, reducing the aspect ratio. They also concluded that CTAB is the better dispersant for nanotubes because it interacts with the outermost nanotubes and modifies their surface characteristics. CNTs containing nanofluids were also tested by Liu et al. (2005). They used ethylene glycol and synthetic engine oil as the fluids. For engine oil they used *N*-hydroxysuccinimide as a surfactant. SE and TE micrographs of their nanotubes showed intense interaction, and a

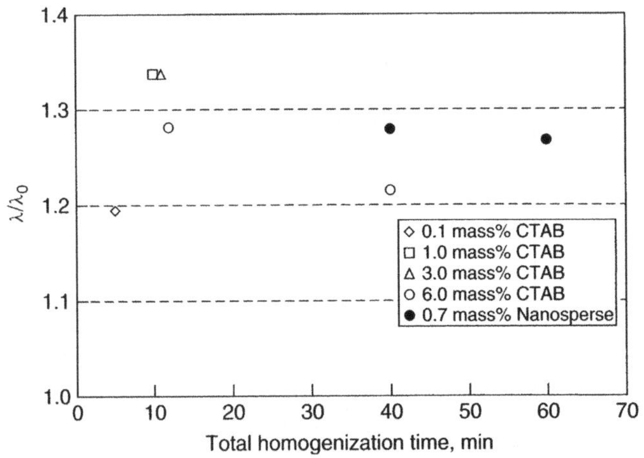

Fig. 3.69 Thermal conductivity enhancement of MWCNTs containing nanofluids (open symbols, group A; closed symbols, group B). [From Assael et al. (2005), with permission from Springer.]

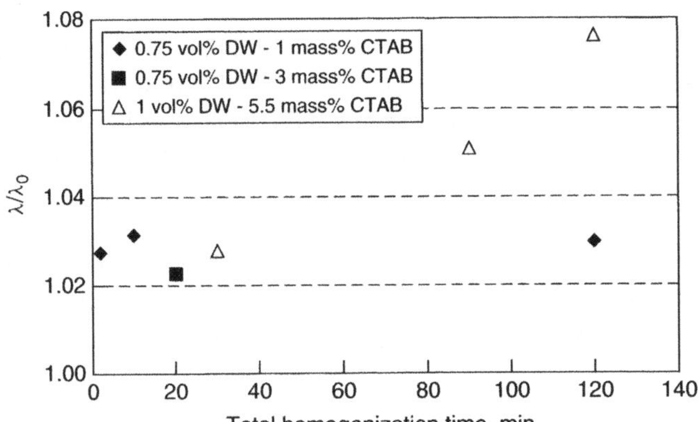

Fig. 3.70 Thermal conductivity of DWCNTs containing nanofluids (group C nanofluids). [From Assael et al. (2005), with permission from Springer.]

comparison of their XRD and TEM results confirm an average tube diameter of 13.4 nm.

Thermal conductivity measurements of nanotubes containing nanofluids were done using the transient hot-wire technique. The results are shown in Figs. 3.71 and 3.72. It was found that in CNT–ethylene glycol nanofluids, the thermal conductivity was enhanced by 12.4% with 1 vol % CNT, while in CNT engine oil nanofluids, the thermal conductivity was enhanced by 30.3% with 2 vol % CNT. Although these enhancements are much higher than the predictions of models such as the Hamilton–Crosser and Maxwell (Fig. 3.73), the enhancement

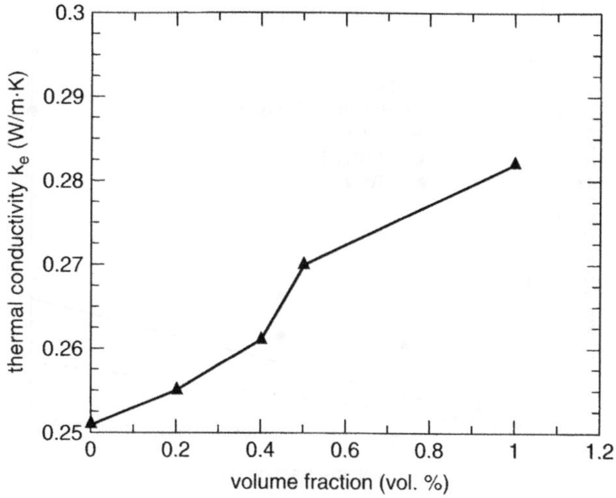

Fig. 3.71 Thermal conductivity of CNT–ethylene glycol nanofluids. [From Liu et al. (2005), with permission from Elsevier.]

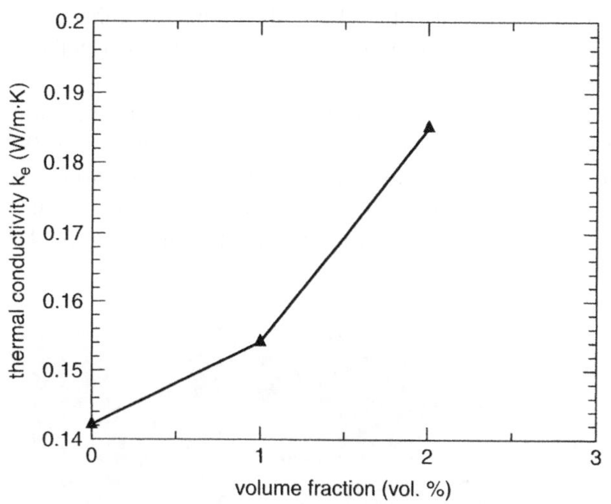

Fig. 3.72 Thermal conductivity of CNT–engine oil nanofluids. [From Liu et al. (2005), with permission from Elsevier.]

reported is much lower than that reported by Choi et al. (2001). The authors argued that the three-dimensional network formed by one-dimensional CNTs are at the root of the conductivity enhancement in CNTs containing nanofluids, as opposed to the Brownian motion in spherical particles. About the deviation from Choi et al. (2001), they cited the different source of CNTs and the presence of surfactant as probable causes, which is not very convincing. They also observed

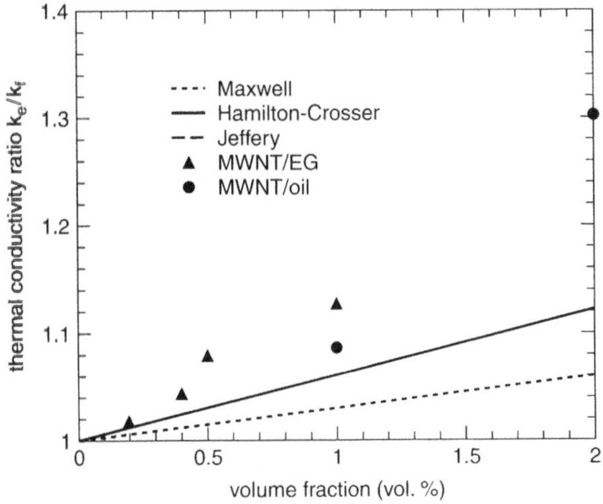

Fig. 3.73 Comparisons of the enhancement ratio of nanofluids relative to the predictive model. [From Liu et al. (2005), with permission from Elsevier.]

that the lower the conductivity of the fluid, the higher the enhancement. However, their data show that at 1% concentration, ethylene glycol–CNT nanofluids show more enhancement.

Although all the studies noted above failed to confirm the anomalously high enhancement of thermal conductivity achieved by Choi et al. (2001), one work showed clear agreement with their trends. This was the work by Ding et al. (2006), who observed that the CNTs as received from commercial products were aggregated and entangled as shown by the SEM image (Fig. 3.74).

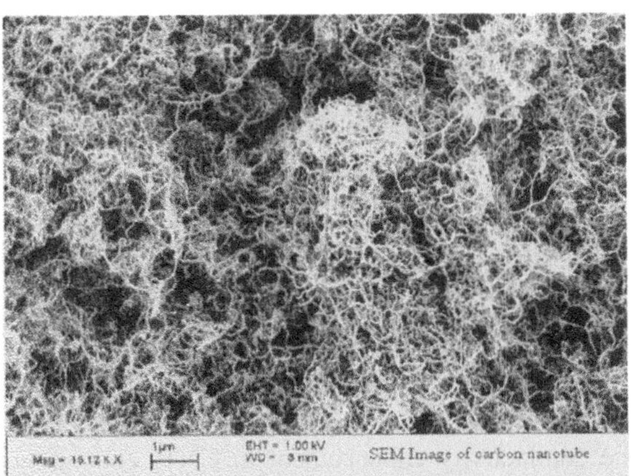

Fig. 3.74 SEM image of as-received CNTs. [From Ding et al. (2006), with permission from Elsevier.]

They used gum arabic (GA) as a dispersant and used a high-speed (24,000-rpm) stator–rotor system following ultrasonication to disperse the CNTs in water. The SEM of the dispersed nanotubes showed good dispersion (Fig. 3.75).

Whereas at room temperature their enhancement was moderate (10 to 25%), with a little elevated temperature (from 20 to 30°C) the enhancement went up to 80%, which in shown in Fig. 3.76. This result falls within the previous results

Fig. 3.75 SEM image of dispersed nanotubes. [From Ding et al. (2006), with permission from Elsevier.]

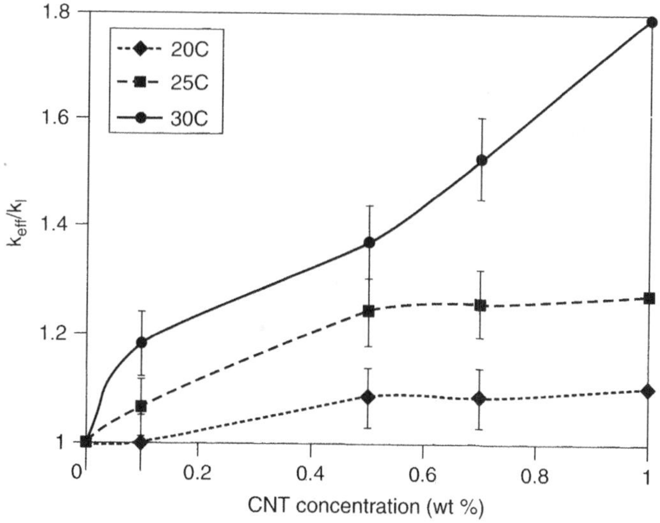

Fig. 3.76 Thermal conductivity enhancement of CNT–water nanofluids with 0.25 vol % gum arabic. [From Ding et al. (2006), with permission from Elsevier.]

of Xie et al. (2003) and Choi et al. (2001). They indicated that Choi et al. (2001) used low-conductivity oil, and hence although the percentage increase in conductivity is large, the absolute increase in not so great.

With the variety of results described above, one conclusion is probably evident: that the thermal conductivity enhancement mechanism in CNTs containing nanofluids is quite different from that of spherical nanoparticles containing nanofluids. The other apparent feature is the fact that the tube aspect ratio and the contact between the tubes play major roles in the thermal conductivity enhancements of CNT-based nanofluids. This was proven conclusively by Yang et al. (2006), who studied the detailed rheological and thermal behavior of CNT–α olefin nanofluids with polyisobutene succinimide as a dispersant. They found that this gives minimum values for both viscosity and thermal conductivity enhancement with respect to the dispersant concentration (this is at about 3 vol % of the dispersant).

Subsequently, they studied the effect of sonication and found that with more dispersing energy, the agglomerates are reduced to smaller sizes. The aspect ratio of the nanotubes is reduced significantly with an increase in the dispersing energy (Fig. 3.77). As a result, the effective conductivity also comes down (Fig. 3.78). which clearly demonstrates that the aspect ratio and tube-to-tube contact are important in enhancing the thermal conductivity of CNT nanofluids. As a direct measure of the relationship of enhancement of thermal conductivity with the aspect ratio, they presented the data in Fig. 3.79, which reconfirms that proposition. Finally, they presented thermal conductivity enhancement with particle loading as shown in Fig. 3.80. Although there is a substantial (about 200%) rise in thermal conductivity with a 0.35 vol % of nanotubes, this enhancement was

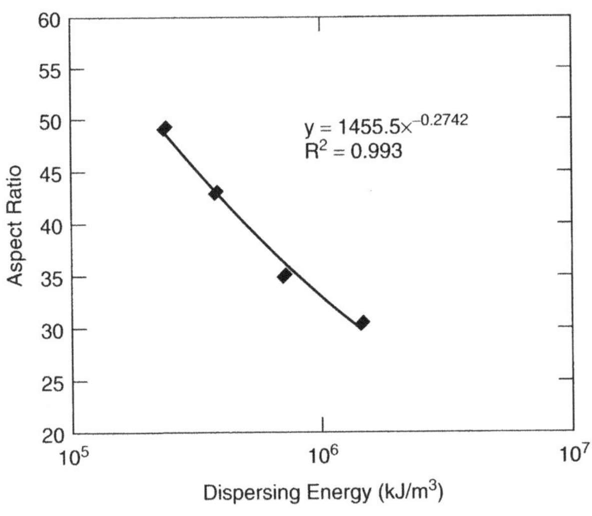

Fig. 3.77 Dispersing energy effect on the aspect ratio of nanotubes. [From Yang et al. (2006), with permission from the American Institute of Physics.]

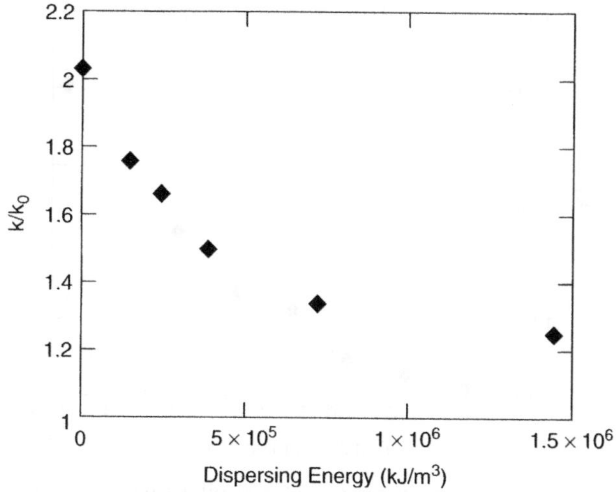

Fig. 3.78 Dispersing energy effect on thermal conductivity of CNT nanofluids. [From Yang et al. (2006), with permission from the American Institute of Physics.]

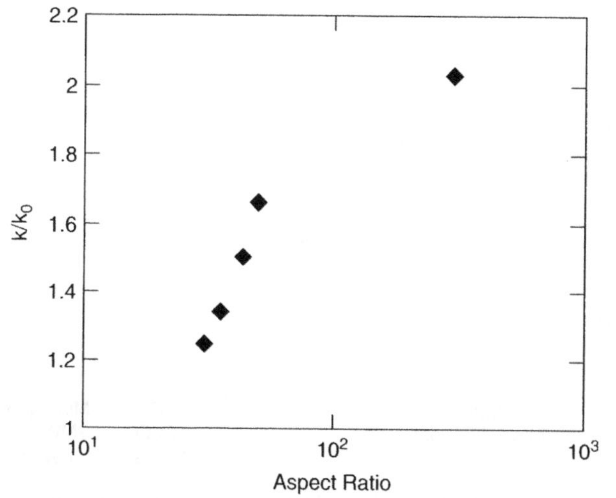

Fig. 3.79 Relationship between thermal conductivity of nanofluid and CNT aspect ratio. [From Yang et al. (2006), with permission from the American Institute of Physics.]

termed "not remarkable," due to the fact that at this volume fraction at the low shear stress of 10 Pa, the viscosity was increased by three orders of magnitude. This requires that a balance between enhancement of thermal conductivity and increase of viscosity be obtained.

Hwang et al. (2006) carried out stability and thermal conductivity studies of a variety of nanofluids with particles such as SiO_2, CuO, and fullerene (C_{60}) and

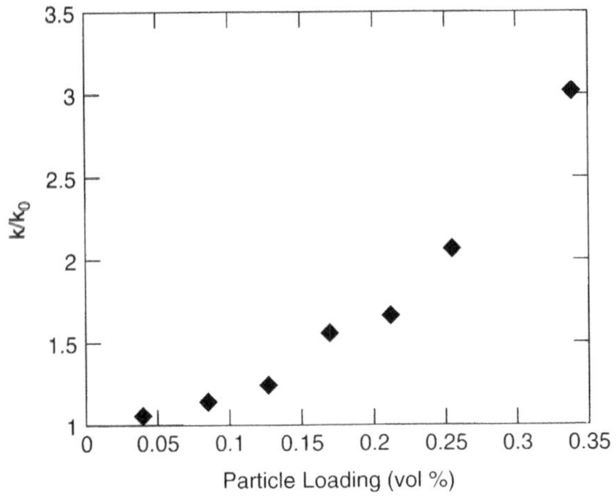

Fig. 3.80 Thermal conductivity increase in CNT–oil suspension. [From Yang et al. (2006), with permission from the American Institute of Physics.]

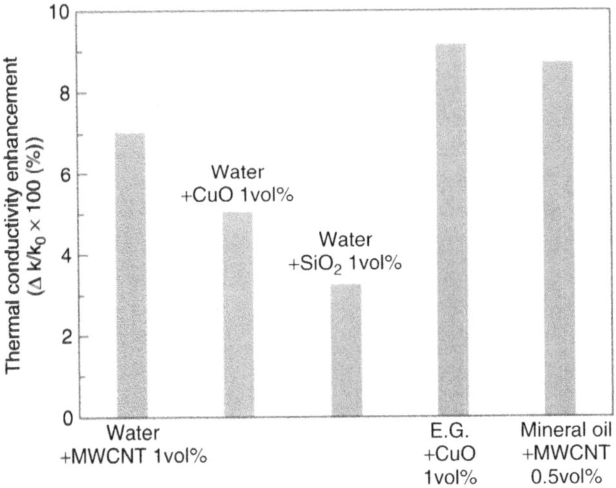

Fig. 3.81 Comparison of the thermal conductivity of various nanofluids. [From Hwang et al. (2005), with permission from Elsevier.]

compared with MWCNTs containing nanofluids. The liquids used were water, ethylene glycol, and oil. The results for thermal conduction measurement are shown in Fig. 3.81. It is evident that MWCNTs containing nanofluids provide much higher thermal conductivity enhancement than do other particles containing nanofluids. However, they found that the stability of CNTs containing nanofluids is poorer, due to agglomeration and that the proper dispersant is necessary to improve it.

The work on CNTs containing nanofluids that is cited above clearly indicates that nanotubes have a higher potential to be used in nanofluids. However, the range of enhancement reported varies from 20 to 200% for about 1 vol % volume of nanotubes and needs to be resolved. Also, any simultaneous rise in viscosity should be monitored and taken into consideration. Finally, the reasons for thermal conductivity enhancement in nanotubes containing nanofluids seem to be quite different from those of nanoparticles containing nanofluids. Arguably, Brownian motion plays a very limited role here in contrast to nanoparticles. However, the intertube connection and as a result, heat percolation through long chains of CNT, may be a dominant mechanism, which is evident from the strong effect of nanofluid conductivity enhancement on the aspect ratio of CNTs.

REFERENCES

Assael, M. J., I. N. Metaxa, J. Arvanitidis, D. Christophilos, and C. Lioutas (2005). Thermal conductivity enhancement in aqueous suspensions of carbon multi-Walled and double-walled nanotubes in the presence of two different dispersants, *Int. J. Thermophys.*, 26: 647–664.

Biercuk, M. J., M. C. Llaguno, M. Radosavljevic, J. K. Hyun, A. T. Johnson, and J. E. Fischer (2002). Carbon nanotube composites for thermal management, *Appl. Phys. Lett.*, 80: 2767–2772.

Bonnecaze, R, T., and J. F. Brady (1990). A method for determining the effective conductivity of dispersions of particles, *Proc. R. Soc. London. A*, 430: 285–313.

Bruggeman, D. A. G. (1935). Berechnung verschiedener physikalischer Konstanten von heterogenen Substanzen: I. Dielektrizitätskonstanten und Leitfähigkeiten der Mischkörper aus isotropen Substanzen, *Ann Phys. Leipzig*, 24: 636–679.

Carslaw, H. S., and J. C. Jaeger (1967). *Conduction of Heat in Solids*, Oxford University Press, New York, pp. 54–56.

Choi, S. U. S. (1995). Enhancing thermal conductivity of fluids with nanoparticles, in *Developments and Applications of Non-Newtonian Flows*, D. A. Singer, and H. P. Wang, Eds., FED 231, ASME, New York, pp. 99–105.

Choi, S. U. S. Z. G. Zhang, W. Yu, F. E. Lockwood, and E. A. Grulke (2001). Anomalous thermal conductivity enhancement in nano-tube suspensions, *Appl. Phys. Lett.*, 79: 2252–2254.

Chon, C. H., and K. D. Kihm (2005). Thermal conductivity enhancement of nanofluids by Brownian motion, *J. Heat Transfer*, 127: 810.

Chopkar, M., P. K. Das, and I. Manna (2006). Synthesis and characterization of nanofluid for advanced heat transfer applications, *Scr. Mater.*, 55: 549–552.

Czarnetzki, W., and W. Roetzel (1995). Temperature oscillation techniques for simultaneous measurement of thermal diffusivity and conductivity, *Int. J. Thermophys*. 16(2): 413–422.

Das, S. K., N. Putra, P. Thiesen, and W. Roetzel (2003). Temperature dependence of thermal conductivity enhancement for nanofluids, *J. Heat Transfer*, 125: 567–574.

Davis, R. H. (1986). Effective thermal conductivity of a composite material with spherical inclusions, *Int. J. Thermophys.*, 7(3): 609–620.

Ding, Y., H. Alias, D. Wen, and R. A. Williams (2006). Heat transfer of aqueous suspensions of carbon nanotubes (CNT nanofluids), *Int. J. Heat Mass Transfer*, 49: 240–250.

Eastman, J. A., S. U. S. Choi, S. Li, L. J. Thompson (1997). Enhanced thermal conductivity through the development of nanofluids, *Proc. Symposium on Nanophase and Nanocomposite Materials II*, Materials Research Society, Boston, MA, 457: 3–11.

Eastman, J. A., S. U. S. Choi, S. Li, W. Yu, and L. J. Thompson (2001). Anomalously increased effective thermal conductivities of ethylene glycol based nanofluids containing copper nanoparticles, *Appl. Phys. Lett.*, 78(6): 718–720.

Fourier, J. B. (1955). *Theorie analytique de la chaleur*, Paris, 1822. (English translation by Freeman, A., Dover Publications, New York.)

Hamilton, R. L., and O. K. Crosser (1962). Thermal conductivity of heterogeneous two component systems, *I & EC Fundam.*, 1(3): 187–191.

Holman, J. P. (1997) *Heat Transfer*, 8th ed., McGraw-Hill, New York.

Hong, T. K., H. S. Yang, and C. J. Choi (2005). Study of the enhanced thermal conductivity of Fe nanofluids, *J. Appl. Phys.*, 97: 064311.

Hwang, J., Y. C. Ahn, H.S. Shin, C. G. Lee, G. T. Kim, H. S. Park, and J. K. Lee (2005). Investigation on characteristics of thermal conductivity enhancement of nanofluids, *Curr. Appl. Phys.*,

Hwang, Y., H. S. Park, J. K. Lee, and W. H. Jung (2006). Thermal conductivity and lubrication characteristics of nanofluids, *Curr. Appl. Phys.*, 6S1: e67–e71.

Incropera, F. P., and D. P. DeWitt (1998). *Fundamentals of Heat Transfer*, 4th ed., Wiley, New York.

Lee, S., S. U. S. Choi, S. Li, and J. A. Eastman (1999). Measuring thermal conductivity of fluids containing oxide nanoparticles, *J. Heat Transfer*, 121: 280–289.

Li, C. H., and G.P. Peterson (2006). Experimental investigation of temperature and volume fraction variations on the effective thermal conductivity of nanoparticle suspensions (nanofluids), *J. Appl. Phys.*, 99: 084314.

Liu, M. S., M. C. C. Lin, I. T. Haung, and C. C. Wang (2005). Enhancement of thermal conductivity with carbon nanotube for nanofluids, *Int. Commun in Heat Mass Transfer*, 32: 1202–1210.

Liu, M., M. Lin, C. Y. Tsai, and C. Wang (2006). Enhancement of thermal conductivity with Cu for nanofluids using chemical reduction method, *Int. J. Heat Mass Transfer*, 49: 3028–3033.

H. Masuda, A. Ebata, K. Teramae, and N. Hishinuma (1993). Alteration of thermal conductivity and viscosity of liquid by dispersing ultra fine particles, *Netsu Bussei*, 4(4): 227–233.

Maxwell, J. C. (1881). *A Treatise on Electricity and Magnetism*, 2nd ed., Vol. 1, Clarendon Press, Oxford.

Murshed S. M. S., K. C. Leong, and C. Yang (2005). Enhanced thermal conductivity of TiO_2—water based nanofluids, *Int. J. Therm. Sci.*, 44: 367–373.

Patel, H. E., S. K. Das, T. Sundararajan, N. A. Sreekumaran, B. George, and T. Pradeep (2003). Thermal conductivities of naked and monolayer protected metal nanoparticle based nanofluids: manifestation of anomalous enhancement and chemical effects, *Appl. Phys. Lett.*, 83(14): 2931–2933.

Wasp, E. J., J. P. Kenny, and R. L. Gandhi (1977). Solid–liquid slurry pipeline transportation, series on bulk material handling, *Trans. Tech. Publications*, Clausthal, Germany.

Xie, H., J. Wang, T. Xi, and Y. Liu (2002a). Thermal conductivity of suspensions containing nanosized SiC particles, *Int. J. Thermophys.*, 23 (2): 571–580.

Xie, H. Q., J. C. Wang, T. G. Xi, Y. Liu, F. Ai, and Q. R. Wu (2002b). Thermal conductivity enhancement of suspensions containing nanosized alumina particles, *J. Appl. Phys.*, 91(7): 4568–4572.

Xie, H., H. Lee, W. Youn, and M. Choi (2003). Nanofluids containing multiwalled carbon nanotubes and their enhanced thermal conductivities, *J. Appl. Phys.*, 94: 4967–4971.

Xuan, Y., and Q. Li (2000). Heat transfer enhancement of nanofluids, *Int. J. Heat Fluid Flow*, 21: 58–64.

Yang, Y., E. A. Grulke, Z. G. Zhang, and G. Wu (2006). Thermal and rheological properties of carbon nanotube-in-oil dispersions *J. Appl Phys.*, 99: 114307.

4 Theoretical Modeling of Thermal Conductivity in Nanofluids

In this chapter we assume that a nanofluid is a mixture consisting of a continuous base fluid component called a *matrix* and a discontinuous solid component called *particles*. The properties of nanofluids depend on the details of their microstructures, such as the component properties, component volume concentrations, particle dimension, particle geometry, particle distribution, particle motion, and matrix–particle interfacial effects. It is impossible to estimate the effective properties of nanofluids unless all the details of their microstructures are known completely. One way to avoid this problem is to attempt to determine upper and lower bounds on the effective properties from partial statistical information on the sample in the form of correlation functions. Many studies have been conducted using this approach, and some of them have been reviewed by Hale (1976) and Torquato (1991). Another way to avoid this problem is to estimate the effective properties based on several reasonable assumptions on the microstructures of the mixtures. Böttcher (1952), van Beek (1967), Landauer (1978), and Nan (1993) provide excellent reviews of this approach. Readers are also referred to other references (Woodside and Messmer, 1961a,b; Davies, 1971, 1974b; Tinga et al., 1973; Fletcher, 1988; Fadale and Taya, 1991; Choi et al., 2004; Eastman et al., 2004) for useful discussions of the effective properties of mixtures. In this chapter we focus on the second approach.

The transport properties of heterogeneous mixtures have been of interest since nearly the time of Maxwell. The reason for this interest is, of course, the enormous variety of physical systems in which inhomogeneities occur. Among various transport properties, the conductivity properties, comprising the dielectric constant, magnetic permeability, electrical conductivity, and thermal conductivity, may be treated together because of the identical form of the conduction laws that control them. In this chapter the terms of the thermal conductivity are used in discussing these properties, which means that the results developed originally for other cases will be transformed into thermal conductivity terms and that with simple changes in nomenclature the results can also be applied to the other cases. Although the effective thermal conductivity of a mixture can vary in a wide range depending on the microstructures of the mixture, it must lie in between the upper

Nanofluids: Science and Technology, By Sarit K. Das, Stephen U. S. Choi, Wenhua Yu, and T. Pradeep

and lower bounds, which correspond to parallel and series particle distributions, respectively, and should reach the bounds under these distributions.

It is virtually impossible in a single chapter to cover all the literature related to the effective conductivities of mixtures because numerous theoretical and experimental studies have been conducted since Maxwell's theoretical work published more than a century ago (Maxwell, 1873). The purpose of this chapter is to review the theoretical studies of the effective conductivities of mixtures, including both important theoretical equations and new developments.

4.1. SIMPLE MIXTURE RULES

Without taking particle dimension, geometry, distribution, and motion into account, the effective thermal conductivity k_e of a mixture consisting of a base fluid matrix and suspended solid particles depends on the matrix thermal conductivity k_m, particle thermal conductivity k_p, and particle volume concentration v_p. In general, the effective thermal conductivity k_e can be expressed as a function of these parameters:

$$k_e = f(k_m, k_p, v_p) \tag{4.1}$$

Obviously, the effective thermal conductivity should give the matrix thermal conductivity at the particle-volume concentration $v_p = 0$ and the particle thermal conductivity at the particle-volume concentration $v_p = 1$. Therefore, the function $f(k_m, k_p, v_p)$ should satisfy

$$f(k_m, k_p, 0) = k_m \quad \text{and} \quad f(k_m, k_p, 1) = k_p \tag{4.2}$$

An empirical equation called the *mixture rule* (Nielsen, 1978; Nan, 1993), which meets the foregoing requirements, is generally employed:

$$k_e^n = (1 - v_p)k_m^n + v_p k_p^n \qquad -1 \le n \le 1 \tag{4.3}$$

which can be rewritten in the alternative form

$$k_e = \left\{1 + v_p\left[\left(\frac{k_p}{k_m}\right)^n - 1\right]\right\}^{1/n} k_m \qquad -1 \le n \le 1 \tag{4.4}$$

For $n = 1$, the empirical equation becomes the *parallel mixture rule* (Wiener, 1912):

$$k_e = (1 - v_p)k_m + v_p k_p = k_m + v_p(k_p - k_m) \tag{4.5}$$

This equation indicates that the effective thermal conductivity of a mixture is simply a linear combination of the matrix and particle thermal conductivities. The parallel mixture rule provides an upper bound for the effective thermal conductivity of a mixture and is valid for describing mixtures with multilayer structures distributed in parallel with respect to the heat propagation direction.

For $n = -1$, the empirical equation becomes the *series mixture rule* (Wiener, 1912):

$$k_e = \frac{1}{(1 - v_p)k_m^{-1} + v_p k_p^{-1}} = k_m + v_p \frac{k_p - k_m}{k_p - v_p(k_p - k_m)} k_m \tag{4.6}$$

The series mixture rule provides a lower bound for the effective thermal conductivity of a mixture and is valid for describing mixtures with multilayer structures distributed in series with respect to the heat propagation direction.

Landau and Lifshitz (1960) considered a mixture as being a homogeneous and isotropic medium and obtained the equation for $n = \frac{1}{3}$ by taking an average over a large volume:

$$k_e = [(1 - v_p)k_m^{1/3} + v_p k_p^{1/3}]^3 = \left\{1 + v_p \left[\left(\frac{k_p}{k_m}\right)^{1/3} - 1\right]\right\}^3 k_m \tag{4.7}$$

For the parameter n approaching zero, the empirical equation becomes the *logarithmic mixture rule:*

$$\ln k_e = (1 - v_p) \ln k_m + v_p \ln k_p \tag{4.8}$$

which also meets the requirements at the particle-volume concentrations $v_p = 0$ and $v_p = 1$ and can be rewritten in the alternative form

$$k_e = k_m^{1-v_p} k_p^{v_p} = \left(\frac{k_p}{k_m}\right)^{v_p} k_m \tag{4.9}$$

The logarithmic mixture rule was proposed by Lichtenecker (1924) based on the experimental data.

For various values of the parameter n, the effective thermal conductivity gradually increases from the lower bound to the upper bound when the parameter n gradually increases from $n = -1$ to $n = 1$. Usually, the simple mixture rules are not capable of accurately predicting the effective thermal conductivity of a mixture without knowing the microstructure details of the mixture. For example, the series and parallel mixture rules are valid only for mixtures with special particle distributions; the logarithmic mixture rule is not valid when the particle thermal conductivity, in comparison to the matrix thermal conductivity, is extremely large.

4.2. MAXWELL'S APPROACH

4.2.1. Maxwell's Equation

Maxwell was one of the first persons to investigate conduction analytically through a suspension particle. Maxwell (1873) considered a very dilute suspension of spherical particles by ignoring interactions among particles. For a mixture containing identical spherical particles of radius r_p in a field of temperature T

and temperature gradient **G**, the governing equation for the steady-state condition is the *Laplace equation:*

$$\nabla^2 T(r) = 0 \tag{4.10}$$

By introducing a large sphere of radius r_0 containing all the spherical particles dispersed in the matrix and being surrounded by the matrix, one can calculate the temperature T outside the sphere r_0 at a distance $r \gg r_0$ in two ways (Stratton, 1941; van Beek, 1967). First, the sphere r_0 is considered to consist of a heterogeneous system with an effective thermal conductivity k_e embedded in a matrix with a thermal conductivity k_m. The temperature T outside the sphere r_0 can therefore be expressed as

$$T(r) = \left(-1 + \frac{k_e - k_m}{2k_m + k_e}\frac{r_0^3}{r^3}\right)\mathbf{G}\cdot\mathbf{r} \tag{4.11}$$

which is obtained by solving the Laplace equation together with the following boundary conditions:

$$\begin{gathered} T(r)|_{r\to\infty} = -\mathbf{G}\cdot\mathbf{r} \qquad T(r)|_{r\to r_0^-} = T(r)|_{r\to r_0^+} \\ k_e\left.\frac{\partial T(r)}{\partial r}\right|_{r\to r_0^-} = k_m\left.\frac{\partial T(r)}{\partial r}\right|_{r\to r_0^+} \end{gathered} \tag{4.12}$$

Second, the temperature T is considered to be produced by all the spherical particles with a thermal conductivity k_p embedded in a matrix with a thermal conductivity k_m, and by following a similar procedure can be calculated from the superposition principle:

$$T(r) = \left(-1 + \frac{k_p - k_m}{2k_m + k_p}\frac{v_p r_0^3}{r^3}\right)\mathbf{G}\cdot\mathbf{r} \tag{4.13}$$

From the equations above, the effective thermal conductivity k_e can readily be obtained after simple algebraic manipulations:

$$k_e = k_m + 3v_p \frac{k_p - k_m}{2k_m + k_p - v_p(k_p - k_m)} k_m \tag{4.14}$$

which for low particle-volume concentrations can be further approximated as

$$k_e = k_m + 3v_p \frac{k_p - k_m}{2k_m + k_p} k_m \tag{4.15}$$

It should be noted that even given its form, Maxwell's equation is only a first-order approximation and applies only to mixtures with low particle-volume concentrations.

4.2.2. Other Extensions

Maxwell's equation has been extended many times since Maxwell's initial investigation. Wiener (1912) proposed replacing Maxwell's equation with a parameter equation:

$$k_e = k_m + (u+1)v_p \frac{k_p - k_m}{uk_m + k_p - v_p(k_p - k_m)} k_m \qquad 0 \le u \le \infty \tag{4.16}$$

For the parameter $u = 2$, the equation becomes Maxwell's equation. Wiener showed that the parameter u is $u = 0$ and $u = \infty$, respectively, for the series and parallel particle distributions, and that the parameter u must be between zero and infinity for any particle aggregate. Realizing that for a mixture the geometrical shape of the dispersed particles might influence the distribution of the temperature field and consequently the effective thermal conductivity of the mixture, Wiener assumed that the parameter u would depend only on the geometrical shape, and accordingly, not on the volume concentration, of the dispersed particles. This assumption is somehow too simplified and is untenable.

When the particle volume concentration is high, the assumption that the large sphere r_0 embedded in the matrix of the thermal conductivity k_m is no longer valid. Instead, Böttcher (1945) assumed that the large sphere r_0 embedded in the matrix of the thermal conductivity k_e rather than k_m and derived the equation

$$(1 - v_p)\frac{k_m - k_e}{2k_e + k_m} + v_p \frac{k_p - k_e}{2k_e + k_p} = 0 \tag{4.17}$$

which was obtained by replacing the denominators $2k_m + k_e$ and $2k_m + k_p$ in the temperature expressions with $3k_e$ and $2k_e + k_p$, respectively. Unlike Maxwell's equation, Böttcher's equation is symmetrical with respect to k_m and k_p. Bruggeman (1935) suggested another way to extend Maxwell's equation to high particle-volume concentrations. Bruggeman employed an integration scheme to take into account the concentration of dispersed particles in the immediate neighborhood of a certain particle. In Bruggeman's scheme the initially low concentration is gradually increased by infinitesimal additions of dispersed component k_p. In the course of this process the thermal conductivity of the medium around a particle changes slowly from k_m to k_e, the finally thermal conductivity of the heterogeneous mixture. Bruggeman applied this idea to the approximation of Maxwell's equation at low particle-volume concentrations and obtained the differential equation

$$\frac{dk}{dv} = \frac{1}{1-v}\frac{3k(k_p - k)}{2k + k_p} \tag{4.18}$$

which upon integration between $v = 0$ $(k = k_m)$ and $v = v_p$ $(k = k_e)$ leads to

$$1 - v_p = \frac{k_p - k_e}{k_p - k_m}\left(\frac{k_m}{k_e}\right)^{1/3} \tag{4.19}$$

Looyenga (1965) assumed that on mixing of two components,

$$k_m = k_e - \Delta k \qquad k_p = k_e + \Delta k \tag{4.20}$$

Inserting these equations into Maxwell's equation gives the variable particle-volume concentration as a function of k_e and Δk:

$$v = \frac{3k_e - \Delta k}{2(3k_e - 2\Delta k)} \approx \frac{1}{2} + \frac{1}{6}\frac{\Delta k}{k_e} \tag{4.21}$$

This assumption indicates that the mixture is composed of components with thermal conductivities $k_m = k_e - \Delta k$ and $k_p = k_e + \Delta k$ in respective variable volume concentrations $1 - v$ (the matrix component) and v (the dispersed particle component). From the superposition principle, the actual particle-volume concentration of the mixture is then

$$v_p(k_e) = (1 - v)v_p(k_e - \Delta k) + v v_p(k_e + \Delta k) \tag{4.22}$$

In equation (4.22) the functions $v_p(k_e - \Delta k)$ and $v_p(k_e + \Delta k)$ represent the actual v_p values for mixture of thermal conductivities $k_e - \Delta k$ and $k_e + \Delta k$, respectively. With Taylor's series expansion, the equation can be expressed as

$$v \approx \frac{1}{2} - \frac{\Delta k}{4}\left(\frac{d^2 v_p}{dk_e^2} \bigg/ \frac{dv_p}{dk_e}\right) \tag{4.23}$$

These equations lead to the differential equation

$$3k_e \frac{d^2 v_p}{dk_e^2} + 2\frac{dv_p}{dk_e} = 0 \tag{4.24}$$

The solution satisfying the boundary conditions $k_e = k_m$ for $v_p = 0$ and $k_e = k_p$ for $v_p = 1$ is

$$k_e = [(1 - v_p)k_m^{1/3} + v_p k_p^{1/3}]^3 = \left\{1 + v_p\left[\left(\frac{k_p}{k_m}\right)^{1/3} - 1\right]\right\}^3 k_m \tag{4.25}$$

which is the same as the Landau and Lifshitz mixture rule for $n = \frac{1}{3}$. Davies (1974a) used Looyenga's approach for a macroscopically isotropic mixture consisting of isotropic fibers embedded in an isotropic matrix, and obtained the same result as equation (4.25).

Maxwell's equation can be rewritten as

$$k_e = k_m + \frac{3v_p[(k_p - k_m)/(2k_m + k_p)]}{(1 - v_p) + 3v_p[k_m/(2k_m + k_p)]} k_m \tag{4.26}$$

To consider the effect of particle clusters and cluster distribution, Wang et al. (2003) replaced the particle thermal conductivity in equation (4.26) with the cluster effective thermal conductivity $k_p(r)$, which was suggested to be estimated with the Bruggeman equation (Bruggeman, 1935), and obtained the following fractal model by taking an average over the cluster radius distribution:

$$k_e = k_m + \frac{3v_p \int_0^\infty \{[k_p(r) - k_m]/[2k_m + k_p(r)]\} n(r)\, dr}{(1 - v_p) + 3v_p \int_0^\infty \{k_m/[2k_m + k_p(r)]\} n(r)\, dr} k_m \tag{4.27}$$

In equation (4.27) the cluster radius distribution function $n(r)$ is estimated approximately by a lognormal distribution of the form

$$n(r) = \frac{1}{\sqrt{2\pi}(\ln \sigma) r} e^{-[\ln(r/r_p)/(\sqrt{2\pi} \ln \sigma)]^2} \tag{4.28}$$

where the standard deviation σ can take the classic value of 1.5 (Wang et al., 2003).

4.3. PARTICLE DISTRIBUTIONS

It is well known that a specification of the particle-volume concentrations alone is not normally sufficient to determine the effective thermal conductivity absolutely. The wide difference between the effective thermal conductivities of the series and parallel particle distributions implies that the particle distributions play an important role in the effective thermal conductivities of mixtures. Many studies have been conducted on the influence of particle distributions on the effective thermal conductivity of a mixture, among which regular and random distributions are two of the simplest and most commanly used distributions.

4.3.1. Regular Distribution

Rayleigh (1892) was the first to analyze the effective thermal conductivity of a mixture with a regular particle distribution. Rayleigh considered the case of a simple cubic array of identical spheres and assumed that the ambient temperature gradient at the center of the reference sphere is equal to the average temperature gradient plus the sum of the fields produced by the surrounding spheres. In essential, Rayleigh solved the Laplace equation for the temperature gradient inside and about a sphere by using the superposition principle to take into account the effects of surrounding spheres on the field in the neighborhood of the central sphere. The resulting final effective thermal conductivity, with two errors corrected, is

$$k_e = k_m + 3v_p \frac{k_p - k_m}{2k_m + k_p - v_p\{1 + 3.939 v_p^{7/3}[(k_p - k_m)/(4k_m + 3k_p)]\}(k_p - k_m)} k_m \tag{4.29}$$

The first correction for the omission of a factor of $1/\pi$ in Rayleigh's derivation was discovered by Runge (1925). The second correction is the omission of a factor of $\frac{5}{2}$ that arises in a partial differentiation (McPhedran and McKenzie, 1978). It is interesting that these two errors compensated for each other to some extent.

Meredith and Tobias (1960) extended Rayleigh's derivation, and by using a different potential function and considering higher terms in the series expansion for the potential in the continuous component obtained an equation that appears to be more satisfactory at high particle-volume concentrations:

$$k_e = k_m + 3v_p\left(1 - 1.227v_p^{7/3}\frac{k_p - k_m}{4k_m + 3k_p}\right) \times \frac{k_p - k_m}{2k_m + k_p - v_p\left[1 + 1.227v_p^{4/3}\frac{(2k_m + k_p) + 2.215v_p(k_p - k_m)}{4k_m + 3k_p}\right](k_p - k_m)} k_m \tag{4.30}$$

The validity of Rayleigh's method has been questioned because it involved the summation of a non-absolutely convergent series (Levine, 1966; Jeffrey, 1973). McPhedran and McKenzie (1978) and O'Brien (1979) have, however, modified Rayleigh's method to overcome this difficulty, and this modified method now has a sound theoretical basis. By means of an extensive application of this technique, McPhedran and McKenzie (1978) and McKenzie et al. (1978) have computed a complete range of v_p and k_p/k_m for the simple, body-centered, and face-centered cubic arrays. Zuzovski and Brenner (1977) devised an alternative method to Rayleigh's original treatment and avoided the convergence problem by formulating methods that rely on the fact that the temperature gradient is periodic. Sangani and Acrivos (1983) further modified Zuzovski and Brenner's method by taking advantage of the symmetry of spherical particles and therefore simplified subsequent calculations significantly. Sangani and Acrivos also derived k_e to the order of $v_p{}^9$ as a function of v_p and k_p/k_m for the simple, body-centered, and face-centered cubic arrays.

It should be noted that the idea of using cubic arrays of identical spherical particles for calculating the effective thermal conductivities of mixtures is not totally out of the box because the cubic array models usually give results similar to those of uniform random distribution models. On the other hand, even with their improvements to the first approximation, the cubic array models, especially low-order explicit solutions, are valid only for low-particle-volume concentrations. When the particle-volume concentrations approach the critical values ($\pi/6$, $\sqrt{3}\,\pi/8$, and $\sqrt{2}\,\pi/6$ for the simple, body-centered, and face-centered cubic arrays, respectively) at which the spheres touch, the low-order explicit solutions no longer hold, and alternative approaches are needed (Keller, 1963; Batchelor and O'Brien, 1977; Bergman, 1979).

4.3.2. Random Distribution

Jeffrey (1973) was the first to calculate the effective thermal conductivity correct to the second order of the particle-volume concentration v_p for a random dispersion of spheres. Assuming a "well-stirred" suspension in which the second sphere occupies all allowable positions with equal probability, Jeffrey was able to calculate the interactions between pairs of spheres and derived the following second-order equation:

$$k_e = k_m + 3v_p \left[1 + v_p \frac{\sigma_1(2k_m + k_p) + (k_p - k_m)}{2k_m + k_p} \right] \frac{k_p - k_m}{2k_m + k_p} k_m \tag{4.31}$$

Davis (1986) also considered the effective thermal conductivity of a mixture containing identical spherical particles randomly dispersed in the mixture and obtained the following second-order equation:

$$k_e = k_m + 3v_p(1 + v_p\sigma_1) \frac{k_p - k_m}{2k_m + k_p - v_p(k_p - k_m)} k_m \tag{4.32}$$

In equations (4.31) and (4.32) the parameter σ_1, depending on the thermal conductivity ratio k_p/k_m, is the sum of a slowly convergent series, requiring the summation of over 100 terms before being correct to three significant figures.

4.3.3. Combination Distribution

Instead of the parallel or series distributions of particles, it is reasonable to use a combination of parallel and series distributions of particles to get a better estimation of the effective thermal conductivity of a mixture. Tsao (1961) used this approach and developed an equation for the effective thermal conductivity of a mixture. In Tsao's approach, a unit cube of a mixture consisting of a continuous component and a discontinuous component is sliced into many thin layers that are perpendicular to the unidirectional heat propagation. Since thermal conductivity in parallel is additive, the discontinuous component of each layer can be rearranged into a rectangle without changing the effective thermal conductivity of each individual layer. The sequence of the layers is then rearranged into a continuous distribution function of the discontinuous component. By assuming a normal distribution of the discontinuous component, the effective thermal conductivity of the mixture can be determined by the integral

$$k_e = \frac{1}{\int_0^1 \left\{ k_m + (k_p - k_m) \int_v^1 (\sqrt{2\pi}\,\sigma)^{-1} e^{-[(v-v_p)/(\sqrt{2}\sigma)]^2} dv \right\}^{-1} dv} \tag{4.33}$$

Unfortunately, it is impossible to get a close form of the integral equation, and experimental data are necessary in determining the standard deviation σ before the effective thermal conductivity can be calculated. Cheng and Vachon (1969) modified Tsao's approach to eliminate the requirement for experimental data. By eliminating all terms except the first two in the Taylor series expansion of the assumed normal distribution of the discontinuous component in the continuous component and by assuming similar shrinkage along the x (parallel to the heat propagation direction) and y (perpendicular to the heat propagation direction) directions from their maximum absolute values $\frac{1}{2}$ and 1, respectively, Cheng and Vachon obtained an approximate parabolic distribution of the discontinuous component:

$$y = \frac{1}{\sqrt{6\upsilon_p}}(3 - 8\upsilon_p x^2) \tag{4.34}$$

With this parabolic distribution of the discontinuous component, the effective thermal conductivity of the mixture can be expressed as

$$k_e = \left\{1 - \sqrt{\frac{3\upsilon_p}{8}}\left[2 - \frac{\sqrt{2/(3\upsilon_p)}k_m}{\sqrt{k_p - k_m}\sqrt{\sqrt{2/(3\upsilon_p)}k_m + (k_p - k_m)}} \times \ln \frac{\sqrt{\sqrt{2/(3\upsilon_p)}k_m + (k_p - k_m)} + \sqrt{k_p - k_m}}{\sqrt{\sqrt{2/(3\upsilon_p)}k_m + (k_p - k_m)} - \sqrt{k_p - k_m}}\right]\right\}^{-1} k_m \tag{4.35}$$

By considering a simple cubic lattice unit of a mixture and assuming that spherical particles are dispersed uniformly in the mixture and are located at the vertexes of the simple cubic lattice, Yu and Choi (2005) obtained the following equation for the effective thermal conductivity from the superposition principle for parallel and series thermal conductivities:

$$k_e = \left\{1 - \sqrt[3]{\frac{3\upsilon_p}{4\pi}}\left[2 - \frac{\sqrt[3]{16/(9\pi\upsilon_p^2)}k_m}{\sqrt{k_p - k_m}\sqrt{\sqrt[3]{16/(9\pi\upsilon_p^2)}k_m + (k_p - k_m)}} \times \ln \frac{\sqrt{\sqrt[3]{16/(9\pi\upsilon_p^2)}k_m + (k_p - k_m)} + \sqrt{k_p - k_m}}{\sqrt{\sqrt[3]{16/(9\pi\upsilon_p^2)}k_m + (k_p - k_m)} - \sqrt{k_p - k_m}}\right]\right\}^{-1} k_m \tag{4.36}$$

This model predicts a unique feature, a nonlinear relationship of the effective thermal conductivity with the particle-volume concentration, whereas the concentration–conductivity curve changes from convex upward to concave upward with increasing particle-volume concentrations. Because it is highly desirable to achieve higher thermal conductivity at smaller particle concentrations for many practical applications of mixtures, this behavior is an attractive feature of structured mixtures that at low particle-volume concentrations would enhance the poor thermal conductivity of base fluids.

4.4. PARTICLE GEOMETRIES

Particle geometries play an important role in the effective thermal conductivity of a mixture. It is well known that elongated particles contact with each other more easily than spherical particles with the same particle-volume concentration. In other words, elongated particles can form longer continuous links of particles and therefore transfer heat better if the thermal conductivity of the particles is higher than that of the matrix, which is usually true for nanofluids.

4.4.1. Depolarization Factors

Ellipsoidal particles are usually chosen in analyzing the influence of particle geometries on the effective thermal conductivity of a mixture. There are two main reasons for choosing ellipsoidal particles. First, a change in the geometrical parameters characterizing them can represent the various types of particles actually used in mixtures. Second, even with some approximations, this is the only case in which a closed expression can be given for the effective thermal conductivity of a mixture. An ellipsoid with semiaxes a, b, and c $(a \geq b \geq c)$ can be expressed as

$$\frac{x^2}{a^2} + \frac{y^2}{b^2} + \frac{z^2}{c^2} = 1 \tag{4.37}$$

whose influences on the effective thermal conductivity of a mixture can be characterized by depolarization factors d_{pi} $(i = a, b, c)$, defined by

$$d_{pi} = \frac{abc}{2} \int_0^{\infty} \frac{1}{(i^2 + w)\sqrt{(a^2 + w)(b^2 + w)(c^2 + w)}}\, dw \tag{4.38}$$

which satisfy

$$d_{pa} + d_{pb} + d_{pc} = 1 \tag{4.39}$$

For a needle-shaped ellipsoid with $a \gg b = c$, d_{pa}, d_{pb}, and d_{pc} tend to 0, $\frac{1}{2}$, and $\frac{1}{2}$, respectively. For a disk-shaped ellipsoid with $a = b \gg c$, d_{pa}, d_{pb}, and d_{pc} tend to 0, 0, and 1, respectively. For a sphere with $a = b = c$, the three factors are $\frac{1}{3}$. In certain conditions, the elliptic integral above can be simplified. For a prolate spheroid with $a > b = c$, d_{pa}, d_{pb}, and d_{pc} are given in terms of the eccentricity $e = \sqrt{1 - (c/a)^2}$:

$$d_{pa} = 1 - 2d_{pb} = 1 - 2d_{pc} = \frac{1 - e^2}{2e^3}\left(\ln \frac{1+e}{1-e} - 2e\right) \tag{4.40}$$

For an oblate spheroid with $a = b > c$, d_{pa}, d_{pb}, and d_{pc} are given by

$$d_{pa} = d_{pb} = \frac{1}{2}(1 - d_{pc}) = \frac{1}{2e^3}\left[\sqrt{1 - e^2}\arcsin e - e(1 - e^2)\right] \tag{4.41}$$

4.4.2. Fricke's Equation

The derivation of the effective thermal conductivity for dispersed ellipsoids proceeds in essentially the same way as for dispersed spheres by inserting the expression for the temperature outside an ellipsoid instead of that for temperature outside a sphere. For ellipsoidal particles with a random distribution of orientations, Fricke (1924) derived an equation by assuming that the ellipsoidal particles are surrounded by a medium of the matrix thermal conductivity k_m:

$$k_e = k_m + \frac{\frac{1}{3}v_p \sum\limits_{i=a,b,c} \dfrac{k_p - k_m}{k_m + d_{pi}(k_p - k_m)}}{(1 - v_p) + \frac{1}{3}v_p \sum\limits_{i=a,b,c} \dfrac{k_m}{k_m + d_{pi}(k_p - k_m)}} k_m \tag{4.42}$$

which for low particle-volume concentrations can be reduced to (Fricke, 1953)

$$k_e = k_m + \frac{1}{3}v_p \sum_{i=a,b,c} \frac{k_p - k_m}{k_m + d_{pi}(k_p - k_m)} k_m \tag{4.43}$$

In obtaining equations (4.42) and (4.43), the average method is used for the effective thermal conductivities along three axes of the ellipsoidal particles with the assumption that the three axes are aligned with the heat propagation direction with equal probability.

With proper values for depolarization factors d_{pi} ($i = a, b, c$), these equations can also be used for the special particle shapes of spheres, spheroids, needles, and disks. For long, thin needles, d_{pa}, d_{pb}, and d_{pc} tend to 0, $\frac{1}{2}$, and $\frac{1}{2}$, respectively, and Fricke's equation is reduced to

$$k_e = k_m + \frac{1}{3}v_p \frac{5k_m + k_p}{k_m + k_p}(k_p - k_m) \tag{4.44}$$

Niesel (1952) applied Bruggeman's integration scheme to equation (4.44) and obtained the differential equation

$$\frac{dk}{dv} = \frac{1}{1-v}\frac{(5k+k_p)(k_p-k)}{3(k+k_p)} \tag{4.45}$$

which upon integration between $v = 0$ $(k = k_m)$ and $v = v_p$ $(k = k_e)$ leads to

$$1 - v_p = \frac{k_p - k_e}{k_p - k_m}\left(\frac{5k_m + k_p}{5k_e + k_p}\right)^{2/5} \tag{4.46}$$

For flat thin disks, d_{pa}, d_{pb}, and d_{pc} tend to 0, 0, and 1, respectively, and Fricke's equation is reduced to

$$k_e = k_m + \frac{1}{3}v_p\frac{k_m + 2k_p}{k_p}(k_p - k_m) \tag{4.47}$$

Applying Bruggeman's integration scheme to equation (4.47), one may obtain the differential equation

$$\frac{dk}{dv} = \frac{1}{1-v}\frac{(k+2k_p)(k_p-k)}{3k_p} \tag{4.48}$$

which upon integration between $v = 0$ $(k = k_m)$ and $v = v_p$ $(k = k_e)$ leads to

$$1 - v_p = \frac{k_p - k_m}{k_p - k_e}\frac{k_e + 2k_p}{k_m + 2k_p} \tag{4.49}$$

Considering a mixture containing carbon nanotube particles with $d_{pa} = 0$ and $d_{pb} = d_{pc} = \frac{1}{2}$ and assuming that $k_p \gg k_m$, Nan et al. (2003) obtained the following approximate equation from Fricke's equation:

$$k_e = \frac{3 + v_p(k_p/k_m)}{3 - 2v_p}k_m \tag{4.50}$$

which can be further simplified for low particle-volume concentrations

$$k_e = k_m + \frac{1}{3}v_p k_p \tag{4.51}$$

This quite simple equation clearly demonstrates the large thermal conductivity enhancement induced by the high thermal conductivity of the carbon nanotubes.

For anisotropic particles with particle thermal conductivities k_{pi} $(i = a, b, c)$ along three axes, one may generalize Fricke's equation and obtain

$$k_e = k_m + \frac{1}{3}v_p \sum_{i=a,b,c}\frac{k_{pi} - k_m}{k_m + d_{pi}(k_{pi} - k_e)}k_m \tag{4.52}$$

By applying Bruggeman's integration scheme, equation (4.52) can be rewritten as the following differential equation:

$$\frac{dk}{dv} = \frac{1}{1-v}\frac{1}{3}\sum_{i=a,b,c}\frac{k(k_{pi}-k)}{k+d_{pi}(k_{pi}-k)} \tag{4.53}$$

which upon integration between $v=0$ ($k=k_m$) and $v=v_p$ ($k=k_e$) leads to

$$\ln(1-v_p) = \int_{k_e}^{k_m}\left[\frac{1}{3}\sum_{i=a,b,c}\frac{k(k_{pi}-k)}{k+d_{pi}(k_{pi}-k)}\right]^{-1} dk \tag{4.54}$$

Using this approach, Gao and Zhou (2006) derived an effective thermal conductivity model for mixtures containing spheroidal particles, which shows a nonlinear dependence of the effective thermal conductivity on the particle-volume concentrations.

4.4.3. Polder and van Santen's Equation

Instead of k_m, Polder and van Santen (1946) and Taylor (1965, 1966) assumed that the ellipsoidal particles are surrounded by a medium of the effective thermal conductivity k_e and obtained the equation

$$k_e = k_m + \frac{1}{3}v_p\sum_{i=a,b,c}\frac{k_p-k_m}{k_e+d_{pi}(k_p-k_e)}k_e \tag{4.55}$$

Even with the assumption, Polder and van Santen (1946) have shown that equation (4.55) only holds for low particle-volume concentrations. However, one may still consider the Polder–van Santen equation as a useful approximation for not very dilute suspensions and prefer the Polder–van Santen equation to Fricke's equation. In general, the Polder–van Santen equation has an irrational expression in k_e. Mandel (1961) derived some approximate expressions of the Polder–van Santen equation that yield, under certain circumstances, rational expressions for k_e.

For long, thin needles, the Polder–van Santen equation is reduced to

$$k_e = k_m + \frac{1}{3}v_p\frac{5k_e+k_p}{k_e+k_p}(k_p-k_m) \tag{4.56}$$

which was also obtained by Davies (1974a) for a macroscopically isotropic mixture consisting of isotropic fibers embedded in an isotropic matrix. For thin, flat disks, the Polder–van Santen equation, after simple algebraic manipulations, is reduced to

$$k_e = k_m + \frac{1}{3}v_p\frac{k_m+2k_p}{k_p-(1/3)v_p(k_p-k_m)}(k_p-k_m) \tag{4.57}$$

which was first obtained by Bruggeman (1935) by means of his integration scheme and is known as the *Bruggeman equation for disks*. It is interesting that this equation is also obtained by interchanging k_m and k_p in Maxwell's equation.

4.4.4. Other Equations

Rayleigh (1892) applied the simple cubic arrays to cylindrical particles whose axes are perpendicular to the heat propagation direction and obtained the equation

$$k_e = k_m + 2v_p \frac{k_p - k_m}{k_m + k_p - v_p\left(1 + 0.3058 v_p^3 \dfrac{k_p - k_m}{k_m + k_p}\right)(k_p - k_m)} k_m \tag{4.58}$$

Perrins et al. (1979) extended Rayleigh's approach to both the simple cubic and hexagonal arrays.

Sillars (1937) showed that the effect of a very long prolate spheroid is negligible when the longest axis is perpendicular to the main field compared with when the longest axis is parallel to the main field. Based on this fact, Sillars derived the following equation for prolate spheroids with their longest axes parallel to the main field:

$$k_e = k_m + v_p \frac{k_p - k_m}{k_m + d_{pa}(1 - v_p)(k_p - k_m)} k_m \tag{4.59}$$

which can be approximated by the following equation for very low particle-volume concentrations:

$$k_e = k_m + v_p \frac{k_p - k_m}{k_m + d_{pa}(k_p - k_m)} k_m \tag{4.60}$$

Brown (1955) used a method analogous to the molecular theory and obtained a rigorous solution in series form that depends on certain statistical properties of the particle geometry. The result up to the second term, without taking statistical properties of the particle geometry into account, can be expressed as

$$k_e = \left\{1 - \frac{1}{3} v_p(1 - v_p)\left[\frac{k_p - k_m}{k_m + v_p(k_p - k_m)}\right]^2\right\} [k_m + v_p(k_p - k_m)] \tag{4.61}$$

which is a correction to the parallel mixture rule.

Based on a wide variety of published experimental data, Pearce (1955) introduced an empirical parameter α and proposed the equation

$$k_e = \left[1 - v_p \frac{\alpha(1 - v_p)}{1 - \alpha v_p} \frac{k_p - k_m}{k_m + v_p(k_p - k_m)}\right] [k_m + v_p(k_p - k_m)] \tag{4.62}$$

which is also a correction to the parallel mixture rule and reduces to the parallel mixture rule for $\alpha = 0$.

Hamilton and Crosser (1962) studied the effective thermal conductivity of a mixture by introducing an empirical shape factor, which can be correlated as a function of the sphericity Ψ of the particle defined as the ratio of the surface area of a sphere, with a volume equal to that of the particle, to the surface area of the particle. Hamilton and Crosser modified Maxwell's equation and obtained the following equation for the effective thermal conductivity of a mixture:

$$k_e = k_m + 3\Psi^{-1} v_p \frac{k_p - k_m}{(3\Psi^{-1} - 1)k_m + k_p - v_p(k_p - k_m)} k_m \tag{4.63}$$

which reduces to Maxwell's equation for spherical particles of the sphericity $\Psi = 1$.

Lu and Lin (1996) determined the effective thermal conductivity of a mixture containing aligned spheroids with the pair interaction taken rigorously into account. Lu and Lin solved a boundary value problem involving two aligned spheroids with a boundary collocation scheme to evaluate the pair interaction and expressed the effective thermal conductivity in the form of expansion in the particle-volume concentration. Lu and Lin derived the explicit expressions of the second-order expansion for the parallel and perpendicular aligned spheroids:

$$k_e = (1 + a_1 v_p + a_2 v_p^2) k_m \tag{4.64}$$

In equation (4.64) the parameters a_1 and a_2 depend on the thermal conductivity ratio k_p/k_m and the aspect ratio of the spheroids. Lu and Lin tabulated the parameters a_1 and a_2 for various thermal conductivity ratios (0, 0.1, 2, 10, 100, 1000, and ∞) and various aspect ratios (1, 10/9, 2, 5, and 10).

Instead of the equal probability average, Xue (2005) introduced a distribution function $P(w)$ and for carbon nanotube–based mixtures derived

$$k_e = k_m + \frac{v_p \int_0^{1/2} \dfrac{k_p - k_m}{k_m + w(k_p - k_m)} P(w)\, dw}{(1 - v_p) + v_p \int_0^{1/2} \dfrac{k_m}{k_m + w(k_p - k_m)} P(w)\, dw} k_m \tag{4.65}$$

In arriving at equation (4.65) it is assumed that the depolarization factor equals 0 and 1/2 when the carbon nanotube particles are parallel and perpendicular to the applied field, respectively. The expression of the effective thermal conductivity can be obtained by choosing proper forms of the distribution function $P(w)$. For a normal-like distribution $P(w) = (2/\pi)/\sqrt{w(1-w)}$, the effective thermal conductivity, with the integral error corrected, is (Xue, 2005)

$$k_e = k_m + \frac{(4/\pi) v_p [(k_p - k_m)/\sqrt{k_m k_p}] \arctan \sqrt{k_p/k_m}}{(1 - v_p) + (4/\pi) v_p (k_m/\sqrt{k_m k_p}) \arctan \sqrt{k_p/k_m}} k_m \tag{4.66}$$

For a uniform distribution $P(w) = 2$, the effective thermal conductivity is (Xue, 2005)

$$k_e = k_m + \frac{2v_p \ln\left(\dfrac{k_p + k_m}{2k_m}\right)}{(1 - v_p) + 2v_p \dfrac{k_m}{k_p - k_m} \ln\left(\dfrac{k_p + k_m}{2k_m}\right)} k_m \tag{4.67}$$

4.5. SYMMETRICAL EQUIVALENT MEDIUM THEORY

A physically very transparent derivation of the effective medium theory was given by Landauer (1952) in terms of the electrical conductivity. Landauer considered a spherical particle of radius r_p and thermal conductivity k_p, which is embedded in an effective medium described by thermal conductivity k_e. For a temperature gradient $\mathbf{G}$ that exists far from the spherical particle, the dipole moment for the spherical particle is $r^3[(k_p - k_e)/(2k_e + k_p)]\mathbf{G}$. If the volume fraction v_p is occupied by such spherical particles, their polarization is

$$\mathbf{P}_1 = v_p \frac{k_p - k_e}{2k_e + k_p} \mathbf{G} \tag{4.68}$$

The rest of the two-component medium is regarded as spherical particles of thermal conductivity k_m embedded similarly in an effective medium of thermal conductivity k_e. Their polarization is then

$$\mathbf{P}_2 = (1 - v_p) \frac{k_m - k_e}{2k_e + k_m} \mathbf{G} \tag{4.69}$$

The effective medium condition is equivalent to stating that the total polarization summed over the two types of components should vanish. Thus, for spherical particles,

$$(1 - v_p) \frac{k_m - k_e}{2k_e + k_m} + v_p \frac{k_p - k_e}{2k_e + k_p} = 0 \tag{4.70}$$

This equation is the same as the Böttcher equation (1945) and was first obtained by Bruggeman (1935). This approximation, which treats the components on a symmetrical basis, has become the most commonly invoked approximation in this field and can easily be generalized to any number of components.

Equation (4.70) can be rewritten as

$$(1 - v_p) \frac{k_m - k_e}{2k_e + k_m} + \frac{1}{3} v_p \frac{k_p - k_e}{k_e + (1/3)(k_p - k_e)} = 0 \tag{4.71}$$

Hence, the term $(k_p - k_e)/(2k_e + k_p)$ is proportional to the polarizability for a sphere, which has a depolarization factor $\frac{1}{3}$. By using the polarizability for an ellipsoid instead of that for a sphere, this equation can be generalized for randomly oriented ellipsoids:

$$(1 - v_p)\frac{k_m - k_e}{2k_e + k_m} + \frac{1}{9}v_p \sum_{i=a,b,c} \frac{k_p - k_e}{k_e + d_{pi}(k_p - k_e)} = 0 \tag{4.72}$$

which was proposed by Granqvist and Hunderi (1977, 1978). Xue (2000) used ellipsoidal polarizability on both matrix and particle components, whose depolarization factors are d_{mi} ($i = a, b, c$) and d_{pi} ($i = a, b, c$), respectively, and further generalized equation (4.72) into

$$(1 - v_p) \sum_{i=a,b,c} \frac{k_m - k_e}{k_e + d_{mi}(k_m - k_e)} + v_p \sum_{i=a,b,c} \frac{k_p - k_e}{k_e + d_{pi}(k_p - k_e)} = 0 \tag{4.73}$$

Stroud (1975) generalized the effective medium approximation to treat, in principle, materials consisting of crystallites of arbitrary shape and conductivity tensors of arbitrary symmetry. Davidson and Tinkham (1976) proposed phenomenological equations for the effective conductivity of microscopically inhomogeneous materials. These equations combine ideas from the effective medium theory and from percolation theory and, for some interesting cases, improve on the effective medium theory in the vicinity of the percolation threshold.

4.6. MATRIX–PARTICLE INTERFACIAL EFFECTS

Because of coated particles, interfacial phenomena, stabilizing agents, adsorbed substances, ordered layers, and other surface effects, the more accurate model for a matrix–particle mixture would be a three-component mixture, of which particles are surrounded with shells having thermal conductivity k_s and volume concentration v_s other than those of the matrix and particles. The shells could be either a thermal barrier or a thermal bridge for heat transfer depending on their nature. However, most of the existing thermal conductivity theories for composites deal with two-component mixtures, although those two-component mixtures are actually three-component mixtures because of the interfacial shells. Therefore, the conventional picture of the matrix–particle mixtures needs to be modified to include the contributions of the shells. It is well known that the surface area/volume ratio of a particle is inversely proportional to the particle dimension. Since the surface area of a particle determines the interfacial shell between the matrix and particle, one may expect that any influence on the mixture properties caused by the interfacial shell must have connections with the particle dimension.

The shell thermal conductivity k_s is dependent on the nature of the matrix–particle interfacial effect. Xie et al. (2005) and Ren et al. (2005) considered a matrix–solidlike liquid layer–spherical particle mixture and reasoned that the thermal conductivity of the solidlike liquid layer should be between the matrix thermal conductivity and particle thermal conductivity. Assuming that the thermal conductivity of the solidlike liquid layer changes linearly from the matrix thermal conductivity at its outside surface to the particle thermal conductivity at its inside surface, Xie et al. (2005) and Ren et al. (2005) obtained the thermal conductivity k_s of the solidlike liquid layer from the averaging equivalent principle:

$$k_s = \frac{(k_p - v_r^{1/3} k_m)^2}{(1 - v_r^{1/3})(k_p - v_r^{1/3} k_m) + v_r^{1/3}(k_p - k_m)\ln[k_p/(v_r^{1/3} k_m)]} \tag{4.74}$$

where the parameter v_r is defined by $v_r = v_p/(v_s + v_p)$.

4.6.1. Two-Component Equation Generalization

There are several ways to tackle mixtures containing three components. The simplest approach is to generalize two-component equations directly into three-component equations. This approach does not require a particle–shell structure and can easily be extended to multicomponent mixtures. The simple mixture rules can be generalized for three-component mixtures:

$$k_e^n = (1 - v_s - v_p)k_m^n + v_s k_s^n + v_p k_p^n \qquad -1 \le n \le 1 \tag{4.75}$$

Applying an analogous method to Maxwell's equation, Brailsford and Major (1964) proposed the following three-component equation:

$$k_e = k_m + \frac{3v_s \dfrac{k_s - k_m}{2k_m + k_s} + 3v_p \dfrac{k_p - k_m}{2k_m + k_p}}{1 - v_s - v_p + 3v_s \dfrac{k_m}{2k_m + k_s} + 3v_p \dfrac{k_m}{2k_m + k_p}} k_m \tag{4.76}$$

The Bruggeman equations can be generalized without difficulty to three-component mixtures; for example, one may use the following equation for spherical particles (Landauer, 1978):

$$(1 - v_s - v_p)\frac{k_m - k_e}{2k_e + k_m} + v_s \frac{k_s - k_e}{2k_e + k_s} + v_p \frac{k_p - k_e}{2k_e + k_p} = 0 \tag{4.77}$$

4.6.2. Complex Particle Approach

In this approach, a complex particle which has an equivalent thermal conductivity k_c and a volume that equals the sum of the particle and shell volumes is constructed to reduce a three-component mixture into a two-component mixture. Miles and Robertson (1932) applied Maxwell's equation to the spherical particle–shell combination and obtained the equivalent thermal conductivity of the complex particles

$$k_c = k_s + 3v_r \frac{k_p - k_s}{2k_s + k_p - v_r(k_p - k_s)} k_s \tag{4.78}$$

where the parameter $v_r = v_p/v_c$ is the ratio of the particle-volume concentration v_p over the complex particle-volume concentration $v_c = v_s + v_p$.

In principle, with the equivalent thermal conductivity and volume concentration of the complex particles, all the two-component equations can be used for three-component mixtures. Applying Maxwell's equation to complex matrix–particle mixtures, one obtains (Kerner, 1956; Pauly and Schwan, 1959; Schwan et al., 1962; Lamb et al., 1978; Benveniste and Miloh, 1991; Yu and Choi, 2003; Wang et al., 2003)

$$k_e = k_m + 3v_c \frac{k_c - k_m}{2k_m + k_c - v_c(k_c - k_m)} k_m \tag{4.79}$$

Ignoring the higher-order terms in equation (4.79), Garboczi et al. (1995) and Schwartz et al. (1995) obtained an approximate equation for low particle-volume concentrations:

$$k_e = k_m + 3v_c \frac{k_c - k_m}{2k_m + k_c} k_m \tag{4.80}$$

By applying Bruggeman's equation to the complex matrix–particle mixture, one obtains (van de Hulst, 1957; Lu and Song, 1996; Xue, 2000)

$$(1 - v_c)\frac{k_m - k_e}{2k_e + k_m} + v_c \frac{k_c - k_e}{2k_e + k_c} = 0 \tag{4.81}$$

This technique can be extended to ellipsoidal particle–shell structures. To calculate the equivalent thermal conductivity of the complex ellipsoidal particles constructed from the ellipsoidal particles and shells, the depolarization factors of the ellipsoidal and complex particles are assumed to be d_{pi} $(i = a, b, c)$ and d_{ci} $(i = a, b, c)$, respectively. Obviously, if the thermal conductivity of the complex ellipsoidal particles is equal to that of the matrix, the temperature gradient and heat flux are unperturbed by the introduction of complex ellipsoidal particles. Based on this argument, Biboul (1969) derived the equivalent thermal conductivity k_{ci} $(i = a, b, c)$ along the axes of the complex ellipsoidal particles as

$$k_{ci} = k_s + v_r \frac{k_p - k_s}{k_s + d_i(k_p - k_s)} k_s \tag{4.82}$$

where the parameters d_i $(i = a, b, c)$ are defined by $d_i = d_{pi} - v_r d_{ci}$.

Applying the complex ellipsoidal particles to Fricke's equation, one obtains

$$k_e = k_m + \frac{1}{3}v_c \sum_{a,b,c} \frac{k_{ci} - k_m}{k_m + d_{ci}(k_{ci} - k_m)} k_m \tag{4.83}$$

Similarly, applying the complex ellipsoidal particles to the Polder–van Santen equation, one obtains

$$k_e = k_m + \frac{1}{3}v_c \sum_{a,b,c} \frac{k_{ci} - k_m}{k_e + d_{ci}(k_{ci} - k_e)} k_e \tag{4.84}$$

For the matrix of ellipsoidal particles, one may use the modified Bruggeman equation and obtain (Xue, 2000, 2003)

$$(1 - v_c) \sum_{i=a,b,c} \frac{k_m - k_e}{k_e + d_{mi}(k_m - k_e)} + v_c \sum_{a,b,c} \frac{k_{ci} - k_e}{k_e + d_{ci}(k_{ci} - k_e)} = 0 \tag{4.85}$$

By applying the average principle to the Hamilton–Crosser equation, Yu and Choi (2004) derived a modified Hamilton–Crosser equation for complex matrix–particle mixtures:

$$k_e = k_m + \Psi^{-1} v_c \sum_{a,b,c} \frac{k_{ci} - k_m}{(3\Psi^{-1} - 1)k_m + k_{ci} - v_c(k_{ci} - k_m)} k_m \tag{4.86}$$

4.6.3. Other Three-Component Equations

Solving the Laplace equation with boundary conditions directly can also be used to obtain the effective thermal conductivity for matrix–shell–particle mixtures. By using this method, Tinga et al. (1973) obtained a self-consistent solution for the effective thermal conductivity of a mixture with confocal ellipsoidal shell–particle inclusions in a rather complicated form. The result, however, can be reduced, for spherical matrix–shell–particle mixtures, into

$$k_e = k_m + 3v_c \frac{k_c - k_m}{2k_m + k_c - v_c(k_c - k_m)} k_m \tag{4.87}$$

which is exactly the same as repeated use of Maxwell's equation.

Lu and Song (1996) considered the effective thermal conductivity of statistically homogeneous mixtures containing identical coated or debonded spherical particles randomly dispersed in the mixtures. Lu and Song adopted an equilibrium hard-sphere model to represent the microstructure of the mixtures, and through solving a boundary value problem involving two coated or debonded spheres with twin spherical expansions, took pair interactions rigorously into account.

Lu and Song derived the effective thermal conductivity of the mixtures in series form and provided the following second-order expression:

$$k_e = k_m + 3v_c \left[1 + v_c \frac{\sigma_2(k_c - k_m)}{2k_m + k_c - v_c(k_c - k_m)}\right] \frac{k_c - k_m}{2k_m + k_c} k_m \tag{4.88}$$

In the equations above, the parameter σ_2 is dependent on the shell-volume concentration v_s, particle-volume concentration v_p, particle radius r_p, and thermal conductivity ratios k_s/k_m and k_p/k_m.

By assuming a simple cubic array of identical spherical particles with shells, Yu and Choi (2005) obtained the effective thermal conductivity of matrix–shell–spherical particle mixtures from the superposition principle for parallel and series thermal conductivities:

$$k_e = \left\{1 - \sqrt[3]{\frac{3v_c}{4\pi}} \left[2 - \frac{\sqrt[3]{16/(9\pi v_c^2)}\, k_m}{k_1 k_2} \ln \frac{(k_2 + k_1)(k_2 - \sqrt[3]{v_r} k_1)}{(k_2 - k_1)(k_2 + \sqrt[3]{v_r} k_1)} \right.\right.$$
$$\left.\left. + \frac{\sqrt[3]{16/(9\pi v_c^2)}\, k_m}{k_3 k_4} \ln \frac{k_4 + \sqrt[3]{v_r} k_3}{k_4 - \sqrt[3]{v_r} k_3}\right]\right\}^{-1} k_m \tag{4.89}$$

where the parameters k_1, k_2, k_3, and k_4 are defined by

$$\begin{aligned} k_1 &= \sqrt{k_s - k_m} \\ k_2 &= \sqrt{\sqrt[3]{16/(9\pi v_c^2)} k_m + (k_s - k_m)} \\ k_3 &= \sqrt{k_p - k_m} \\ k_4 &= \sqrt{\sqrt[3]{16/(9\pi v_c^2)} k_m + (k_s - k_m) + \sqrt[3]{v_r^2}(k_p - k_s)} \end{aligned} \tag{4.90}$$

Based on the mean-field concept of Mori and Tanaka (1973), Benveniste et al. (1990) considered the effective thermal conductivity of mixtures containing coated cylindrical fibers with a circular cross section and for isotropic matrix, coating, and particles with completely random particle distribution obtained

$$k_e = k_m + \frac{1}{3} v_c \frac{(4k_r - 1) + (1 - v_r)(k_s/k_m) + v_r(k_p/k_m)}{1 - (2/3)v_c k_r} k_m \tag{4.91}$$

where the parameter k_r is defined by

$$k_r = \frac{(k_s - k_m)(k_p + k_s) + v_r(k_s + k_m)(k_p - k_s)}{(k_s + k_m)(k_p + k_s) + v_r(k_s - k_m)(k_p - k_s)} \tag{4.92}$$

4.7. INTERFACIAL THERMAL RESISTANCE

It is well known that interfacial thermal resistance is present even at ideal interfaces of different components of a mixture. This interfacial thermal resistance is known as the *Kapitza resistance* after the discovery by Kapitza (1941) of the temperature discontinuity at the interface between liquid helium and copper due to heat flow. This effect can have a substantial influence on the effective thermal conductivity of mixtures. Interfacial thermal resistance is a special particle shell that has zero volume concentration, and its effect can be viewed as a reduction on the particle thermal conductivity of a matrix–particle mixture.

Benveniste (1987) considered a dispersal of spherical particles of radius r_p embedded in a matrix with interfacial thermal resistance R, or the reciprocal of the interfacial heat transfer coefficient called the *skin constant*, between the matrix and particles. Benveniste generalized the self-consistent scheme and the Mori and Tanaka theory (1973) to allow for the phenomenon of interfacial thermal resistance at matrix–particle interfaces and obtained

$$k_e = k_m + 3v_p \frac{k_p^R - k_m}{2k_m + k_p^R - v_p(k_p^R - k_m)} k_m \tag{4.93}$$

which is Maxwell's equation with k_p replaced by $k_p^R = k_p/(1 + Rk_p/r_p)$. Hasselman and Johnson (1987) adopted equivalent inclusion concept and derived exactly the same equation. Equation (4.93) can be approximated for low-particle-volume concentrations as

$$k_e = k_m + 3v_p \frac{k_p^R - k_m}{2k_m + k_p^R} k_m \tag{4.94}$$

By applying the Bruggeman integration scheme to equation (4.94), one obtains the differential equation

$$\frac{dk}{dv} = \frac{1}{1-v} \frac{3k(k_p^R - k)}{2k + k_p^R} \tag{4.95}$$

which upon integration between $v = 0$ ($k = k_m$) and $v = v_p$ ($k = k_e$) leads to

$$1 - v_p = \frac{k_p^R - k_e}{k_p^R - k_m} \left(\frac{k_m}{k_e}\right)^{1/3} \tag{4.96}$$

which is reduced to Bruggeman's equation for $R = 0$. Every et al. (1992) suggested this approach but obtained an incorrect result, due to an error in deriving the differential equation.

Chiew and Glandt (1987) extended the calculation to second order in the particle-volume concentration for a dispersion having the structure of equilibrium hard spheres:

$$k_e = k_m + 3v_p\left[1 + \frac{1}{3}v_p\frac{\sigma_3(2k_m + k_p^R)^2 - 3(k_p^R - k_m)^2}{(2k_m + k_p^R)(k_p^R - k_m)}\right] \times \frac{k_p^R - k_m}{2k_m + k_p^R - v_p(k_p^R - k_m)}k_m \tag{4.97}$$

where the parameter σ_3 is defined in a series function form of the particle-volume concentration v_p, thermal conductivity ratio k_p/k_m, and the Biot number $(Rk_p/r_p)^{-1}$.

Ni et al. (1997) applied the perturbation expansion method to nonlinear mixtures containing low-particle-volume concentrations of cylindrical particles with interfacial thermal resistance and derived formulas for linear and nonlinear effective thermal conductivities. The results for linear effective thermal conductivity can be expressed as

$$k_e = k_m + 2v_p\frac{k_p^R - k_m}{k_m + k_p^R}k_m \tag{4.98}$$

which is the first-order approximation of Rayleigh's cylindrical equation with the modification of the particle thermal conductivity k_p replaced by $k_p^R = k_p/(1 + Rk_p/r_p)$.

For a matrix–ellipsoidal particle mixture, one may introduce the concept of equivalent particle thermal conductivities k_{pi}^R $(i = a, b, c)$, defined as the combination effects of particle thermal conductivities and matrix–particle interfacial thermal resistance along the three ellipsoidal particle axes. Applying this concept to Fricke's equation for low volume concentrations, one may obtain

$$k_e = k_m + \frac{1}{3}v_p\sum_{i=a,b,c}\frac{k_{pi}^R - k_m}{k_m + d_{pi}(k_{pi}^R - k_m)}k_m \tag{4.99}$$

For a needle-shaped ellipsoid with $a \gg b = c$, d_{pa}, d_{pb}, and d_{pc} tend to 0, $\frac{1}{2}$, and $\frac{1}{2}$, respectively, and equation (4.99) reduces to

$$k_e = k_m + \frac{1}{3}v_p\left(\frac{k_{pa}^R - k_m}{k_m} + 2\frac{k_{pb}^R - k_m}{k_{pb}^R + k_m} + 2\frac{k_{pc}^R - k_m}{k_{pc}^R + k_m}\right)k_m \tag{4.100}$$

Nan et al. (2004) proposed the following approximate equation obtained from equation (4.100) under the condition that k_{pa}^R, k_{pb}^R, and k_{pc}^R are much larger than k_m:

$$k_e = k_m + \frac{1}{3}v_p\frac{k_p}{1 + Rk_p/a} \tag{4.101}$$

In arriving at this equation, Nan et al. applied the simple series rule for the equivalent particle thermal conductivities k_{pi}^R $(i = a, b, c)$:

$$k_{pi}^R = \frac{1}{R/i + 1/k_p} = \frac{k_p}{1 + Rk_p/i} \tag{4.102}$$

Even though the model predicts the experimental data of carbon nanotube suspensions reasonably well, the assumption of a series distribution along the longitudinal direction lacks a physical basis.

Ju and Li (2006) and Xue (2006) also suggested models for the effective thermal conductivities of carbon nanotube–based mixtures with an interfacial thermal resistance effect.

4.8. DYNAMIC MODELS OF THERMAL CONDUCTIVITY IN NANOFLUIDS

4.8.1. Background of Dynamic Models

The classical continuum models of the thermal conductivity of solid–liquid suspensions are based on the central assumption that the heat transport in each phase is governed by the diffusion equation. As a result, continuum models such as the Fourier model can adequately explain the effective thermal conductivity of conventional suspensions of millimeter- or micrometer-sized particles. However, they simply fail to explain the new thermal transport phenomena in nanofluids, such as high thermal conductivity at low volume fractions of nanoparticles, and strongly temperature- and size-dependent conductivities. Discovery of these new transport phenomena has created a great need to understand new thermal transport mechanisms at small length scales. A number of new mechanisms have been proposed to explain the enhanced thermal conductivity of nanofluids.

Wang et al. (1999) were first to propose new static and dynamic mechanisms behind enhanced thermal transport in nanofluids. They attributed enhanced conductivity to the microscopic motions of nanoparticles and fluids, which are induced by microscopic forces acting on a nanoparticle such as the van der Waals force, the electrostatic force resulting from the electric double layer at the particle surface, the stochastic force that gives rise to the Brownian motion of particles, and the hydrodynamic force. Although they did not develop a dynamic model because these microscopic forces cannot be calculated accurately, they suggested for the first time that nanoparticle size is important in enhancing the thermal conductivity of nanofluids. They also attributed enhanced conductivity to the chain structure of nanoparticle clusters.

Xuan and Li (2000) suggested several possible mechanisms for enhanced thermal conductivity of nanofluids, such as the increased surface area of nanoparticles, particle–particle collisions, and the dispersion of nanoparticles. Years later, Keblinski et al. (2002) proposed four possible microscopic mechanisms for the anomalous increase in the thermal conductivity of nanofluids: Brownian motion of the nanoparticles, molecular-level layering of the liquid at the liquid–particle interface, the ballistic rather than diffusive nature of heat conduction in the nanoparticles, and the effects of nanoparticle clustering.

Modeling for the thermal conductivity of nanofluids typically falls into two broad categories: extension of existing conduction models and the development of new models. Briefly, structural models such as nanolayer, fractal, or percolation structures and dynamic models such as Brownian motion-based collision of nanoparticles belong to the first category. Nanoconvection induced by Brownian motion of nanoparticles and near-field radiation belong to the second category. A number of investigators have proposed both static (or structural) and dynamic mechanisms and models in both categories to account for the anomalously high thermal conductivity enhancements reported in recent measurements.

In previous sections we have discussed static or structural models of the effective thermal conductivity of nanofluids, assuming that the nanoparticles are static when there is no bulk motion of the nanofluids. However, nanoparticles in nanofluids are engaged in relentless random thermal motion. Therefore, nanofluids are dynamical systems and the effective thermal conductivity of nanofluids depends not only on the nanostructures of the suspensions but also the dynamics of nanoparticles in liquids. Therefore, the motion of nanoparticles, including the interactions between Brownian nanoparticles or between Brownian nanoparticles and liquid molecules, should be considered to develop more realistic models. It should be noted that the shape of nanoparticles is critical in determining the dominant mechanism of heat transport in nanofluids. For example, it seems that dynamic mechanisms such as Brownian motion play a key role in nanofluids containing spherical nanoparticles, but static mechanisms such as percolation are dominant in nanofluids containing CNTs. In some nanofluids there may be a synergistic effect of static and dynamic mechanisms.

The Brownian motion of nanoparticles was considered as a most probable mechanism early in developing theoretical models of nanofluids. However, Wang et al. (1999) pointed out that Brownian motion does not contribute significantly to energy transport in nanofluids. Later, Keblinski et al. (2002) showed clearly that Brownian motion is not a significant mechanism of enhanced heat conduction, based on the results of a time-scale study. However, it is important to understand that the Brownian motion mechanism explored by both Wang et al. (1999) and Keblinski et al. (2002) is heat conduction through particle–particle collisions caused by Brownian motion of nanoparticles.

4.8.2. Dynamic Models

Even though it was stated earlier that the direct Brownian motion contribution to heat conduction in nanofluids is negligible (Wang et al., 1999; Keblinski et al., 2002), a number of investigators have not dropped the idea that the Brownian motion of nanoparticles is a likely mechanism for modeling one of the most important thermal phenomena in nanofluids: the strongly temperature-dependent thermal conductivity of nanofluids (Das et al., 2003; Patel et al., 2003). Furthermore, Xie et al. (2002) measured the thermal conductivity of aqueous Al_2O_3

nanofluids with varying particle sizes and showed for the first time that the thermal conductivity of nanofluids depends strongly on particle size. A number of dynamic models, all of which incorporate the concept of Brownian motion of nanoparticles in enhancing the thermal conductivity of nanofluids, have been developed to address the temperature- and size-dependent thermal conductivity of nanofluids. In fact, one of the key concepts used in most promising dynamic models is that nanoconvection induced by nanoparticle motion is essential to enhanced energy transport in nanofluids.

Xuan et al. (2003) developed a dynamic model into which the effects of Brownian motion of nanoparticles and the aggregation structure of nanoparticle clusters (i.e., fractals) are taken. Predictions from the model agree with experimental data for copper–water nanofluids when the effect of nanoparticle aggregation is taken into account. Their model simulations are probably the first to show that the thermal conductivity of nanofluids depends on fluid temperature and the structure of nanoparticle clusters. Although their model is one of the earliest models to include the effect of Brownian motion, it cannot correctly predict the strongly temperature-dependent thermal conductivity data obtained by Das et al. (2003) and Patel et al. (2003) because the dependence suggested by them is too weak ($\sim T^{1/2}$).

Yu et al. (2003) have developed a simplified one-dimensional drift velocity model of a nanofluid thermal conductivity. First, they have shown that the collision of particles and the drift velocity in the presence of a temperature gradient can account for a very small part of the enhancement. They then assumed that a pair of thermophoretically drifting nanoparticles with correlated Brownian motion drag a modest amount of the surrounding fluid with them to create nanoconvection. Although not yet knowing the exact form of the nanoconvection, they estimated an order-of-magnitude improvement over conventional models that do not take the effect of nanoconvection into account. However, this type of nanoconvection model failed to show the effect of nanoparticle size and so does not present a complete, accurate model. But this work is significant in proposing a new mechanism of nanoconvection induced by a pair of drifting nanoparticles and in modeling the enhanced property of nanofluids on fundamental physics without fitting parameters.

Bhattacharya et al. (2004) computed the thermal conductivity of nanofluids using Brownian dynamics simulations and the equilibrium Green–Kubo method. Their simulation results are in agreement with experimental data for alumina and copper nanofluids. However, in their Brownian dynamics simulations they introduced an interparticle potential with a range of the order of one light-to reproduce experimental data (Keblinski et al., 2005).

Jang and Choi (2004) have developed for the first time a dynamic model that takes into account convection induced by a single Brownian nanoparticle. They derived a general expression for the thermal conductivity of nanofluids involving four modes of energy transport, as shown in Fig. 4.1. The first mode is collision between the base fluid molecules (i.e., the thermal conductivity of the base fluid). The second mode is the thermal diffusion in nanoparticles in fluids

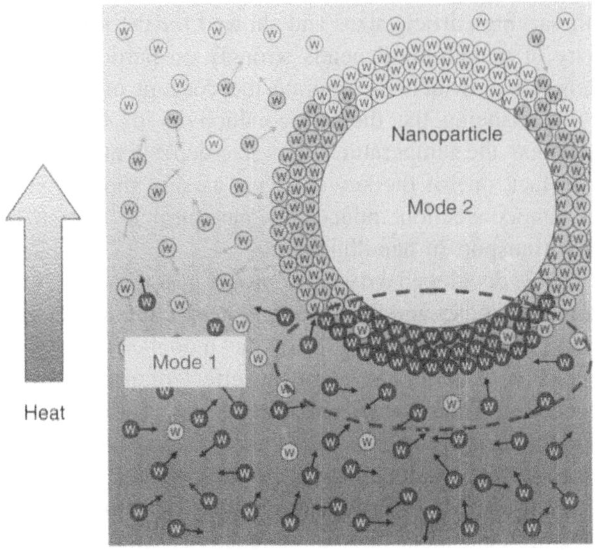

Fig. 4.1 Modes of energy transport in nanofluids. The first mode is collision between base fluid molecules; the second mode is the thermal diffusion in nanoparticles suspended in fluids; the third mode is collision between nanoparticles (not shown); and the fourth mode is thermal interactions of dynamic or dancing nanoparticles with base fluid molecules. [From Jang and Choi (2004), with permission from the American Institute of Physics.]

(i.e., the thermal conductivity of suspended nanoparticles involving the Kapitza resistance). The third mode is collision between nanoparticles due to Brownian motion, which can be neglected because Brownian diffusion of nanoparticles is a very slow process compared to thermal diffusion (Keblinski et al., 2002). The last mode is thermal interactions of dynamic nanoparticles with base fluid molecules. This mode, which had been overlooked by earlier researchers, is now known to be a key to temperature- and size-dependent conductivity.

Most important, they introduced the new idea that a Brownian nanoparticle produces a convectionlike effect at the nanoscale. The effective thermal conductivity of nanofluids k_{eff} for their model is given by (Jang and Choi, 2004)

$$k_{eff} = k_{Bf}(1 - f) + k_{nano} f + 3C_1 \frac{d_{Bf}}{d_{nano}} k_{Bf} \mathrm{Re}_{d_{nano}}^2 \mathrm{Pr} f \qquad (4.103)$$

where k_{BF}, f, k_{nano}, C_1, d_{BF}, d_{nano}, and Pr are the base fluid conductivity, the volume fraction of nanoparticles, the thermal conductivity of the suspended nanoparticles including the interface thermal resistance, an empirical constant, the diameter of the base fluid molecule, and the diameter of a nanoparticle, respectively. $\mathrm{Re}_{d_{nano}}$ is the Reynolds number, defined by

$$\mathrm{Re}_{d_{nano}} = \frac{\overline{C}_{R.M.} d_{nano}}{\nu} \qquad (4.104)$$

where ν is the dynamic viscosity of the base fluid. The random motion velocity of nanoparticles, $\overline{C}_{R.M.}$ can be defined by

$$\overline{C}_{R.M.} = \frac{D_0}{l_{Bf}} \tag{4.105}$$

where l_{BF} is the mean free path of a base fluid molecule and D_0 is the nanoparticle diffusion coefficient, given by

$$D_0 = \frac{k_b T}{3\pi\mu d_{nano}} \tag{4.106}$$

where μ is the viscosity of a base fluid, T is the temperature of the base fluid molecules, and $k_b = 1.3807 \times 10^{-23}\ J/K$ is the Boltzmann constant.

Figure 4.2 shows that the present model predictions (solid curve) are in excellent agreement with temperature-dependent conductivity data for water-based nanofluids containing 38.4-nm Al_2O_3 nanoparticles. In contrast, conventional theories such as Maxwell's with motionless nanoparticles (horizontal dashed line)

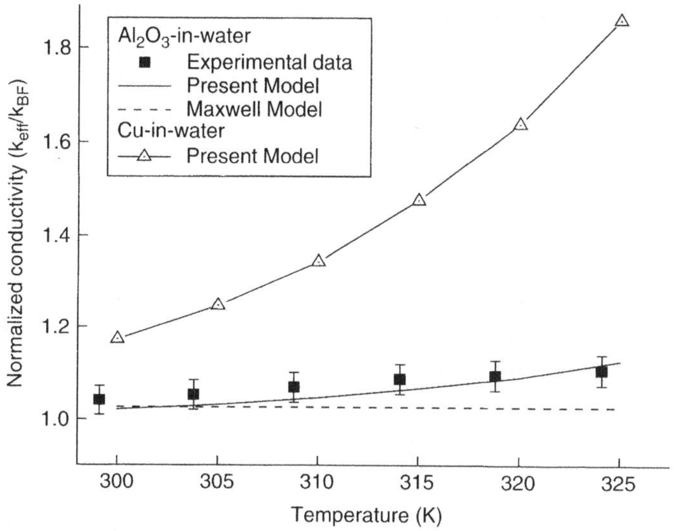

Fig. 4.2 Temperature-dependent thermal conductivities of nanofluids at a fixed nanoparticle concentration of 1 vol %, normalized to the thermal conductivity of the base fluid. The present model predictions (solid curve) are in excellent agreement with experimental data from Das et al. (2003) (solid squares) for water-based nanofluids containing 38.4-nm Al_2O_3 nanoparticles. In contrast, the experimental data are not predicted by conventional theories such as Maxwell's with motionless nanoparticles (horizontal dashed line). Incredibly, the model predicts that water-based nanofluids containing 6-nm Cu nanoparticles (curve with open triangles) are much more temperature sensitive than those containing 38.4-nm Al_2O_3 particles, with an increase in conductivity of nearly a factor of 2 at 325 K. [From Jang and Choi (2004), with permission from the American Institute of Physics.]

fail to capture the temperature-dependent conductivity data. This new dynamic model, which accounts for the fundamental role of convection, not only captures the concentration- and temperature-dependent conductivity, but also predicts for the first time strongly size-dependent conductivity. Thus, they showed that local, short-lived convection induced by Brownian motion of the nanoparticles is a key nanoscale mechanism governing the thermal conductivity of nanofluids. This study, which gives a new direction in theoretical modeling of nanofluids by adding to the pure conduction mechanism Brownian-motion-induced nanoconvection as a new fundamental mechanism of energy transport in nanofluids, could be a turning point in the century-old research on thermal properties of suspensions. Furthermore, Jang and Choi (2004) discovered a fundamental difference between solid–solid composites and solid–liquid suspensions in size-dependent conductivity due to different thermal transport mechanisms at small length scales. The nanofluid conductivity increases with decreasing particle size due to Brownian-motion-induced convection. In contrast, solid–solid composites have a reverse size effect (i.e., decreasing conductivity with decreasing particle size), due to boundary scattering of phonons. However, the new concepts introduced for submodels and simplifying assumptions made to develop a simplified theoretical model have not been validated.

Kumar et al. (2004) presented a comprehensive model that consists of the stationary particle model and the moving particle model. The *stationary particle model* accounts for the particle size effect through increased surface area with decreasing particle size. It assumes two parallel paths of heat flow through the suspension, one through the liquid particles and the other through the nanoparticles. The stationary particle model shows the linear dependence of thermal conductivity enhancement on particle concentration and the inverse dependence of thermal conductivity enhancement on particle size. The *moving particle model*, developed from the Stokes–Einstein formula, accounts for the temperature effect. In this model, the effective thermal conductivity of particles is modeled by drawing a parallel to the kinetic theory of gases. The effective thermal conductivity of nanofluids $\boldsymbol{k}_{eff}$ for their model is expressed as (Kumar et al., 2004)

$$k_{eff} = k_m \left[1 + \frac{k_p \varepsilon r_m}{k_m (1 - \varepsilon) r_p} \right] \tag{4.107}$$

where k_m, ε, r_m, and r_p are the base fluid conductivity, nanoparticle volume fraction, liquid particle radius, and nanoparticle radius, respectively. The thermal conductivity of the nanoparticle, k_p, is given as

$$k_p = c\overline{u}_p \tag{4.108}$$

where the mean velocity of the nanoparticle has been calculated using the Stokes–Einstein formula,

$$\overline{u}_p = \frac{2k_b T}{(\pi \eta d_p^2)} \tag{4.109}$$

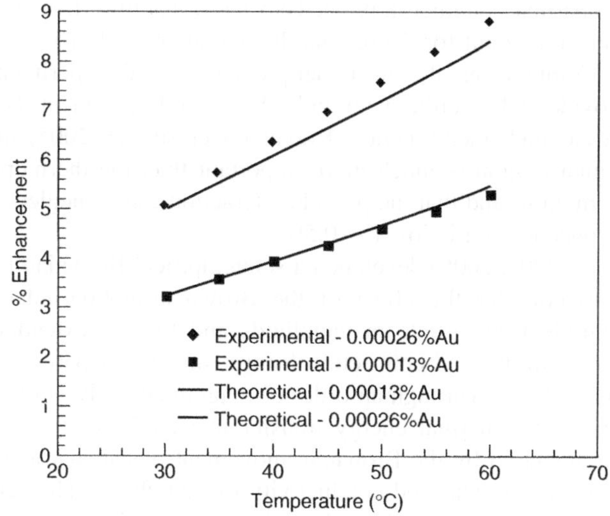

Fig. 4.3 Comparison of experimental and theoretical enhancement against temperature for a Au nanofluid of 17 nm mean diameter. [From Kumar et al. (2004), with permission from the American Physical Society.]

where T, η, and d_p are the fluid temperature, the dynamic viscosity of the fluid, and the nanoparticle diameter, respectively.

Their combined model accounts for the dependence of thermal conductivity of nanofluids on particle size, concentration, and temperature. Figure 4.3 shows that predictions from the combined model are in excellent agreement with experimental data for gold nanofluids with vanishingly small concentration. However, several researchers made arguements to the contrary at the conceptual level (Bastea, 2005; Keblinski and Cahill, 2005). For example, as Bastea (2005) pointed out, one problem with their stationary particle model is that if the particle radius is much larger than the liquid molecules, the thermal conductivity of the nanofluid will be the same as that of the base fluid, which is unrealistic. In the moving particle model, to match their data they made an unphysical assumption that the mean free path of a nanoparticle in nanofluid is on the order of 1 cm (Keblinski and Cahill, 2005). This dynamic model is probably one of the most hotly debated theoretical models in nanofluids literature (Bastea, 2005; Das et al., 2005a,b; Keblinski and Cahill, 2005). Nonetheless, the issues involved in the arguments have not been completely resolved, and further studies are needed to settle debates on the pros and cons of the new dynamic model.

Extending the idea of Yu et al. (2003) that a pair of drifting nanoparticles drag some of the surrounding fluid between them to create nanoconvection, Koo and Kleinstreuer (2004) assumed that a randomly moving nanoparticle drags the surrounding liquid to form a moving nanoparticle–liquid cell, which may interact with its neighboring cell and create micromixing. Then they developed for the first time a micromixing model that takes into account the effect of Brownian motion of nanoparticles on micromixing. They show that the effect of interparticle potential is very important for dense nanofluids (nanoparticle volume concentrations $> 1\%$). Comparisons of their model predictions with experimental data for a number of oxide and metallic nanofluids show good agreement. However, the model includes an undefined function. Koo and Kleinstreuer (2005) show that the role of Brownian motion is much more important than the thermophoretic and osmophoretic motions and that the particle interaction can be neglected when the nanoparticle concentration is low ($< 0.5\%$).

Prasher et al. (2005, 2006) developed a semiempirical Brownian model. They are the first to consider the effect of the Brownian-motion-induced convection from multiple nanoparticle in nanofluids and thus to extend the concept of Brownian-motion-induced convection from a single nanoparticle in the Jang and Choi model (2004). Through an order-of-magnitude analysis of various possible mechanisms for thermal energy transfer in nanofluids, they showed that local convection caused by the Brownian movement of nanoparticles is primarily responsible for the enhanced conductivity of nanofluids. The semiempirical model for the normalized thermal conductivity of nanofluids k/k_f is expressed as (Prasher et al., 2006)

$$\frac{k}{k_f} = (1 + A \cdot \mathrm{Re}^m \mathrm{Pr}^{0.333} \phi) \frac{[k_p(1 + 2\alpha) + 2k_m] + 2\phi k_p(1 - \alpha) - k_m]}{[k_p(1 + 2\alpha) + 2k_m] - \phi[k_p(1 - \alpha) - k_m]} \quad (4.110)$$

where A and m are empirical constants, and ϕ and α are the volume fraction of nanoparticles and the nanoparticle Biot number, respectively. Figure 4.4 shows that predictions from the model are in good agreement with temperature-dependent conductivity data for oxide nanofluids for particular values of constants.

Ren et al. (2005) developed a model that takes into account kinetic-theory-based microconvection and liquid layering in addition to conduction through both particles and liquid. They also considered a fixed nanolayer thickness of 2 nm and determined the thermal conductivity of the nanolayer as the volume-averaged thermal conductivity of the base liquid and particles. The model accounts for the enhanced thermal conductivity of nanofluids with respect to nanoparticle concentration, particle size, and temperature. Predictions from the model for oxide nanofluids agree quite well with experiment.

Patel et al. (2005) developed a microconvection model for evaluation of thermal conductivity of the nanofluid by taking into account nanoconvection induced by Brownian nanoparticles and the specific surface area of nanoparticles. They modeled microconvection with empiricism in the Nusselt number definition. The model works well over a wide range of nanofluid parameters.

Chon et al. (2005) developed a completely empirical model for alumina nanofluids by fitting a curve through a linear regression analysis to the existing experimental data. To model microconvection around nanoparticles, they derived the Brownian velocity of nanoparticles using the mean free path of liquid molecules as a characteristic length. Their empirical model clearly shows the key

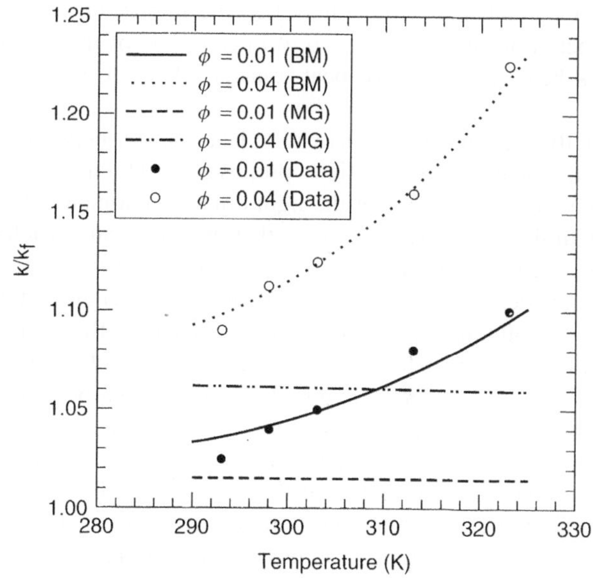

Fig. 4.4 Comparison of the semiempirical Brownian model with experimental data for 38-nm Al_2O_3 nanoparticles in water from Das et al. (2003) for varying temperatures. [From Prasher et al. (2005), with permission from the American Physical Society.]

role of temperature and nanoparticle size for thermal conductivity enhancement. Also, they are probably the first to validate experimentally that the Brownian motion of nanoparticles constitutes a key mechanism of the thermal conductivity enhancement with increasing temperature and decreasing nanoparticle size.

Xu et al. (2006) were probably the first to develop a fractal–convection model which takes into account the fractal size distribution of nanoparticles and convection caused by Brownian motion of nanoparticles. Their fractal–convection model accounts for the dependence of thermal conductivity of nanofluids on particle concentration, average size, fractal dimension, and temperature. Interestingly, the model shows that the contribution of Brownian-motion-induced convection reaches a maximum value at a critical concentration of 12.6 vol %. Predictions from the fractal model agree with available experimental data for oxide nanofluids.

4.9. NEAR-FIELD RADIATION MODEL

Recently, Domingues et al. (2005) proposed a new physical mechanism based on near-field heat transfer. When the volume fraction exceeds a few percent, the mean distance between particles in nanofluids is on the order of the particle diameter. This distance is much lower than the dominant wavelength of far-field radiation (i.e., when photons are emitted or absorbed), and near-field radiation (i.e., Coulomb interaction) may become important. They showed that the near-field heat transfer becomes two to three orders of magnitude more efficient than bulk heat conduction or heat transfer when the nanoparticles are in contact.

Now we briefly describe the mechanism of heat transfer between two nanoparticles through the near-field interactions proposed by Domingues et al. (2005). Considering a mixture consisting of the liquid matrix and many identical spherical particles with the radius R, we assume that the spherical particles of thermal conductivity k_p are uniformly dispersed in the liquid matrix of thermal conductivity k_m and are located at the vertexes of the simple cubic lattice. For most well-suspended nanofluids, this assumption is reasonable. Based on the assumption above, the particle center-to-center distance d for particle-volume concentration c is given by $d = \sqrt[3]{4\pi/3c}R$. Figure 4.5 shows the particle center-to-center distance as a function of the particle-volume concentration. It can be seen from the figure that the particle center-to-center distance d is on the order of $10R$ to $4R$ for a particle-volume concentration c of 0.4 to 6.5% and becomes smaller with an increase in the particle-volume concentration. Because of the very small size of the nanoparticles in nanofluids, it is possible for the particle center-to-center distance to be reduced to the nanometer range at particle-volume concentration levels that are practical for engineering applications. Therefore, it is important to study the heat transfer mechanism involved at a much smaller dimension than the traditional heat transfer range.

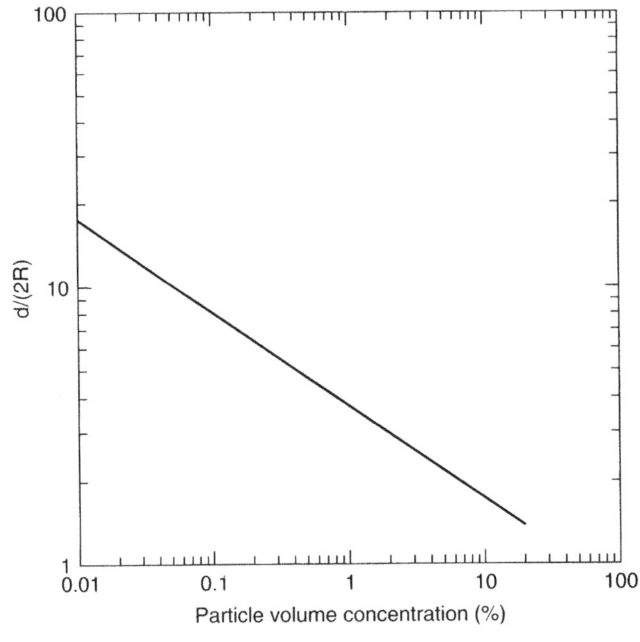

Fig. 4.5 Particle center-to-center distance as a function of particle-volume concentration.

How energy is exchanged between two objects just before contact is still an open question. The conventional mechanism is based on the radiation heat transfer and reaches limits at contact. This mechanism gives heat transfer flux due to the emission and absorption of photons in the far field proportional to d^{-2} and is not enough to explain the drastic enhancement of thermal conductivities of nanofluids. In their molecular dynamics study, Domingues et al. (2005) used the BKS interaction potential (van Beest, 1990) to provide a full physical picture of long-range electromagnetic and repulsive–attractive short-range interactions. Accordingly, the BKS potential can be decomposed into a Coulomb potential and a Buckingham potential. The Buckingham part includes an exponential term to describe the short-range van der Waals attractive forces. The Coulomb potential takes into account the interatomic electrostatic force. Figure 4.6 shows their results of conductance as a function of the particle center-to-center distance. There are some unusual characteristics that should be noted from Fig. 4.6. First, the conductance obtained from the molecular dynamics technique is much higher than that from the conventional far-field emission and absorption model. Second, the conductance increases dramatically when the particle center-to-center distance is less than $4R$, due to multipolar contributions. Finally, the contact conductance (gray points) is two to three orders of magnitude lower than the conductance just before contact, which suggests that the thermal conductance of a chain of particles might be larger than the conductance of a continuous rod. This phenomenon can be explained as follows. Since the Buckingham contribution is negligible before contact, the conductance is due only to the autocorrelation of the Coulomb

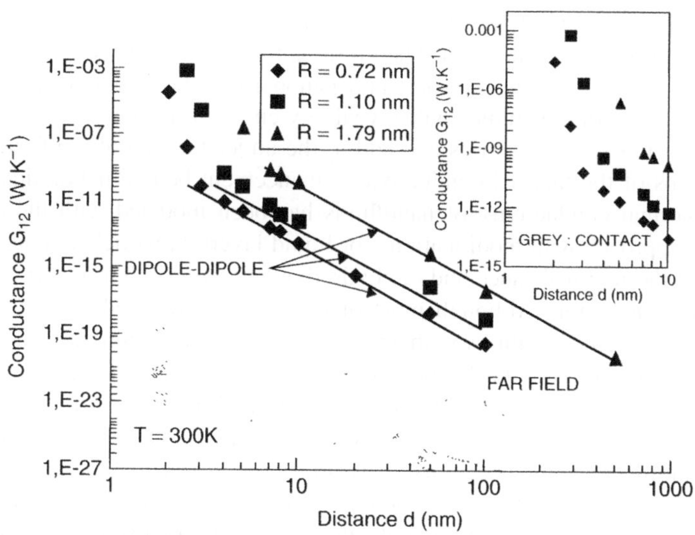

Fig. 4.6 Thermal conductance G12 as a function of particle center-to-center distance d. R corresponds to the nanoparticle radius and N is the number of atoms in each particle. [From Domingues et al. (2005), with permission from the American Physical Society.]

contribution. At contact, the Coulomb contribution does not vary much, but three other terms appear: the autocorrelation of the Buckingham contribution and two cross terms between the Coulomb contribution and the Buckingham contribution. The calculation shows that the cross terms are negative and on the order of the autocorrelation of the Buckingham contribution. Therefore, the final sum is lower than the autocorrelation of the Coulomb contribution. The origin of this decay is still an open question. It might be possible that the contact produces a correlation of positions of the atoms of both particles that results in a smaller fluctuation between two particles. These results show that the traditional separation between conduction and radiation is no longer meaningful at small length scales.

Near-field radiation could be a new mechanism of energy transport between the nanoparticles in nanofluids, due to photon tunneling through evanescent modes. A combined experimental and theoretical work is needed, although direct measurement of near-field thermal radiation in nanofluids is very challenging.

4.10. FUTURE RESEARCH

Thermal scientists have made great efforts to understand new mechanisms and develop new models of the enhanced thermal conductivity of nanofluids. The three main categories of new mechanisms include conduction, nanoscale convection, and near-field radiation. It should be noted that the validity of most of the mechanisms discussed in earlier sections remains a subject of debate. At present, there is no agreement in the nanofluids community about the mechanisms of the anomalous thermal behavior of nanofluids. Furthermore, there are no systematic experimental studies of fundamental mechanisms of energy transport in nanofluids at the nanoscale level. Therefore, in the future we need to explore both structure-enhanced energy transport and nanoparticle-mobility-enhanced energy transport. These future studies will reveal key energy transport mechanisms that are missing in existing theories and add to the understanding of the fundamental mechanisms of the thermal conductivity enhancement behind nanofluids.

The thermal conductivity of nanofluids has been modeled with the classical diffusion approach with modifications for liquid layering, particle aggregation, or interfacial thermal resistance and the hydrodynamic approach to incorporate new nanoscale convection. Although most of the models proposed seem to provide reasonable agreement with experimental data, one or two empirical constants are used to reproduce the data. Numerical simulations such as molecular dynamics simulations are needed to understand the origin and nature of fitting parameters employed in the models proposed. However, at present they lack the ability to model nanoscale phenomena realistically. Furthermore, each of the many assumptions and concepts used in new models need a fundamental proof. Therefore, systematic and controlled experiments are needed to establish the validity of the new models and their assumptions made to model thermal transport mechanisms at small length and time scales. Most important, a more theoretical basis for modeling the thermal transport phenomena in nanofluids is needed to model the intriguing thermal phenomena in nanofluids on fundamental physics and chemistry without adjustable parameters.

Understanding the fundamentals of energy transport in nanofluids is important for developing extremely energy-efficient nanofluids for a range of heat transfer applications. To our knowledge, no fundamental studies have been carried out on the energy transport in nanofluids at the nanoscale. At present, little is known about the interfacial thermal resistance or extremely fast Brownian particle dynamics in nanofluids at small scales in space and time. Thus, to develop such a basic understanding we need integrated experimental, simulation, theoretical, and modeling studies.

REFERENCES

Bastea, S. (2005). Comment on "Model for heat conduction in nanofluids," *Phys. Rev. Lett.*, 95 (1): 019401.

Batchelor, G. K., and R. W. O'Brien (1977). Thermal or electrical conduction through a granular material, *Proc. R. Soc. London A*, 355: 313–333.

Benveniste, Y. (1987). Effective thermal conductivity of composites with a thermal contact resistance between the constituents: nondilute case, *J. Appl. Phys.*, 61: 2840–2843.

Benveniste, Y., and T. Miloh (1991). On the effective thermal conductivity of coated short-fiber composites, *J. Appl. Phys.*, 69: 1337–1344.

Benveniste, Y., T. Chen, and G.J. Dvorak (1990). The effective thermal conductivity of composites reinforced by coated cylindrically orthotropic fibers, *J. Appl. Phys.*, 67: 2878–2884.

Bergman, D. J. (1979). The dielectric constant of a simple cubic array of identical spheres, *J. Phys. C Solid State Phys.*, 12: 4947–4960.

Bhattacharya, P., S. K. Saha, A. Yadav, P. E. Phelan, and R.S. Prasher (2004). Brownian dynamics simulation to determine the effective thermal conductivity of nanofluids, *J. Appl. Phys.*, 95: 6492–6494.

Bilboul, R. R. (1969). A note on the permittivity of a double-layer ellipsoid, *Br. J. Appl. Phys.* (Ser. 2), 2: 921–923.

Böttcher, C. J. F. (1945). The dielectric constant of crystalline powders, *Recl. Trav. Chim. Pays-Bas*, 64: 47–51.

Böttcher, C. J. F. (1952). Theory of Electric Polarization, Elsevier, Amsterdam; The Netherlands.

Brailsford, A. D., and K.G. Major (1964). The thermal conductivity of aggregates of several phases, including porous materials, *Br. J. Appl. Phys.*, 15: 313–319.

Brown, W. F., Jr. (1955). Solid mixture permittivities, *J. Chem. Phys.*, 23: 1514–1517.

Bruggeman, D. A. G. (1935). Berechnung verschiedener physikalisher Konstanten von heterogenen Substanzen: I. Dielektrizitätskonstanten und Leitfähigkeiten der Mischkörper aus isotropen Substanzen, *Ann. Phys.*, 24: 636–664.

Cheng, S. C., and R.I. Vachon (1969). The prediction of the thermal conductivity of two and three phase solid heterogeneous mixtures, *Int. J. Heat Mass Transfer*, 12: 249–264.

Chiew, Y. C., and E. D. Glandt. (1987). Effective conductivity of dispersions: the effect of resistance at the particle surfaces, *Chem. Eng. Sci.*, 42: 2677–2685.

Choi, S. U. S., Z. G. Zhang, and P. Keblinski (2004). Nanofluids. In *Encyclopedia of Nanoscience and Nanotechnology*, vol. 6, H. S. Nalwa; Ed., American Scientific Publishers, city, pp. 757–773.

Chon, C. H., K. D. Kihm, S. P. Lee, and S.U.S. Choi (2005). Empirical correlation finding the role of temperature and particle size for nanofluid (Al^2O^3) thermal conductivity enhancement, *Appl. Phys. Lett.*, 87 (15): 153107.

Das, S. K., N. Putra, P. Thiesen, and W. Roetzel (2003). Temperature dependence of thermal conductivity enhancement for nanofluids, *J. Heat Transfer*, 125: 567–574.

Das, S. K., T. Sundararajan, T. Pradeep, and H.E. Patel (2005a). Comment on "Model for heat conduction in nanofluids"—Reply, *Phys. Rev. Lett.*, 95 (1): 019402.

Das, S. K., T. Sundararajan, T. Pradeep, and H.E. Patel (2005b). Comment on "Model for heat conduction in nanofluids"—Reply, *Phys. Rev. Lett.*, 95 (20): 209402.

Davidson, A., and M. Tinkham (1976). Phenomenological equations for the electrical conductivity of microscopically inhomogeneous materials, *Phys. Rev. B*, 13: 3261–3267.

Davies, W. E. A. (1971). The theory of composite dielectrics, *J. Phys. D Appl. Phys.*, 4: 318–328.

Davies, W. E. A. (1974a). The dielectric constants of fibre composites, *J. Phys. D Appl. Phys.*, 7: 120–130.

Davies, W. E. A. (1974b). The dielectric constants of axially isotropic composite materials, *J. Phys. D Appl. Phys.*, 7: 1016–1029.

Davis, R. H. (1986). The effective thermal conductivity of a composite material with spherical inclusions, *Int. J. Thermophys.*, 7: 609–620.

Domingues, G., S. Volz, K. Joulain, and J.-J. Greffet (2005). Heat transfer between two nanoparticles through near field interaction, *Phys. Rev. Lett.*, 94: 085901.

Eastman, J. A., S. R. Phillpot, S. U. S. Choi, and P. Keblinski (2004). Thermal transport in nanofluids, *Annu. Rev. Mater. Sci.*, 34: 219–246.

Every, A. G., Y. Tzou, D. P. H. Hasselman, and R. Raj. (1992). The effect of particle size on the thermal conductivity of ZnS/diamond composites, *Acta Metall. Mater.*, 40: 123–129.

Fadale, T. D., and M. Taya (1991). Effective thermal conductivity of composites with fibre–matrix debonding, *J. Mater. Sci. Lett.*, 10: 682–684.

Fletcher, L. S. (1988). Recent developments in contact conductance heat transfer, *J. Heat Transfer*, 110: 1059–1070.

Fricke, H. (1924). A mathematical treatment of the electric conductivity and capacity of disperse systems: I. The electric conductivity of a suspension of homogeneous spheroids, *Phys. Rev.*, 24: 575–587.

Fricke, H. (1953). The Maxwell–Wagner dispersion in a suspension of ellipsoids, *J. Phys. Chem.*, 57: 934–937.

Gao, L., and X.F. Zhou (2006). Differential effective medium theory for thermal conductivity in nanofluids, *Phys. Lett. A* 348: 355–360.

Garboczi, E. J., L. M. Schwartz, and D.P. Bentz (1995). Modeling the influence of the interfacial zone on the d.c. electrical-conductivity of mortar, *Adv. Cement Based Mater.*, 2: 169–181.

Granqvist, C. G., and O. Hunderi (1977). Optical properties of ultrafine gold particles, *Phys. Rev. B*, 16: 3513–3534.

Granqvist, C. G., and O. Hunderi (1978). Conductivity of inhomogeneous materials: effective-medium theory with dipole–dipole interaction, *Phys. Rev. B*, 18: 1554–1561.

Hale, D. K. (1976). The physical properties of composite materials, *J. Mater. Sci.*, 11: 2105–2141.

Hamilton, R. L., and O.K. Crosser (1962). Thermal conductivity of heterogeneous two-component systems, *Ind. Eng. Chem. Fundam.*, 1: 187–191.

Hasselman, D. P. H., and L. F. Johnson. (1987). Effective thermal conductivity of composites with interfacial thermal barrier resistance, *J. Compos. Mater.*, 21: 508–515.

Jang, S. P., and S.U.S. Choi (2004). Role of Brownian motion in the enhanced thermal conductivity of nanofluids, *Appl. Phys. Lett.*, 84 (21): 4316–4318.

Jeffrey, D. J. (1973). Conduction through a random suspension of spheres, *Proc. R. Soc. London A*, 335: 355–367.

Ju, S., and Z.Y. Li (2006). Theory of thermal conductance in carbon nanotube composites, *Phys. Lett. A*, 353: 194–197.

Kapitza, P. L. (1941). The study of heat transfer in helium: II, *J. Phys.*, 4: 181–210.

Keblinski, P., and D.G. Cahill (2005). Comment on Model for heat conduction in nanofluids, *Phys. Rev. Lett.*, 95 (20): 209401.

Keblinski, P., S. R. Phillpot, S. U. S. Choi, and J.A. Eastman (2002). Mechanisms of heat flow in suspensions of nano-sized particles (nanofluids), *Int. J. Heat Mass Transfer*, 45 (4): 855–863.

Keblinski, P., J. A. Eastman, and D.G. Cahill (2005). Nanofluids for thermal transport, *Mater. Today*, June, pp. 36–44.

Keller, J. B. (1963). Conductivity of a medium containing a dense array of perfectly conducting spheres or cylinders or nonconducting cylinders, *J. Appl. Phys.*, 34: 991–993.

Kerner, E. H. (1956). The electrical conductivity of composite media, *Proc. Phys. Soc.*, B 69: 802–807.

Koo, J., and C. Kleinstreuer (2004). A new thermal conductivity model for nanofluids, *J. Nanopart. Res.*, 6 (6): 577–588.

Koo, J., and C. Kleinstreuer (2005). Impact analysis of nanoparticle motion mechanisms on the thermal conductivity of nanofluids, *Int. Commun. Heat Mass Transfer*, 32 (9): 1111–1118.

Kumar, D. H., H. E. Patel, V. R. R. Kumar, T. Sundararajan, T. Pradeep, and S.K. Das (2004). Model for heat conduction in nanofluids, *Phys. Rev. Lett.*, 93 (14): 144301.

Lamb, W., D. M. Wood, and N. W. Ashcroft (1978). Optical properties of small particle composites: theories and applications, in *Electrical Transport and Optical Properties of Inhomogeneous Media*, J. C. Garland and D. B. Tanner, American Institute of Physics, Eds., New York, pp. 240–255.

Landau, L. D., and E.M. Lifshitz (1960). Electrodynamics of Continuous Media, translated by J. B. Sykes and J. S. Bell, Pergamon Press, Oxford.

Landauer, R. (1952). The electrical resistance of binary metallic mixtures, *J. Appl. Phys.*, 23: 779–784.

Landauer, R. (1978). Electrical conductivity in inhomogeneous media, in *Electrical Transport and Optical Properties of Inhomogeneous Media*, J. C. Garland and D. B. Tanner, American Institute of Physics, Eds., New York, pp. 2–43.

Levine, H. (1966). The effective conductivity of a regular composite medium, *J. Inst. Math. Appl.*, 2: 12–28.

Lichtenecker, K. von (1924). Der elektrische Leitungswiderstand künstlicher und natürlicher Aggregate, *Phys. Z.*, 25: 225–233.

Looyenga, H. (1965). Dielectric constants of heterogeneous mixtures, *Physica*, 31: 401–406.

Lu, S.-Y., and H.-C. Lin (1996). Effective conductivity of composites containing aligned spheroidal inclusions of finite conductivity, *J. Appl. Phys.*, 79: 6761–6769.

Lu, S.-Y., and J.-L. Song (1996). Effective conductivity of composites with spherical inclusions: effective of coating and detachment, *J. Appl. Phys.*, 79: 609–618.

Mandel, M. (1961). The dielectric constant and Maxwell–Wagner dispersion of suspensions of oriented prolate spheroids, *Physica*, 27: 827–840.

Maxwell, J. C. (1873). Treatise on Electricity and Magnetism, Clarendon Press, Oxford.

McKenzie, D. R., R. C. McPhedran, and G.H. Derrick (1978). The conductivity of lattices of spheres: II. The body centred and face centred cubic lattice, *Proc. R. Soc. London A*, 362: 211–232.

McPhedran, R. C., and D.R. McKenzie (1978). The conductivity of lattices of spheres: I. The simple cubic lattice, *Proc. R. Soc. London A*, 359: 45–63.

Meredith, R. E., and C.W. Tobias (1960). Resistance to potential flow through a cubical array of spheres, *J. Appl. Phys.*, 31: 1270–1273.

Miles, J. B., Jr., and H.P. Robertson (1932). The dielectric behavior of colloidal particles with an electric double-layer, *Phys. Rev.*, 40: 583–591.

Mori, T., and K. Tanaka (1973). Average stress in matrix and average elastic energy of materials with misfitting inclusions, *Acta Metall.*, 21: 571–574.

Nan, C.-W. (1993). Physics of inhomogeneous inorganic materials, *Prog. Mater. Sci.*, 37: 1–116.

Nan, C.-W., Z. Shi, and Y. Lin (2003). A simple model for thermal conductivity of carbon nanotube-based composites, *Chem. Phys. Lett.*, 375: 666–669.

Nan, C.-W., G. Liu, Y. Lin, and M. Li (2004). Interface effect on thermal conductivity of carbon nanotube composites, *Appl. Phys. Lett.*, 85: 3549–3551.

Ni, F., G. Q. Gu, and K.M. Chen (1997). Effective thermal conductivity of nonlinear composite media with contact resistance, *Int. J. Heat Mass Transfer*, 40: 943–949.

Nielsen, L. E. (1978). *Predicting the Properties of Mixtures: Mixture Rules in Science and Engineering*, Marcel Dekker, New York.

Niesel, W. von (1952). Die Dielektrizitätstanen heterogener Mischkörper aus isotropen und anisotropen Substanzen, *Ann. Phys.*, 10: 336–348.

O'Brien, R. W. (1979). A method for the calculation of the effective transport properties of suspensions of interacting particles, *J. Fluid Mech.*, 91: 17–39.

Patel, H. E., S. K. Das, T. Sundararajan, N. A. Sreekumaran, B. George, and T. Pradeep (2003). Thermal conductivities of naked and monolayer protected metal nanoparticle based nanofluids: manifestation of anomalous enhancement and chemical effects, *Appl. Phys. Lett.*, 83 (14): 2931–2933.

Patel, H. E., T. Sundararajan, T. Pradeep, A. Dasgupta, N. Dasgupta, and S.K. Das (2005). A micro-convection model for thermal conductivity of nanofluids, *Pramana J. Phys.*, 65: 863–869.

Pauly, H., von and H. P. Schwan (1959). Über die Impedanz einer Suspension von kugelförmigen Teilchen mit einer Schale, *Z. Naturforsch.*, 146: 125–131.

Pearce, C. A. R. (1955). The permittivity of two phase mixtures, *Br. J. Appl. Phys.*, 6: 358–361.

Perrins, W. T., D. R. McKenzie, and R.C. McPhedran (1979). Transport properties of regular arrays of cylinders, *Proc. R. Soc. London A*, 369: 207–225.

Polder, D., and J.H. van Santen (1946). The effective permeability of mixtures of solids, *Physica*, 12: 257–271.

Prasher, R., P. Bhattacharya, and P.E. Phelan (2005). Thermal conductivity of nanoscale colloidal solutions (nanofluids), *Phys. Rev. Lett.*, 94 (2): 025901.

Prasher, R., P. Bhattacharya, and P.E. Phelan (2006). Brownian-motion-based convective–conductive model for the efffective thermal conductivity of nanofluids, *J. Heat Transfer*, 128: 588–595.

Rayleigh, L. (1892). On the influence of obstacles arranged in rectangular order upon the properties of a medium, *Philo. Mag.*, 34: 481–502.

Ren, Y., H. Xie, and A. Cai (2005). Effective thermal conductivity of nanofluids containing spherical nanoparticles, *J. Phys. D Appl. Phys.*, 38: 3958–3961.

Runge, I. von (1925). Zur elektrischen Leitfähigkeit metallischer Aggregate, *Z. Tech. Phys.*, 6: 61–68.

Sangani, A. S., and A. Acrivos (1983). The effective conductivity of a periodic array of spheres, *Proc. R. Soc. London A*, 386: 263–275.

Schwan, H. P., G. Schwarz, J. Maczuk, and H. Pauly (1962). On the low-frequency dielectric dispersion of colloidal particles in electrolyte solution, *J. Phys. Chem.*, 66: 2626–2635.

Schwartz, L. M., E. J. Garboczi, and D.P. Bentz (1995). Interfacial transport in porous media: application to d.c. electrical conductivity of mortars, *J. Appl. Phys.*, 78: 5898–5908.

Sillars, R. W. (1937). The properties of a dielectric containing semi-conducting particles of various shapes, *J. Inst. Electr. Eng.*, 80: 378–394.

Stratton, J. A. (1941). Electromagnetic Theory, McGraw-Hill, New York.

Stroud, D. (1975). Generalized effective-medium approach to the conductivity of an inhomogeneous material, *Phys. Rev. B*, 12: 3368–3373.

Taylor, L. S. (1965). Dielectrics properties of mixtures, *IEEE Trans. Antennas Propagation*, AP- 13: 943–947.

Taylor, L. S. (1966). Dielectrics loaded with anisotropic materials, *IEEE Trans. Antennas Propagation*, AP- 14: 669–670.

Tinga, W. R., W. A. G. Voss, and D.F. Blossey (1973). Generalized approach to multiphase dielectric mixture theory, *J. Appl. Phys.*, 44: 3897–3902.

Torquato, S. (1991). Random heterogeneous media: microstructure and improved bounds on effective properties, *Appl. Mech. Rev.*, 44: 37–76.

Tsao, G. T.-N. (1961). Thermal conductivity of two-phase materials, *Ind. Eng. Chem.*, 53: 395–397.

Van Beest, B. W. H., G. J. Kramer, and R.A. van Santen (1990). please supply titlePhys. Rev. Lett., 64: 1955–1958.

Van Beek, L. K. H. (1967). Dielectric behaviour of heterogeneous system, *Progr. Dielectr.*, 7: 69–114.

Van de Hulst, H. C. (1957). Light Scattering by Small Particles, Wiley, New York.

Wang, B.-X., L.-P. Zhou, and X.-F. Peng (2003). A fractal model for predicting the effective thermal conductivity of liquid with suspension of nanoparticles, *Int. J. Heat Mass Transfer*, 46: 2665–2672.

Wang, X., X. Xu, and S.U.S. Choi (1999). Thermal conductivity of nanoparticle–fluid mixture, *J. Thermophys. Heat Transfer*, 13 (4): 474–480.

Wiener, O. (1912). Die Theorie des Mischkörpers für das Feld der stationären Strömung: 1. Abhandlung: Die Mittelwertsätze für Kraft, Polarisation und Energie, *Abh. Math.-Phys. Kl. Koniglich Saechsis. Ges. Wiss.*, 32: 507–604.

Woodside, W., and J.H. Messmer (1961a). Thermal conductivity of porous media: I. Unconsolidated sands, *J. Appl. Phys.*, 32: 1688–1699.

Woodside, W., and J.H. Messmer (1961b). Thermal conductivity of porous media: II. Consolidated rocks, *J. Appl. Phys.*, 32: 1699–1706.

Xie, H. Q., J. C. Wang, T. G. Xi, Y. Liu, F. Ai, and Q.R. Wu (2002). Thermal conductivity enhancement of suspensions containing nanosized alumina particles, *J. Appl. Phys.*, 91 (7): 4568–4572.

Xie, H., M. Fujii, and X. Zhang (2005). Effect of interfacial nanolayer on the effective thermal conductivity of nanoparticle–fluid mixture, *Int. J. Heat Mass Transfer*, 48: 2926–2932.

Xu, J., B. Yu, M. Zou, and P. Xu (2006). A new model for heat conduction of nanofluids based on fractal distributions of nanoparticles, *J. Phys. D Appl. Phys.*, 39: 4486–4490.

Xuan, Y., and Q. Li (2000). Heat transfer enhancement of nano-fluids, *Int. J. Heat Fluid Flow*, 21: 58–64.

Xuan, Y., Q. Li, and W. Hu (2003). Aggregation structure and thermal conductivity of nanofluids, *AIChE J.*, 49 (4): 1038–1043.

Xue, Q. (2000). Effective-medium theory for two-phase random composite with an interfacial shell, *J. Mater. Sci. Technol.*, 16: 367–369.

Xue, Q.-Z. (2003). Model for effective thermal conductivity of nanofluids, *Phys. Lett. A*, 307: 313–317.

Xue, Q. Z. (2005). Model for thermal conductivity of carbon nanotube–based composites, *Physica B*, 368: 302–307.

Xue, Q. Z. (2006). Model for the effective thermal conductivity of carbon nanotube composites, *Nanotechnology*, 17: 1655–1660.

Yu, W., and S.U.S. Choi (2003). The role of interfacial layers in the enhanced thermal conductivity of nanofluids: a renovated Maxwell model, *J. Nanopart. Res.*, 5: 167–171.

Yu, W., and S.U.S. Choi (2004). The role of interfacial layers in the enhanced thermal conductivity of nanofluids: a renovated Hamilton–Crosser model, *J. Nanopart. Res.*, 6: 355–361.

Yu, W., and S.U.S. Choi (2005). An effective thermal conductivity model of nanofluids with a cubic arrangement of spherical particles, *J. Nanosci. Nanotechnol.*, 5: 580–586.

Yu, W., J. H. Hull, and S.U.S. Choi (2003). Stable and highly conductive nanofluids: experimental and theoretical studies, Paper. TED-AJ03-384, *Proc. 6th ASME-JSME Thermal Engineering Joint Conference*, Hawaiian Islands, Mar. 16–23, 2003, ASME, New York.

Zuzovski, M., and H. Brenner (1977). Effective conductivities of composite materials composed of cubic arrangements of spherical particles embedded in an isotropic matrix, *(J. Appl. Math. Phys).*, *Z. Angew. Math. Phys.* 28: 979–992.

5 Convection in Nanofluids

The discussion in earlier chapters revealed fascinating facts about the structure, composition, and thermal conductivity of nanofluids. The primary early interest in nanofluids from a technological viewpoint was the possibility of using these fluids for cooling purposes. Although the higher conductivity is an encouraging phenomenon, it is by no means conclusive evidence of the cooling capabilities of such fluids. For that, it is necessary to have definitive proof of the performance of these fluids under a convective environment. It is also important not only to reveal the convective behavior of nanofluids but also to bring out comprehensively the fluid dynamics and heat transfer theories of nanofluids. Untill now, convective studies of nanofluids have been very limited compared to experimental and theoretical studies on conduction. In this chapter we present the limited understanding of convective heat transfer in nanofluids that has been developed in recent years. Before doing so, it is important to describe some of the fundamentals of convective heat transfer, particularly for interdisciplinary readers who have not undertaken detailed studies of convective heat transfer.

5.1. FUNDAMENTALS OF CONVECTIVE HEAT TRANSFER

Convection is the mode of heat transfer in which the transport of heat from a solid wall is effected by a fluid flowing adjacent to the wall. The transfer of heat can also be in the opposite direction (i.e., from the fluid to the wall). Convection can be divided broadly into two types: forced convection and free (or natural) convection. When the fluid is "made to flow" by external agents such as a pump, fan, compressor, or blower on the heat-dissipating (or heat-accepting) surface, the convection is called *forced convection*. When flow is generated by buoyancy force during heating or cooling a fluid, it is called *free* or *natural convection*. While heating a fluid in a container on the oven, we see fluid movement before it starts boiling. This is due to natural convection.

The fundamental law of convection was proclaimed by Newton even before Fourier's conduction law was proposed. This is called the *Newton's law of cooling*, given by

$$Q = hA(T_w - T_f) \tag{5.1}$$

Nanofluids: Science and Technology, By Sarit K. Das, Stephen U. S. Choi, Wenhua Yu, and T. Pradeep

Here Q is the amount of heat transfer between the wall and the fluid that is in motion, A the solid–liquid interface area, T_w the wall temperature, and T_f the bulk fluid temperature (which is the temperature far away from the wall). The equation can be written in flux form as

$$q = \frac{Q}{A} = h(T_w - T_f) \tag{5.2}$$

This gives the essence of *Newton's law of cooling*, which simply states that convective heat flux is proportional to the temperature difference between the wall and the fluid. Here, h is merely the constant of proportionality called the *heat transfer coefficient*, with the unit W/m^2· K in the SI system. It must be noted unlike thermal conductivity, which is a material property, that h is not a constant. On the contrary, h depends on a number of factors, such as fluid and its properties, flow geometry, flow regime (laminar or turbulent), flow velocity, and thermal boundary condition at the wall.

As a result, convective heat transfer is closely related to fluid mechanics. Depending on the problem at hand, convective study may involve the solution of fluid dynamic and thermal transport equations successively or simultaneously. We do not go into the details of such solutions because they are available in standard textbooks on heat transfer, but the fundamental transport equations and resulting solutions that are used in subsequent sections to describe heat convection in nanofluids are presented.

5.1.1. Governing Equations of Fluid Flow and Heat Transfer

When fluid flows, two general conservation principles are usually satisfied: conservation of mass and conservation of momentum, which are the pillars of Newtonian mechanics. For convective heat transfer in a flowing fluid, the energy transport equation has to be satisfied in addition to the flow equations. Thus, the resulting equations for the general case of three-dimensional flow with convective heat transfer in Cartesian coordinates is given by the equations that follow. These equations can be derived by making mass, momentum, or energy balance over the elemental fluid volume shown in Fig. 5.1, which also shows mass fluxes.

1. The conservation of mass or continuity equation is given by

$$\frac{\partial \rho}{\partial t} + \frac{\partial(\rho u)}{\partial x} + \frac{\partial(\rho v)}{\partial y} + \frac{\partial(\rho w)}{\partial z} = 0 \tag{5.3}$$

where, u, v, and w are the velocities in the x, y, and z coordinate directions, respectively, ρ is the local density, and t is the time. This general continuity equation can be simplified for incompressible (liquids can be treated as incompressible) and steady flow:

$$\frac{\partial u}{\partial x} + \frac{\partial v}{\partial y} + \frac{\partial w}{\partial z} = 0 \tag{5.4}$$

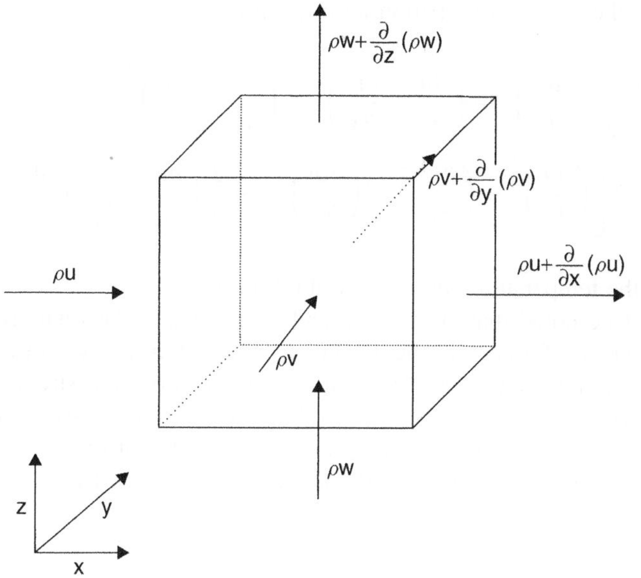

Fig. 5.1 Fluid control volume in Cartesian coordinates.

2. The momentum equations derived from Newton's second law of motion for fluid flow, popularly known as the *Navier–Stokes equations*, are given by

$$\rho\left[\frac{\partial u}{\partial t}+u\frac{\partial u}{\partial x}+v\frac{\partial u}{\partial y}+w\frac{\partial u}{\partial z}\right]=\frac{\partial \sigma_{xx}}{\partial x}+\frac{\partial \tau_{yx}}{\partial y}+\frac{\partial \tau_{zx}}{\partial z}+F_x \qquad (5.5)$$

$$\rho\left[\frac{\partial v}{\partial t}+u\frac{\partial v}{\partial x}+v\frac{\partial v}{\partial y}+w\frac{\partial v}{\partial z}\right]=\frac{\partial \sigma_{yy}}{\partial y}+\frac{\partial \tau_{xy}}{\partial x}+\frac{\partial \tau_{zy}}{\partial z}+F_y \qquad (5.6)$$

$$\rho\left[\frac{\partial w}{\partial t}+u\frac{\partial w}{\partial x}+v\frac{\partial w}{\partial y}+w\frac{\partial w}{\partial z}\right]=\frac{\partial \sigma_{zz}}{\partial z}+\frac{\partial \tau_{yz}}{\partial x}+\frac{\partial \tau_{xz}}{\partial x}+F_z \qquad (5.7)$$

Here σ and τ are normal and shear stresses on the fluid and F is the body force in the particular direction.

Finally, the energy equation is given as

$$\rho C_p\left[\frac{\partial T}{\partial t}+u\frac{\partial T}{\partial x}+v\frac{\partial T}{\partial y}+w\frac{\partial T}{\partial z}\right]=\frac{\partial}{\partial x}\left(k\frac{\partial T}{\partial x}\right)+\frac{\partial}{\partial y}\left(k\frac{\partial T}{\partial y}\right)+\frac{\partial}{\partial z}\left(k\frac{\partial T}{\partial z}\right)+\left(\frac{\partial P}{\partial t}+u\frac{\partial P}{\partial x}+v\frac{\partial P}{\partial y}+w\frac{\partial P}{\partial z}\right)+\mu\Phi \qquad (5.8)$$

where Φ is the viscous dissipation term given by

$$\Phi = \left(\frac{\partial u}{\partial y} + \frac{\partial v}{\partial x}\right)^2 + \left(\frac{\partial u}{\partial z} + \frac{\partial w}{\partial x}\right)^2 + \left(\frac{\partial v}{\partial z} + \frac{\partial w}{\partial y}\right)^2 + 2\left[\left(\frac{\partial u}{\partial x}\right)^2 + \left(\frac{\partial v}{\partial y}\right)^2 + \left(\frac{\partial w}{\partial z}\right)^2\right] - \frac{2}{3}\left(\frac{\partial u}{\partial x} + \frac{\partial v}{\partial y} + \frac{\partial w}{\partial z}\right)^2 \tag{5.9}$$

Here T is the temperature and p is the fluid pressure.

It should be noted that since mass and energy are scalar quantities, there is only one equation for each of them, whereas momentum is a vector quantity and hence there are three equations for them in each direction. The stresses appearing in equations (5.5) to (5.7) can be expressed in terms of fluid pressure and viscosity when the fluid flow is laminar. Whether a flow is laminar or turbulent depends on a dimensionless parameter, the Reynolds number, given by

$$\mathrm{Re} = \frac{\rho v D}{\mu} \tag{5.10}$$

where v is a characteristic fluid velocity, D a characteristic flow dimension, and μ the fluid viscosity. This velocity and dimension can be different for different cases. For example, usually for flow over a flat plate, the length of the plate is taken as the length and the free stream velocity far away from the plate is taken as the velocity. On the other hand, for flow inside a tube the tube diameter is taken as the dimension and the mean velocity is the characteristic velocity. For different geometries the transition from laminar to turbulent flow takes place at different Reynolds numbers. For a flat plate it is usually taken to be 5×10^5, and for a smooth tube, 2300. However, this transition is not abrupt but takes place over a range of Reynolds numbers.

For laminar flow, after substituting expressions for the stress terms σ and τ, the momentum equation reduces to the following form for x directions [equation (5.5)]:

$$\rho\left[\frac{\partial u}{\partial t} + u\frac{\partial u}{\partial x} + v\frac{\partial u}{\partial y} + w\frac{\partial u}{\partial z}\right] = -\frac{\partial P}{\partial x} + 2\frac{\partial}{\partial x}\left(\mu\frac{\partial u}{\partial x}\right) + \frac{\partial}{\partial y}\left[\mu\left(\frac{\partial u}{\partial y} + \frac{\partial v}{\partial x}\right)\right] + \frac{\partial}{\partial z}\left[\mu\left(\frac{\partial w}{\partial x} + \frac{\partial u}{\partial z}\right)\right] - \frac{2}{3}\frac{\partial}{\partial x}\left[\mu\left(\frac{\partial u}{\partial x} + \frac{\partial v}{\partial y} + \frac{\partial w}{\partial z}\right)\right] + F_x \tag{5.11}$$

Similar equations can be written for the y and z directions. A simple, often useful form of this equation is the case when the flow is two-dimensional, steady, and incompressible and body forces are absent:

$$\rho\left[u\frac{\partial u}{\partial x}+v\frac{\partial u}{\partial y}\right]=-\frac{\partial P}{\partial x}+\mu\left(\frac{\partial^2 u}{\partial x^2}+\frac{\partial^2 u}{\partial y^2}\right) \tag{5.12}$$

$$\rho\left[u\frac{\partial v}{\partial x}+v\frac{\partial v}{\partial y}\right]=-\frac{\partial P}{\partial y}+\mu\left(\frac{\partial^2 v}{\partial x^2}+\frac{\partial^2 v}{\partial y^2}\right) \tag{5.13}$$

5.1.2. Boundary Layer Approximation

The Navier–Stokes equation in its original form [equation (5.11)] cannot be solved analytically. Ludwig Prandtl, the German scientist, introduced an important approximation which works quite well for a class of fluid flow and heat transfer problems. He observed that in a simple flow situation the effect of viscosity and wall is limited to a thin layer adjacent to the wall called the *boundary layer*. He proposed that this layer is of small thickness compared to the other flow dimensions. With an order-of-magnitude analysis, this assumption can simplify the Navier–Stokes equation considerably. Figure 5.2 shows the hydrodynamic boundary layer where the initial part of the layer is laminar, followed by a transition zone and finally, a turbulent zone.

It can be noted that the velocity profile within the boundary layers depends on whether the flow is laminar or turbulent. Laminar flow is viscosity-dominated flow, whereas turbulent flow is vortex-dominated. Under the boundary layer assumption for two-dimensional steady flow, the y momentum equation drops out and only three simple equations of continuity, momentum, and energy prevail:

$$\frac{\partial u}{\partial x}+\frac{\partial v}{\partial y}=0 \tag{5.14}$$

$$\rho\left(u\frac{\partial u}{\partial x}+v\frac{\partial u}{\partial y}\right)=-\frac{dP}{dx}+\mu\frac{\partial^2 u}{\partial y^2} \tag{5.15}$$

$$\rho C_p\left(u\frac{\partial T}{\partial x}+v\frac{\partial T}{\partial y}\right)=k\left(\frac{\partial^2 T}{\partial x^2}+\frac{\partial^2 T}{\partial y^2}\right) \tag{5.16}$$

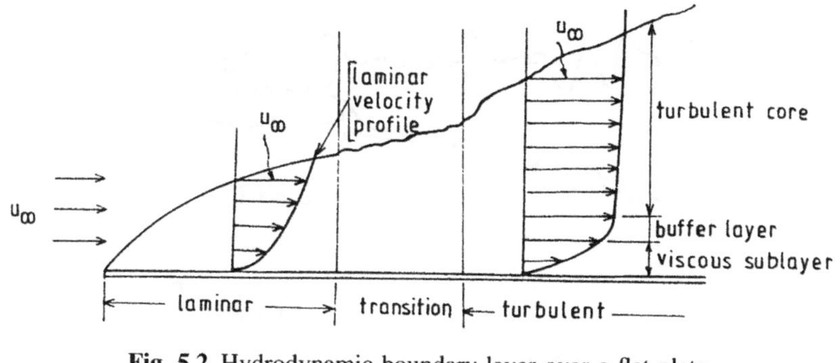

Fig. 5.2 Hydrodynamic boundary layer over a flat plate.

It must be mentioned here that the hydrodynamic boundary layer is different from the thermal boundary layer. The thermal boundary layer is a layer in which wall heat transfer affects the fluid temperature. The relative thickness of the hydrodynamic and thermal boundary layers is given by the ratio of momentum diffusivity (or kinematic viscosity) to thermal diffusivity, known as the *Prandtl number*:

$$\mathrm{Pr} = \frac{\nu}{\alpha} = \frac{\mu C_p}{k} \tag{5.17}$$

5.1.3. Turbulent Flow

Turbulent flow is one of the most complex phenomena in fluid mechanics. When the inertial forces are much higher than the viscous forces, flow does not remain in the form of undisturbed layers but starts fluctuating about the mean value. The mean flow may be steady or time dependent, as shown in Fig. 5.3. The hydrodynamic boundary layer over a flat plate depends on the eddies and mixing in the flow. Modeling turbulent flow is a century-old challenge which even after tremendous research effort remains far short of being satisfactory. However, in simpler commonly used geometries, a number of well-accepted features and correlations have emerged over the past few decades. Usually, the instantaneous quantities in a turbulent flow can be split into mean and functioning quantities, such as

$$\begin{aligned} u &= \overline{u} + u' \qquad v = \overline{v} + v' \\ P &= \overline{P} + P' \qquad T = \overline{T} + T' \end{aligned} \tag{5.18}$$

Using these quantities, the original boundary layer equations, (5.14) to (5.16), can be reduced to the form

$$\frac{\partial \overline{u}}{\partial x} + \frac{\partial \overline{v}}{\partial y} = 0 \tag{5.19}$$

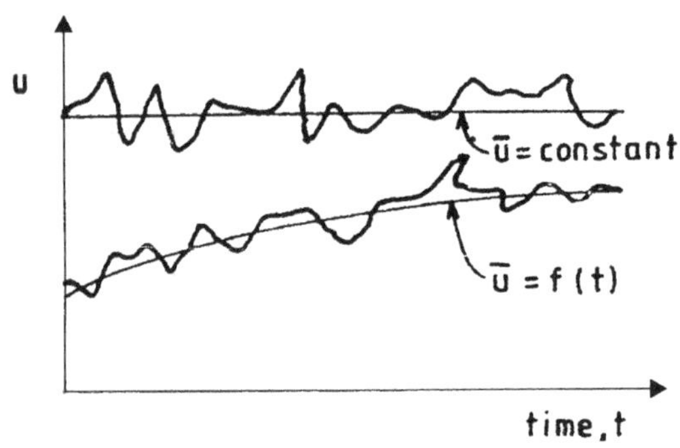

Fig. 5.3 Turbulent velocity fluctuation.

$$\rho\left(\overline{u}\frac{\partial \overline{u}}{\partial x}+\overline{v}\frac{\partial \overline{u}}{\partial y}\right)=-\frac{\partial \overline{P}}{\partial x}+\frac{\partial}{\partial y}\left(\mu\frac{\partial \overline{u}}{\partial y}-\rho\overline{v'u'}\right) \tag{5.20}$$

$$\rho C_p\left(\overline{u}\frac{\partial \overline{T}}{\partial x}+\overline{v}\frac{\partial \overline{T}}{\partial y}\right)=-\frac{\partial}{\partial y}\left(-k\frac{\partial \overline{T}}{\partial y}+\rho C_p\overline{v'T'}\right)+\mu\left(\frac{\partial^2 \overline{u}}{\partial y^2}\right)^2 \tag{5.21}$$

where the terms with overbars are averaged over time. It is interesting to note that these equations are same as the laminar flow equations except for the turbulent stress and turbulent heat flux terms, which are given by

$$\tau_{\text{turbulent}}=-\rho\overline{v'u'} \tag{5.22}$$

$$q_{\text{turbulent}}=\rho C_p\overline{v'T'} \tag{5.23}$$

Modeling these terms for different flow conditions still remains a challenge for the scientific community. The best one can do is to assume an equation similar to that for laminar flow:

$$\tau_{\text{turbulent}}=\mu_t\frac{\partial \overline{u}}{\partial y} \tag{5.24}$$

$$q_{\text{turbulent}}=-k_t\frac{\partial \overline{T}}{\partial y} \tag{5.25}$$

where μ_t and k_t are turbulent viscosity and turbulent thermal conductivity, respectively. This does not solve the problem anyway because unlike molecular viscosity and conductivity of laminar equations, the turbulent quantities are not fluid properties but functions of the flow conditions. The effort to model this has been going on starting with Prandtl and extending to the current fluid mechanics community. Prandtl tried to model it by drawing an analogy between kinetic theory of gases and turbulence. Hence, he proposed a *mixing length* which is analogous to the mean free path of kinetic theory and suggested that

$$\mu_t=\rho l_m^2\left|\frac{\partial \overline{u}}{\partial y}\right| \tag{5.26}$$

where l_m in the mixing length, which depends on geometry and flow conditions. For some known cases this model works quite well and l_m is the well-known length scale

$$l_m=\begin{cases}0.75\times\text{half jet width (plane jet)}\\0.07\times\text{half the layer width (plane mixing layer)}\end{cases}$$

However, for more complex situations, more sophisticated equations are needed. The most popular of them is to solve transport equations for turbulent kinetic energy (κ) and its dissipation rate (ε) and to combine them to give μ_t, known as the κ−ε *model*. However, it must be said here that the success of analytical or even numerical treatment of turbulence is still very limited, and hence many fluid mechanics and heat transfer correlations for turbulent flow are purely experimental.

5.1.4. Natural Convection

In natural convection, flow is generated due to the temperature variation within the fluid arising out of the buoyancy force. Let us consider the vertical plate shown in Fig. 5.4. Heat conduction from the plate to the fluid (in the case of a hot plate) will create a density difference within the fluid. This will bring buoyancy into play, and the fluid adjacent to the wall will rise and the cold fluid from an adjacent area will replenish it. Thus, a convective current will set in which will remove heat form the wall continuously without any external flow-maintaining device such as a fan or pump. In case of natural convection, the buoyancy force, which is a body force, is dominant, and hence the laminar boundary layer equation in two dimensions is reduced to

$$\frac{\partial u}{\partial x} + \frac{\partial v}{\partial y} = 0 \tag{5.27}$$

$$\rho\left(u\frac{\partial u}{\partial x} + v\frac{\partial u}{\partial y}\right) = g\beta(T - T_\alpha) + \mu\frac{\partial^2 u}{\partial y^2} \tag{5.28}$$

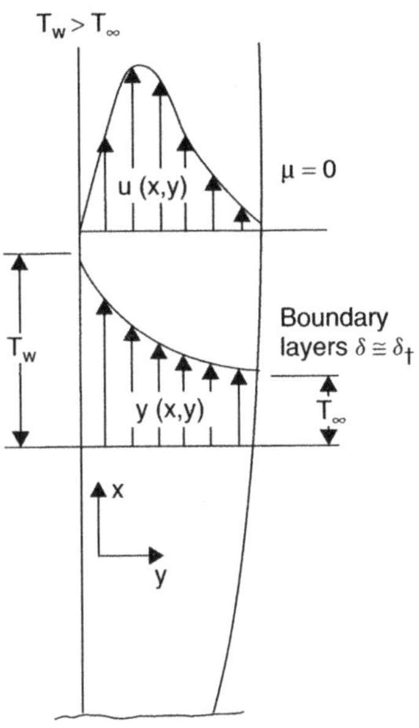

Fig. 5.4 Temperature and velocity profiles in natural convective boundary layer on a vertical plate.

$$\rho C_p \left(u \frac{\partial T}{\partial x} + v \frac{\partial T}{\partial y} \right) = k \frac{\partial^2 T}{\partial y^2} \tag{5.29}$$

Here we have used an approximation known as the *Boussinesq approximation*, which treats all properties, including density, as a constant excepting the density term in the body force term [first-right-hand-side term of equation (5.28)] where density variation with temperature is considered. Instead of the Reynolds number, a dimensionless number that indicates whether a natural convective flow is laminar or turbulence is the *Grashof number*, given by

$$\mathrm{Gr} = \frac{g\beta(T_w - T_f)L^3}{\nu^2} \tag{5.30}$$

where β is the volume compressibility of the fluid, T_{w} and T_f are wall and bulk fluid temperatures, L is the characteristic length, and ν is the kinematic viscosity. In recent times it has been more popular to use another dimensionless number, the *Rayleigh number*, to indicate a transition from laminar to turbulent natural convection:

$$\mathrm{Ra} = \mathrm{Gr} \cdot \mathrm{Pr} \tag{5.31}$$

For a vertical plate, $\mathrm{Ra} > 10^8$ is usually the turbulent flow regime.

5.1.5. Fluid Flow and Heat Transfer Correlations in Convection

Based on the fundamentals of fluid and energy transport presented in Section 5.1.4, many heat and fluid flow correlations have been developed either by solving these equations analytically or numerically or by conducting experiments over a wide range of parameters. It is impossible to present even a selective collection of such correlations. Hence, here we present only those correlations that are important for convective heat transfer studies in nanofluids. Before taking up this job, some important definitions need to be stated. The most important parameter in convective heat transfer is the *heat transfer coefficient*, defined by Newton's law of cooling [equation(5.1)]. To present more generalized results, this quantity is nondimensionalized and called the *Nusselt number*, defined as

$$\mathrm{Nu} = \frac{hL}{k} \tag{5.32}$$

where L is a characteristic dimension and k is the thermal conductivity of the fluid. Locally, heat transfer coefficient can be defined (at location x) as

$$h_x = \frac{q_x}{T_w - T_\infty} \tag{5.33}$$

where T_w and T_α are the local wall and bulk fluid temperatures and q_x is the local heat flux at the wall. The mean value of Nusselt number is based on the mean heat transfer coefficient as

$$Nu_m = \frac{h_m L}{k} \qquad \text{where} \quad h_m = \frac{1}{L}\int_0^1 h_x\,dx \tag{5.34}$$

On the fluid flow side for the flat plate, the local fluid frictional drag coefficient is defined as

$$C_{f.x} = \frac{\tau_{w.x}}{\frac{1}{2}\rho u_\infty^2} \tag{5.35}$$

where $\tau_{w..x}$ is the local fluid shear stress at the wall and u_α is the fluid free stream velocity (far away from the wall). The drag coefficient can be calculated as

$$C_{f.m} = \frac{1}{L}\int_0^L C_{f.x}\,dx \tag{5.36}$$

For flow inside tubes and channels, the fluid friction causes a pressure drop. This pressure drop is related conventionally to a *friction factor*, f. For practical purposes f is defined as

$$\Delta P = f\frac{L}{D}\left(\frac{1}{2}\rho u_m^2\right) \tag{5.37}$$

Hence for calculating pressure drop the friction factor must be known either from the solution of momentum equations or experimentally.

Based on these definitions it is now possible to describe the heat transfer correlations. However, before that, one needs to understand that when arriving at these correlations the boundary conditions under which the solutions were obtained from the equations described earlier need to be specified. For fluid flow the layers next to the wall usually stick to the wall under the influence of viscosity. Called the *no-slip condition*, this is satisfied in most practical cases except for the flow of rarefied gas, where a "slip" may occur at the boundary. On the thermal side, at the wall a variety of thermal conditions may be present. The two limiting cases that are important are (1) the constant wall temperature condition and (2) the constant wall heat flux condition. There can also be a variable temperature or heat flux at the wall. With the definitions given above we are now in a position to present the heat transfer and fluid friction correlations for convection.

Flow over a Flat Plate First let us consider the flat plate shown in Fig. 5.2. The frictional drag coefficient at a distance from the leading edge for laminar flow is given by

$$C_{f.x} = \frac{0.664}{(\text{Re}_x)^{1/2}} \qquad \text{where} \quad \text{Re}_x = \frac{\rho u_\alpha x}{\nu} \tag{5.38}$$

The mean value of the drag coefficient over a length L of the plate is given by

$$C_{f,m} = 2C_{f,L} = \frac{1.328}{(\mathrm{Re}_L)^{1/2}} \qquad \text{where} \quad \mathrm{Re}_L = \frac{\rho u_\alpha L}{\nu} \tag{5.39}$$

For heat transfer, the local Nusselt number for this case is given by

$$\mathrm{Nu}_x = \frac{h_x x}{k} = 0.332\mathrm{Re}_x^{1/2}\mathrm{Pr}^{1/3} \tag{5.40}$$

The mean value is given by

$$\mathrm{Nu}_m = \frac{h_m L}{k} = 2\mathrm{Nu}_L = 0.664\mathrm{Re}_L^{1/2}\mathrm{Pr}^{1/3} \tag{5.41}$$

It is important to note that in this case, the fluid properties are to be taken at the *mean film temperature*, given by

$$T_f = \frac{T_w + T_\infty}{2} \tag{5.42}$$

For turbulent flow, the correlations for local drag and heat transfer coefficient are given by

$$C_{f,x} = \frac{0.0592}{\mathrm{Re}_x^{0.2}} \tag{5.43}$$

$$\mathrm{Nu}_x = 0.029\mathrm{Re}_x^{0.8}\mathrm{Pr}^{1/3} \tag{5.44}$$

Whitaker (1972) suggested a little adjustment of the constants to make it more agreeable to experimental results:

$$\mathrm{Nu} = 0.0296\mathrm{Re}_x^{0.8}\mathrm{Pr}^{1/3} \tag{5.45}$$

Flow over a Tube Flow over a tube is of importance in applications such as heat exchangers, automotive radiators, and refrigeration tubes. Flow over a cylinder and tube is quite complex since flow separation takes place and a phenomenon known as *vortex shedding* occurs, shown schematically in Fig. 5.5. As a result, the drag coefficient C_D is also different. For different ranges of Reynolds number,

$$C_D = \begin{cases} 10.41\mathrm{Re}_D^{-0.6872} & 0.1 < \mathrm{Re}_D < 4 \\ 5.67\mathrm{Re}_D^{-0.2511} & 4 < \mathrm{Re}_D < 1000 \\ 0.31\mathrm{Re}_D^{-0.1375} & 5000 < \mathrm{Re}_D < 10^4 \\ 1.1 & 10^4 < \mathrm{Re}_D < 10^5 \end{cases}$$

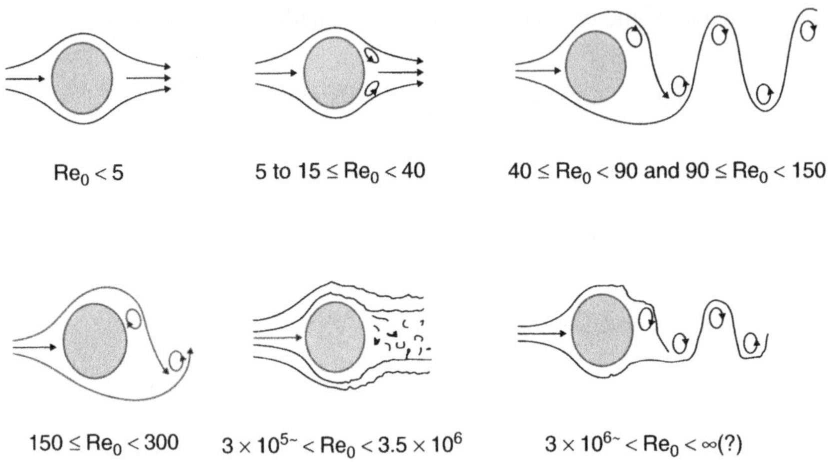

Fig. 5.5 Flow over a cylinder and vortex shedding.

The Nusselt numbers for the case with a constant cylinder wall temperature is given by Whitaker's (1972) experimental correlation:

$$\mathrm{Nu}_D = (0.4\mathrm{Re}^{1/2} + 0.06\mathrm{Re}^{2/3})\mathrm{Pr}^{0.4}\left(\frac{\mu_\infty}{\mu_w}\right)^{0.25} \tag{5.46}$$

where μ_∞ and μ_w are the values of viscosity at the free stream and wall temperatures respectively. A more general correlation was proposed by Churchill and Bernstein (1977):

$$\mathrm{Nu}_D = 0.3 + \frac{0.62\mathrm{Re}^{1/2}\mathrm{Pr}^{1/3}}{\left[1 + (0.4/\mathrm{Pr})^{2/3}\right]^{1/4}}\left[1 + \left(\frac{\mathrm{Re}}{282{,}000}\right)^{5/8}\right]^{4/5} \qquad 100 < \mathrm{Re} < 10^5 \tag{5.47}$$

There are many correlations for turbulent flow which are not relevant in the present context.

Flow Inside Tubes Flow and heat transfer for a fluid flowing inside a tube is of special importance in cooling applications. Inside the tube a special case occurs. The boundary layers emerging from all the sides (both hydrodynamic and thermal boundary layers) merge and fill the entire tube. From that time the velocity profile remains unchanged and is called a *fully developed flow*. The region prior to this is called the *entry length* or *developing flow region*. Hence, in pipe flow, laminar flow remains laminar throughout the pipe and turbulent flow remains turbulent, as shown in Fig. 5.6. The thermal boundary layer developed is not the same as the hydrodynamic boundary layer. Hence, the temperature profile changes continuously, due to heat transfer. Thus, flow inside a tube is

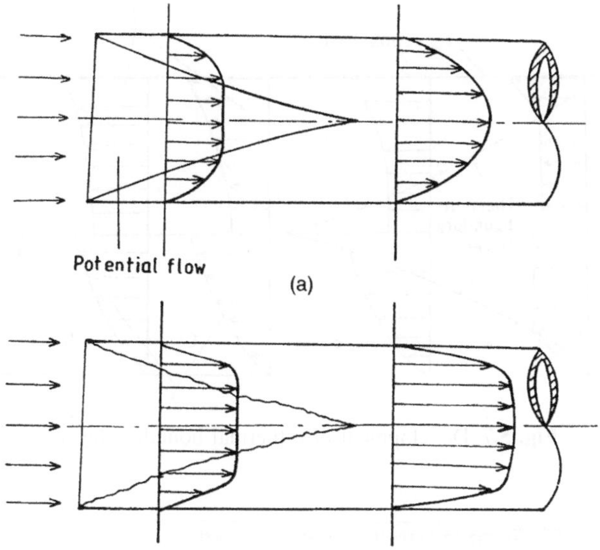

Fig. 5.6 Flow development in a tube: (a) laminar; (b) turbulent.

called *thermally developed* if the dimensionless temperature (Θ) profile remains unchanged, where

$$\Theta = \frac{T_w - T}{T_w - T_m} \tag{5.48}$$

The mean temperature at a particular section in given by

$$T_m = \frac{\int_A \rho u C_p T \, dA}{m C_p} \tag{5.49}$$

where m is the mass flow rate of the fluid. The real and dimensionless temperatures for flow inside a tube with a wall at a higher temperature are shown in Fig. 5.7.

Using the boundary layer–approximated Navier–Stokes equation in cylindrical coordinates, the friction factor for flow inside the tube can be found as follows: for lalminar flow,

$$f = \frac{64}{\mathrm{Re}_D} \tag{5.50}$$

and for turbulent flow,

$$f = \begin{cases} \dfrac{0.316}{\mathrm{Re}_D^{0.25}} & \mathrm{Re} < 2 \times 10^4 \quad (5.51) \\ \dfrac{0.184}{\mathrm{Re}} & \mathrm{Re} > 2 \times 10^4 \quad (5.52) \end{cases}$$

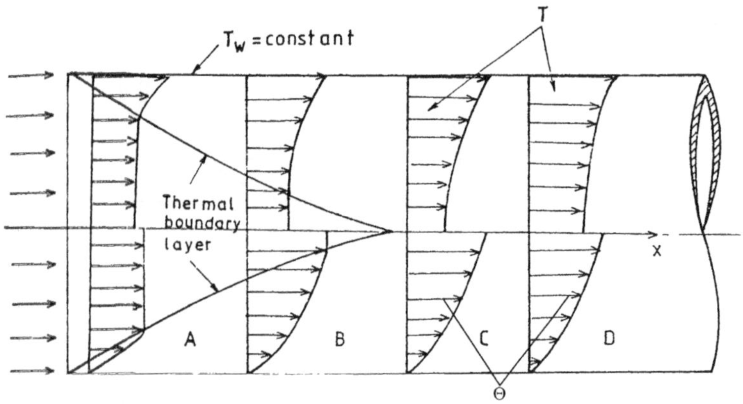

Fig. 5.7 Development of a thermal boundary layer.

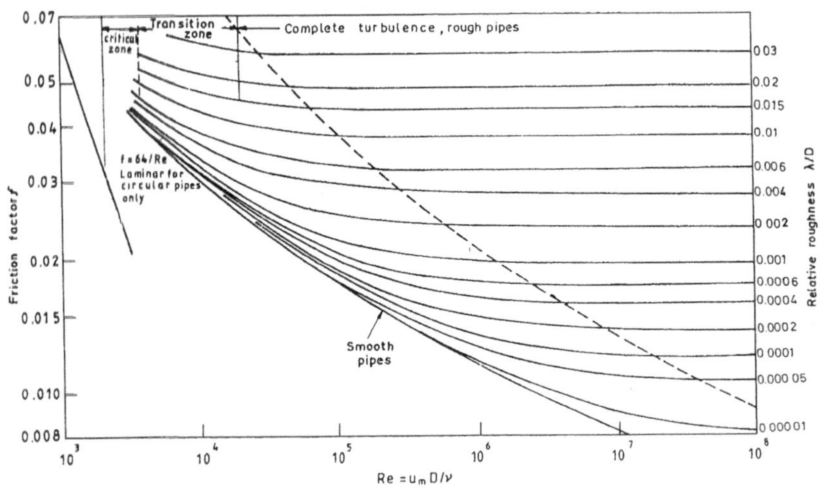

Fig. 5.8 Moody's diagram for a friction factor inside a tube.

However, it is more convenient to use a the Moody diagram (Fig. 5.8) to obtain the friction factor not only for smooth tubes but also for rough tubes in which the friction factor increases significantly in the turbulent regime with increasing tube roughness. With heat transfer it is interesting that for fully developed laminar flow inside channels, the Nusselt number is a constant number that depends only on the shape of the channel and the thermal boundary conditions.

For circular tubes the Nusselt number for two standard boundary conditions is

$$\mathrm{Nu}_D = \begin{cases} \dfrac{hD}{k} = 4.364 & \text{for constant heat flux} \quad (5.53) \\ 3.66 & \text{for constant wall temperature} \quad (5.54) \end{cases}$$

Table 5.1 Nusselt Number for Hydrodynamically and Thermally Developed Laminar Flow in Ducts of Various Cross Sections (Constant Heat Flux)

Geometry ($L/D_h > 100$)		Nu_{III}
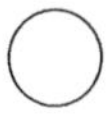		4.364
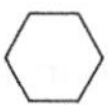		2.002
$2b$ 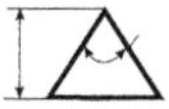 $2a$	$\frac{2b}{2a}=\frac{\sqrt{3}}{2}$	3.111
$2a$  $2a$	$\frac{2b}{2a}=1$	3.608
$2b$  $2a$	$\frac{2b}{2a}=\frac{1}{2}$	4.123
$2b$  $2a$	$\frac{2b}{2a}=\frac{1}{4}$	5.099
$2b$  $2a$	$\frac{2b}{2a}=\frac{1}{8}$	6.490
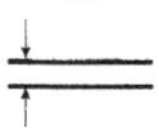	$\frac{2b}{2a}=0$	8.325
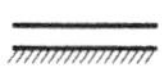	$\frac{b}{a}=0$	5.385

For other geometries, Table 5.1 gives the Nusselt number for constant wall heat flux conditions. However, for turbulent flow the Nusselt number depends on the Reynolds and Prandtl numbers. One experimental correlation, known as the *Dittus–Boelter equation*, is very popular in pipe flow:

$$\mathrm{Nu}_D = 0.023\mathrm{Re}_D^{0.8}\mathrm{Pr}^n \qquad (5.55)$$

where, n is 0.3 for cooling of the fluid and 0.4 for heating of the fluid. This correlation can be used for $0.7 < \mathrm{Pr} < 100$ and L/D for a the tube larger than 60. Here the properties are to be evaluated at the bulk mean temperature between inlet and outlet:

$$T_m = \frac{T_{\text{in}} + T_{\text{out}}}{2} \qquad (5.56)$$

This equation can also be used for a noncircular cross section with the diameter replaced by hydraulic diameter D_h:

$$D_h = \frac{4 \times \text{flow cross-sectional area}}{\text{wetted perimeter}} \qquad (5.57)$$

The Dittus–Boelter equation is not recommended for large property variation, due to the temperature difference. In such cases the *Sieder–Tate* (1936) *equation* is recommended:

$$\mathrm{Nu} = 0.027\mathrm{Re}^{0.8}\mathrm{Pr}^{1/3}\left(\frac{\mu_b}{\mu_w}\right)^{0.14} \tag{5.58}$$

for $0.7 < \mathrm{Pr} < 16{,}700$ and $\mathrm{Re} \geq 10{,}000$. There are two more important experimental correlations which are often used, due to their greater accuracy. The first is the *Petukhov* (1970) *correlation*:

$$\mathrm{Nu} = \frac{(f/8)\mathrm{RePr}}{1.07 + 12.7\sqrt{f/8}(\mathrm{Pr}^{2/3} - 1)}\left(\frac{\mu_b}{\mu_w}\right)^{n} \tag{5.59}$$

where, n is 0.11 for uniform $T_w > T_b$, 0.25 for uniform $T_w < T_b$, and 0 for constant heat flux or for gases. This is applicable for $10^4 < \mathrm{Re} < 5 \times 10^6$, $0.5 < \mathrm{Pr} < 200$ with 5% error, and $200 < \mathrm{Pr} < 2000$ with 10% error. This equation was modified as the *Gnielinski* (1976) correlation

$$\mathrm{Nu} = \frac{f/8(\mathrm{Re} - 1000)\mathrm{Pr}}{1 + 12.7\sqrt{f/8}(\mathrm{Pr}^{2/3} - 1)}\left[1 + \left(\frac{d}{L}\right)^{2/3}\right] \tag{5.60}$$

The friction factor was given as

$$f = \frac{1}{(1.82\log_{10}\mathrm{Re} - 1.64)^2} \tag{5.61}$$

For thermally developing flow with a constant wall temperature, *Hausen's* (1943) *correlation* is used:

$$\mathrm{Nu}_D = 3.66 + \frac{0.0668\mathrm{Gz}}{1 + 0.04\mathrm{Gz}^{2/3}} \tag{5.62}$$

where Gz, the *Graetz number*, $= \mathrm{Re}\cdot\mathrm{Pr}(D/L)$

Often, for developing flow, *Sieder and Tate's* (1936) *correlation* is used:

$$\mathrm{Nu} = 1.86\mathrm{Re}^{1/3}\mathrm{Pr}^{1/3}\left(\frac{D}{L}\right)^{1/3}\left(\frac{\mu_b}{\mu_w}\right)^{0.14} \tag{5.63}$$

For a thermal entrance region, *Shah's correlation* for the local Nusselt number is very popular:

$$\mathrm{Nu} = 1.953\left(\mathrm{Re}\cdot\mathrm{Pr}\frac{D}{x}\right)^{1/3} \quad \text{for} \quad \mathrm{Re}\cdot\mathrm{Pr}\frac{D}{x} \geq 33.3 \tag{5.64}$$

Natural Convection The correlations in natural convection are critically dependent on the orientation of the heat transfer surface with respect to gravity. For flow on a vertical plate, the local Nusselt number is given by

$$\mathrm{Nu}_x = 0.508\mathrm{Pr}^{1/2}\frac{1}{(0.952+\mathrm{Pr})^{1/4}}\mathrm{Gr}^{1/4} \quad \text{where} \quad \mathrm{Gr}_x = \frac{g\beta(T_w - T_\infty)x^3}{\nu^2} \tag{5.65}$$

In terms of the Rayleigh number, the Nusselt number can be written as

$$\mathrm{Nu}_x = 0.508\ \mathrm{Ra}_x^{1/4}\left(\frac{\mathrm{Pr}}{(0.952+\mathrm{Pr})^{1/4}}\right)^{1/4} \tag{5.66}$$

The mean Nusselt number over the plate can be written as

$$\mathrm{Nu}_m = 0.667\ \mathrm{Ra}_L^{1/4}\left(\frac{\mathrm{Pr}}{0.952+\mathrm{Pr}}\right)^{1/4} \tag{5.67}$$

A more versatile and experimentally verified passed correlation was suggested by Churchill and Chu (1975):

$$\mathrm{Nu}_m = 0.68 + \frac{0.67\ \mathrm{Ra}^{1/4}}{[1+(0.492/\mathrm{Pr})^{9/16}]^{4/9}} \tag{5.68}$$

for $0 < \mathrm{Ra} < 10^9$ and $0 < \mathrm{Pr} < \infty$, and for $\mathrm{Ra} > 10^9$,

$$\mathrm{Nu}_m = \frac{0.15\ \mathrm{Ra}^{1/3}}{[1+(0.492/\mathrm{Pr})^{9/16}]^{16/27}} \tag{5.69}$$

There are wide varieties of correlation for horizontal plates. The one by McAdams (1954) for constant wall temperature is very popular:

$$\mathrm{Nu}_m = \begin{cases} 0.54\ \mathrm{Ra}_L^{1/4}, & 10^5 < \mathrm{Ra}_L < 3\times 10^7 \\ 0.14\ \mathrm{Ra}_L^{1/3}, & 2\times 10^7 < \mathrm{Ra}_L < 3\times 10^{10} \end{cases} \tag{5.70}$$

5.2. CONVECTION IN SUSPENSIONS AND SLURRIES

Before entering the domain of convection in nanofluids it will be useful to have a brief look at slurries and suspensions with respect to their hydrodynamics and resulting heat transfer. The flow of solid particles with fluids is given various names, such as particle-laden flow, pneumatic particle transport, particulate flow, fluidized flow, slurry flow, and hydrotransport. Whereas the last two types of flow occur in liquids with solid particles, the others refer to different types of gases with solid particles. Usually, the term *particulate flow* refers to both of these suspensions.

Mathematical treatment of particulate flow and heat transfer behavior is quite complex and forms the subject matter of a different field of study in which a large number of investigations have been carried out in the past few decades. There is also no guarantee that these methods can be useful for the very special types of suspensions such as nanofluids. However, these methods of modeling, as well as the resulting heat transfer correlations, can be very good starting points for the analytical and experimental treatment of convection in nanofluids. For this purpose we first discuss different modeling approaches to hydrodynamics and heat transport in suspensions and then describe several interesting studies from the literature.

5.2.1. Hydrodynamics of Suspensions

To ascertain the hydrodynamics of suspensions, one first has to choose the frame of reference for the analysis. This frame can be Eulerian (static frame) or Lagrangian (frame moving with the particle). However, it is always convenient to model the fluid flow in the Eulerian frame. Hence, the following combinations are possible.

Eulerian–Eulerian Approach In this approach, the fluid and the particle are both treated in the Eulerian frame. In this approach two methods are popular.

Single-Fluid Approach In this approach the entire suspension is treated as a single fluid whose properties are in between the solid and fluid properties that form the suspensions. Defining these properties is often tricky and may need experimental input. Hence, the continuity, momentum, and energy equations can be written. The continuity equation is

$$\frac{\partial}{\partial t}(\rho_m) + \frac{\partial(\rho_m u_m)}{\partial x} + \frac{\partial(\rho_m v_m)}{\partial y} + \frac{\partial(\rho_m w_m)}{\partial z} = 0 \tag{5.71}$$

Here, u_m, v_m, and w_m are mass-averaged velocities of the single fluid in the x, y, and z directions respectively, and ρ_m is the average density of the medium. For example, u_m is given by

$$u_m = \frac{\varepsilon_p \rho_p u_p + \varepsilon_f \rho_f u_f}{\rho_m} \tag{5.72}$$

where ε_p, ρ_p, and u_p are the volume fraction, density, and velocity of the particle, and ε_f, ρ_f, and u_f are the same for the fluid. The average density of the medium (or assumed single fluid) is given by

$$\rho_m = \varepsilon_p \rho_p + \varepsilon_f \rho_f \tag{5.73}$$

The momentum equation for incompressible suspension (solid in liquid) is given by

$$\frac{\partial}{\partial t}(\rho_m u_m) + u_m \frac{\partial u_m}{\partial x} + v_m \frac{\partial u_m}{\partial y} + w_m \frac{\partial u_m}{\partial z}$$
$$= -\frac{\partial P}{\partial x} + (\mu_m + \mu_t)\left(\frac{\partial^2 u_m}{\partial x^2} + \frac{\partial^2 u_m}{\partial y^2} + \frac{\partial^2 u_m}{\partial z^2}\right) + \rho_m g + F_x + \frac{\partial}{\partial x}(\varepsilon_p \rho u_{pd}^2) \tag{5.74}$$

where u_{pd} is the particle drift velocity, given by $u_{pd} = u_p - u_m$ and μ_t is the turbulent viscosity of the suspension. The medium viscosity is to be taken from experimental data on the viscosity of the suspension. For an evaluation of turbulent viscosity we need a turbulence model. A wide variety of models have been suggested in the literature. A significant contribution in particulate flow with heat transfer is that of Michaelides (1986), who assumed a *mixing length model* for turbulence. However, a more popular model recently has been the $k-\varepsilon$ model (section 5.1.3), in which additional transport equations for turbulent kinetic energy (k) and its dissipation rate (ε) are solved and the turbulent viscosity is calculated as

$$\mu_t = \rho C_\mu \frac{k^2}{\varepsilon} \tag{5.75}$$

where C_μ is a constant (usually, $\approx$0.09).

Viscosity is difficult to model in particulate flow. Often, it is taken as the sum of collisional viscosity, kinetic viscosity, and frictional viscosity, and each of these terms is evaluated by empirical correlations available in the literature. Due to the stable nature of nano-fluids, the best option seems to be to use an experimental value. For dilute suspensions ($\varepsilon_p < 2\%$) the famous Stokes–Einstein formula for viscosity can be a good approximation for nanofluids:

$$\mu = \mu_L(1 + 2.5\varepsilon_p) \tag{5.76}$$

The momentum equation for the y and z directions can be written similarly. The drift velocity can be correlated with the interfacial drag and particle acceleration. When there is no slip between the particle and the fluid (i.e., the particle moves with the fluid), the last term of equation (5.74) drops out.

The energy equation for this case is

$$\frac{\partial}{\partial t}(\varepsilon_p \rho_p h_p + \varepsilon_f \rho_f h_f) + \varepsilon_p \rho_p \left(u_p \frac{\partial h_p}{\partial x} + v_p \frac{\partial h_p}{\partial y} + w_p \frac{\partial h_p}{\partial z}\right)$$
$$+ \varepsilon_f \rho_f \left(u_f \frac{\partial h_f}{\partial x} + v_f \frac{\partial h_f}{\partial y} + w_f \frac{\partial h_f}{\partial z}\right) = k_{eff}\left(\frac{\partial^2 T}{\partial x^2} + \frac{\partial^2 T}{\partial y^2} + \frac{\partial^2 T}{\partial z^2}\right) \tag{5.77}$$

Here h is the enthalpy of the corresponding phase, which can be replaced by C_pT (where C_p is the specific heat) and k_{eff} is the effective conductivity of the medium (see Chapter 3 for a detailed discussion).

It must be mentioned here that all these complex equations are required only if the liquid and solid move with different velocities. When the particles move with the same velocity as the fluid, they can be modeled simply as single-phase flow, as described in Section 5.1, with the effective properties of the medium replacing the liquid properties. Due to interaction of the particles and the particle-fluid, in single-fluid formulation a thermal dispersion term that acts as an additional virtual conduction is usually added to the energy equation. The single-fluid equation with dispersion term can be written as

$$(\rho C_p)_m\left(\frac{\partial T_m}{\partial t}+u_m\frac{\partial T_m}{\partial x}+v_m\frac{\partial T_m}{\partial y}+w_m\frac{\partial T_m}{\partial z}\right)$$
$$=(k_{eff}+k_d)\left(\frac{\partial^2 T_m}{\partial x^2}+\frac{\partial^2 T_m}{\partial y^2}+\frac{\partial^2 T_m}{\partial z^2}\right) \tag{5.78}$$

Here T_m is the mean medium temperature (which is assumed to be in thermal equilibrium between particle and fluid), k_{eff} the effective medium conductivity, and k_d the dispersive equivalent conductivity of the medium.

The Pure Eulerian Approach The Eulerian model is a more detailed model in which both the particle and fluid phases are modeled in the Eulerian frame. The interfacial heat transfer is accounted for and other types of forces, such as the lift force, are accounted for in the momentum equation. Usually, suspensions with a high volume of particle loading (typically, more that 10%) are modeled using this approach. Since nanofluids are of low-particle-volume fraction, this model does not seem relevant here. Also, more computer time is required in this type of analysis.

The Eulerian–Lagrangian Approach In this model the particle is tracked in the Lagrangian frame and the fluid in the Eulerian frame. The Lagrangian particle momentum equation can be written

$$\frac{du_p}{dt}=F_D(u-u_p)+\frac{g_x(\rho_p-\rho)}{\rho_p}+F_x \tag{5.79}$$

where F_D is the drag force, given by

$$F_D=\frac{18\mu}{\rho_p d_p^2}\,\frac{C_D\cdot \mathrm{Re}}{24} \tag{5.80}$$

Here u_p is the particle velocity, u the fluid velocity, g_x the component of g in the x direction, and d_p the particle diameter. Re is the relative Reynolds number, given by

$$\mathrm{Re} = \frac{\rho d_p |u_p - u|}{\mu}$$

There are various models of the drag coefficient C_D for nanoparticles. Stoke's law seems to be appropriate:

$$F_D = \frac{18\mu}{d_p^2 \rho_p C_c} \qquad \text{where} \quad C_c = 1 + \frac{2\lambda}{d_p}(1.257 + 0.4e^{-(1.1d_p/2\lambda)}) \tag{5.81}$$

λ being the molecular mean free path. In this model the particle energy balance equation is formulated as

$$hA_p(T - T_p) = \rho_p V_p C_p \frac{dT_p}{dt} \tag{5.82}$$

where A_p is the particle surface area and V_p is the particle volume. The interfacial heat transfer equation is given in terms of the Nusselt number. Often the following equations are used:

$$\mathrm{Nu} = \begin{cases} 1 + (1 + \mathrm{Re}_p\mathrm{Pr})^{1/3} & \text{for} \quad 0 \le \mathrm{Re}_p \le 1 \qquad (5.83) \\ \mathrm{Re}_p^{0.41}\left(1 + \dfrac{1}{\mathrm{Re}_p\mathrm{Pr}}\right)^{1/3}\mathrm{Pr}^{1/3} + 1 & \text{for} \quad 1 \le \mathrm{Re}_p \le 100 \qquad (5.84) \\ 0.752\mathrm{Re}_p^{0.472}\left(1 + \dfrac{1}{\mathrm{Re}_p\mathrm{Pr}}\right)^{1/3}\mathrm{Pr}^{1/3} + 1 & \text{for} \quad 100 \le \mathrm{Re}_p \le 2000 \qquad (5.85) \end{cases}$$

Here the particle Reynolds number is given by

$$\mathrm{Re}_p = \frac{\rho d_p U}{\mu} \qquad \text{where} \quad U = \sqrt{(u - u_p)^2 (v - v_p)^2 + (w - w_p)^2}$$

It goes without saying in all these approaches the final solution requires tedious numerical computer calculations, and the chances of getting analytical solutions are very remote. However, for simple cases of single-fluid treatment or with large number of assumptions, some analytical solutions are possible, but the applicability of these solutions is doubtful. Hence, quite often we depend on experimental correlations such as that by Kim and Lee (2001):

$$\mathrm{Nu} = 1.8\mathrm{Re}_p^{0.43}\mathrm{Pr}^{1/3}\left(\frac{d_p}{D}\right)^{0.74}\varepsilon_p^{0.04} \tag{5.86}$$

were D is the tube diameter. However, it must be kept in mind that such equations are exclusively for one system of fluid and particle and cannot be extended to other systems—not to speak of nanofluids. For example, equation (5.86) is only for the flow of large glass beads with water in a vertical tube.

Thus, it is clear that the equations, methods, and correlations presented above may not be directly applicable for convective heat transfer with nanofluids, but they will certainly be helpful in understanding the analysis and data presented by a large number of investigators working on convection in nanofluids.

5.2.2. Special Features of Particulate Flow

Apart from the effects of fluid–particle interaction, particle–particle collision, and wall–particle collision, and under certain circumstances, some special effects become significant in particulate flows. These effects may be important for nanofluids and are discussed here.

Thermophoresis The solid particles suspended in a fluid experience a force in the direction opposite to the temperature gradient imposed. A number of studies, such as those by Talbot et al. (1980) and Yamamoto and Ishihara (1988), explained and analyzed this phenomenon. The thermophoric force on a particle is given by

$$F_T = \frac{6\pi\mu^2 C_s(K_r + 2.18K_r)}{\rho(1 + 3 \times 1.14K_r)(1 + 2K_r + 4.36K_r)} \frac{1}{m_p T} \frac{\partial T}{\partial x} \tag{5.87}$$

This equation can be written in diffusionlike form as

$$F_T = -D_T \frac{1}{m_p T} \frac{\partial T}{\partial x} \tag{5.88}$$

where D_T is the thermophoric diffusion coefficient, m_p the mass of the particle, K_r the ratio of the thermal conductivity of the fluid and particle, $C_s = 1.17$, and μ is the fluid viscosity. However, this equation is derived for the suspension of solid particles in ideal gases, A modification may be necessary to use it for solid–liquid suspensions such as nanofluids.

Shear Lift Force The lift force due to shear described by Li and Ahmadi (1992) was generalized by Saffman (1965), hence is also called *Saffman's lift force*:

$$\vec{F} = \frac{2K\nu^{1/2}\rho d_{ij}}{\rho_p d_p (d_{kl} d_{kl})^{1/4}} (\vec{v} - \vec{v}_p) \tag{5.89}$$

where $K = 2.594$ and d_{ij} is the deformation tensor. This force is only of importance in submicrometre particles and hence can be important with nanofluids.

Brownian Motion In Chapter 3 it was shown that a large number of investigators consider Brownian motion of the particles to be important for nanofluids. Its link with the temperature effect on conductivity is particularly well established. The Brownian force is a periodic force with a spectral intensity of

$$S_{i,j} = S_0 \delta_{ij} \tag{5.90}$$

where $S_0 = 216\gamma k_B T / \pi^2 \rho d_p^5 (\rho_p/\rho)^2 C_c$, δ_{ij} is the Kronecker delta, and k_B is Boltzmann's constant. C_c is a constant that depends on the particle, diameter and mean free path of the particles, known as the *Cunningham correction factor* for Stoke's drag force. The amplitude of the Brownian force is given by

$$F_{bi} = \zeta_i \sqrt{\frac{\pi S_0}{\Delta t}} \tag{5.91}$$

where ζ_i is a Gaussian random number.

Soret and Dufour Effects When a temperature gradient is applied to a liquid mixture, the components separate, creating a concentration gradient with respect to the temperature gradient. A salt solution contained in a tube with two ends at different temperatures does not remain uniform in composition. The salt is more concentrated near the cold end than near the hot end of the tube. For the x component of the mass flux of the reference chemical compound in a binary mixture,

$$J_x = -\rho D \frac{\partial c}{\partial x} - \rho D_T (1 - C_0) \frac{\partial T}{\partial x} \tag{5.92}$$

where the first term on the right-hand side is Fick's law of diffusion, with C the mass fraction of the reference component and D the molecular diffusion coefficient. The second term describes the *Soret* or *thermodiffusion effect*. The Soret coefficient is defined as

$$S_T = \frac{D_T}{D} \tag{5.93}$$

Some researchers propose the use of a binary mixture instead of a pure liquid as a base fluid to make a binary nanofluid. Such a mixture may be useful in absorption refrigeration, as a solution in electro or electroless plating or as a transfer medium in medical treatment. The *Dufour effect* is a reciprocal phenomenon to the Soret effect: the occurrence of a heat flux due to a chemical potential gradient.

5.3. CONVECTION IN NANOFLUIDS

5.3.1. Forced Convection

With our introduction to single-phase heat transfer and particulate flow and heat transfer, we are now in a position to enter the domain of forced convection and related heat transfer in nanofluids. Although the number of studies available in this area is limited compared to that in the area of thermal conductivity of nanofluids, the views and approaches used in convective studies are quite divergent and hence require careful examination. Convective heat transfer is closely related to the viscosity of suspensions, and hence in many of these studies the viscosity variation is discussed before taking up convective issues. Here we also examine the nature of viscosity variation of nanofluids first. Subsequently, we look at the experimental results, followed by various approaches to analyzing phenomena leading to numerical work on convection in nanofluids.

5.3.2. Viscosity Variation in Nanofluids

The first question to be asked with respect to the viscosity of nanofluids is whether nanofluids are Newtonian fluids or whether a shear thinning process (i.e, shear-rate-dependent viscosity) is important to them. The first convective study on nanofluids, by Pak and Cho (1998), considered this question. If the viscosity of a fluid is independent of shear rate, it is said to be a Newtonian fluid that follows

$$\tau = \mu \frac{du}{dy} \tag{5.94}$$

Using the standard Brooksfield viscometer they found that for γAl_2O_3 and TiO_2 particles of 13 and 27 nm average size suspended in water, the suspensions are Newtonian at very low particle-volume fractions and start showing shear thinning behavior (i.e, decrease in viscosity with shear rate) with an increase in particle-volume fraction.

However, they found a difference between the two nanofluids. Whereas water–Al_2O_3 nanofluid started showing shear thinning behavior at 3% particle volume, water–TiO_2 nanofluid showed it at 10% particle volume onward. A large number of studies are available on the theoretical modeling of suspension viscosity. The model by Batchelor (1977) is applicable to dilute fluids dispersed with a solid sphere, given by

$$\mu_r = 1 + 2.5\varepsilon_p + 6.2\varepsilon_p^2 \tag{5.95}$$

where μ_r is the relative viscosity of the suspension and ε_p is the particle-volume fraction. Often, a simplified form of this equation known as the *Einstein equation* is used to estimate the viscosity of dilute suspensions:

$$\mu = \mu_L(1 + 2.5\varepsilon_p) \tag{5.96}$$

where μ_L is the liquid viscosity.

Pak and Cho (1998) observed substantial increases in the viscosity of the two nanofluids and showed that the Batchelor model fails completely for these fluids, although the volume fraction of the particles are in the range of the applicability of Batchelor's equation. This is shown in Fig. 5.9. Pak and Cho found that with

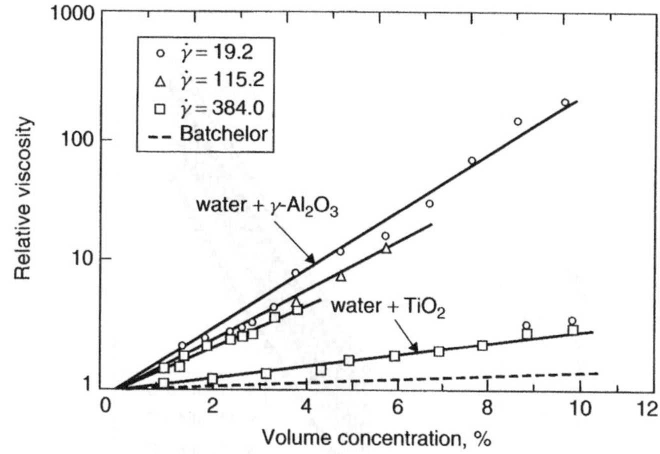

Fig. 5.9 Comparison of relative viscosity of nanofluids with Batchelor's equation. [From Pak and Cho (1998), with permission from Taylor & Francis.]

an increase in temperature, the viscosity of nanofluids decreases, following the same trend as that for a base liquid, although the viscosity value is much higher.

Lee et al. (1999) measured the viscosity of water- and ethylene glycol–based nanofluids with Al_2O_3 nanoparticles and made the similar observation that the relative viscosity of nanofluids increases substantially, which can offset the advantage in heat transfer. Das et al. (2003) measured the viscosity of water–Al_2O_3 nanofluid and showed a shear-rate-independent viscosity. This follows Newtonian theory, results for which are given in Fig. 5.10 for particle concentrations up to 4%. They also demonstrated the effect of temperature on viscosity (Fig 5.11).

Carbon nanotubes (CNTs) containing nanofluids behave quite differently not only with respect to thermal conductivity but also with respect to viscosity. The study by Ding et al. (2006) on aqueous CNTs containing nanofluids showed interesting the linear shear thinning behavior of the nanofluid at lower shear rates. Since they used gum arabic as the stabilizing agent, they also measured the viscosity of the base water with gum arabic, which showed nonlinear behavior different from that of CNT–water nanofluid. Figrue 5.12 shows the viscosity of CNTs containing nanofluids.

Note that traditional viscometers are not designed for nanosuspension, and hence viscosity values measured with them may not give a correct picture. Also, during dynamic flow, particle collisions yield a collision viscosity that cannot be measured by a traditional viscometer.

5.3.3. Experimental Works on Convection in Nanofluids

For the use of nanofluids in heat transfer devices, the higher thermal conductivity is an encouraging feature but not conclusive proof of their applicability. There needs to be hard evidence about the performance of these fluids under convective

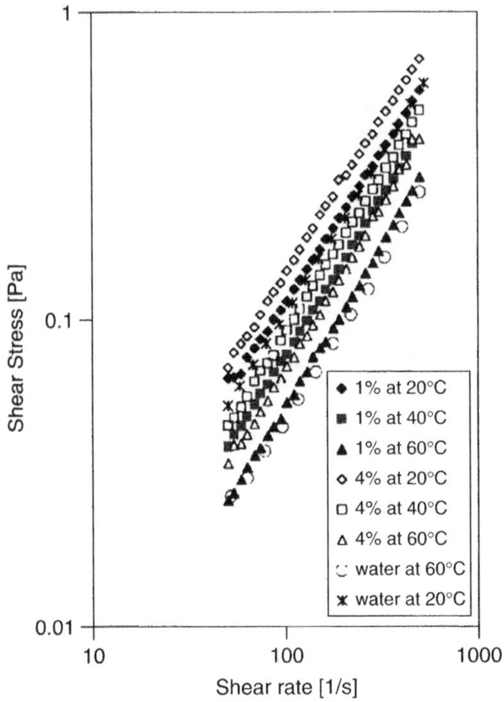

Fig. 5.10 Rheological behavior of nanofluids at 1% and 4% concentration and pure water.

conditions. This is why we first present the experimental findings related to convective heat transfer in nanofluids.

The first work on convective flow and heat transfer of nanofluids, inside a 10.66-mm-diameter tube, was presented by Pak and Cho (1998) before Choi and his group introduced the term *nano-fluids*. Pak and Cho referred to a "dispersed fluid with submicron particles." However, these particles were small enough (13 and 27 nm) for the suspension to be called a nanofluid. The first observation they made was a substantial increase in the heat transfer coefficient in the turbulent flow regime. The experiment showed the applicability of the Dittus–Boelter equation(5.55) for pure water flow and a substantial increase in the heat transfer coefficient for flow with the particles suspended. The increase was 45% with 1.34% Al_2O_3 particles and 75% with 2.78% Al_2O_3 particles. One can readily find that this increase is more than an increase in conductivity alone, and hence the enhanced convective heat transfer cannot be attributed to the increased conductivity of the nanofluid alone. The increase in the heattransfer coefficient is shown in Fig. 5.13.

It is interesting to note that like them, many investigators prefer to plot a dimensional heat transfer coefficient rather than a dimensionless Nusselt number for nanofluids because Nu contains the nanofluid's conductivity and hence may not be an indicator of the increased heat transfer coefficient. They also proposed

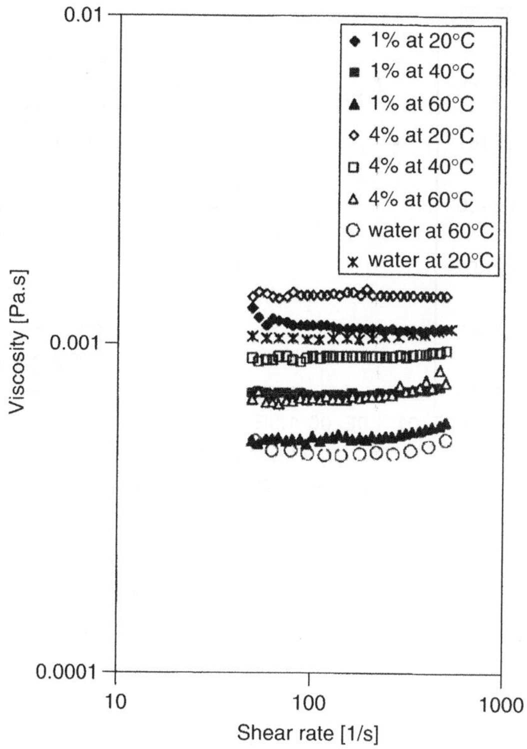

Fig. 5.11 Dynamic viscosity of nanofluid and pure water at different temperatures.

a modified Dittus–Boelter correlation (1930) for their data:

$$\mathrm{Nu} = 0.021\mathrm{Re}^{0.8}\mathrm{Pr}^{0.5} \tag{5.97}$$

However, it must be understood that this equation is not of much practical significance because it does not account for particle-volume fraction or particle size separately (but through Re, nanofluid conductivity, and Pr). As a result, extending these results to other particles is questionable.

Also, their overall depiction is much more gloomy. They found that the Darcy friction factor follows the Kays correlation (shown in Fig. 5.14). Thus, due to a rise in viscosity, there is a substantial rise in frictional pressure-drop. This means that although the heat transfer coefficient increases in nanofluids, the pressure-drop penalty is substantial. In convection heat transfer applications there is always competition between enhancement of heat transfer and the resulting pressure penalty. Any enhancement method, such as creating high turbulence or interruption of the boundary layer, has an associated pressure penalty, and this results in a higher pumping power requirement, which may offset the advantage of heat transfer enhancement. Often, we compare heat transfer enhancement at the same pumping power, which gives a better picture. In the case of nanofluids,

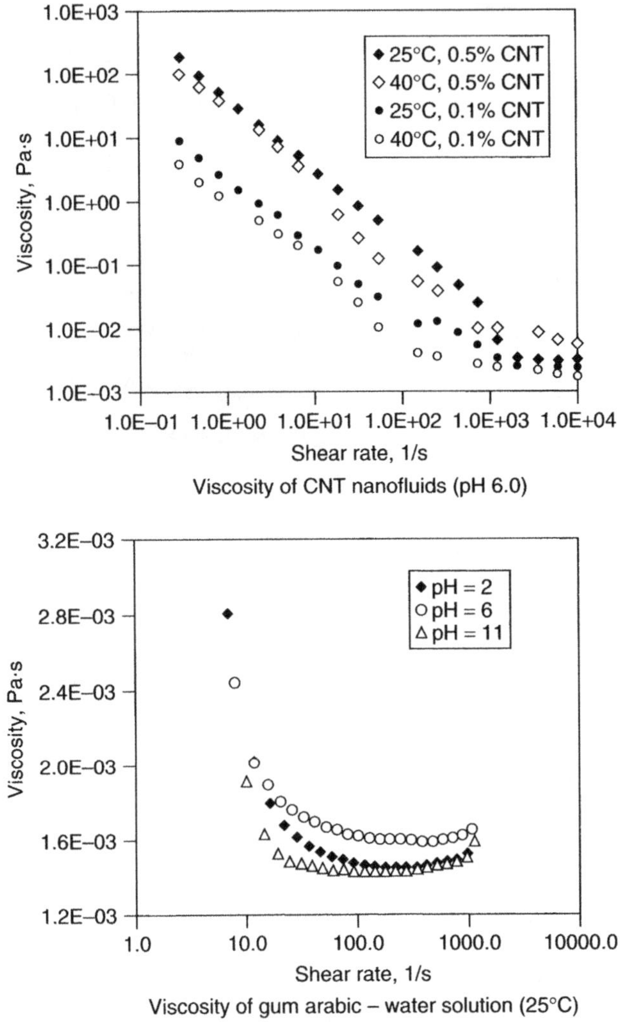

Fig. 5.12 Viscosity of CNTs containing nanofluid. [From Ding et al. (2006), with permission from Elsevier.]

Pak and Cho (1998) claimed that for a constant average velocity there was actually a 3 to 12% decrease in heat transfer coefficient.

This picture changed substantially with the result of Xuan and Li (2003). One major difference was that their particles were pure copper particles but slightly larger (≈100 nm). They had a well-designed test loop (Fig. 5.15). The dimensional heat transfer coefficient measured against the flow velocity in this experimental setup is shown in Fig 5.16. This shows a dramatic increase in convective heat transfer with nanofluids, a result that contradicts the conclusion of Pak and Cho (1998) that for constant average velocity, the heat transfer coefficient

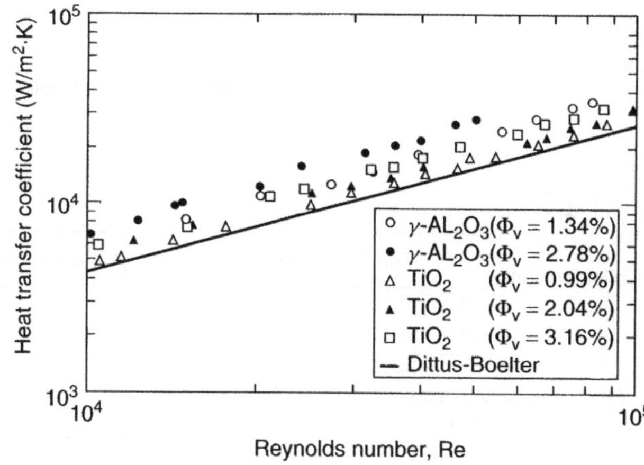

Fig. 5.13 Heat transfer coefficient versus Reynolds number for nanofluids. [From Pak and Cho (1998), with permission form Taylor & Francis.]

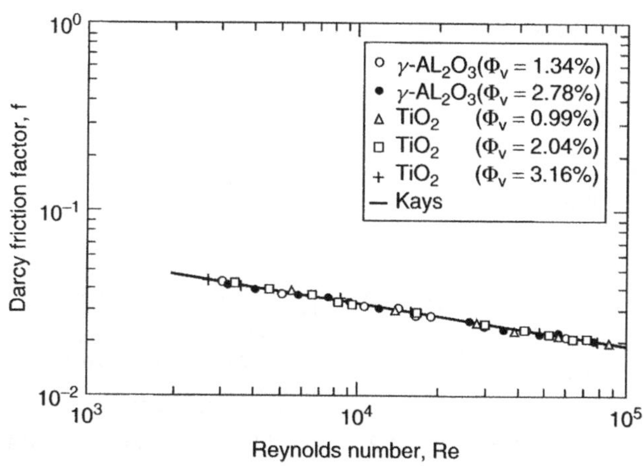

Fig. 5.14 Friction factor for nanofluids. [From Pak and Cho (1998), with permission from Taylor & Francis.]

decreases as much as 12% with nanofluids. On the contrary, Xuan and Li (2003) showed an increase of as much as 40% at the same velocity. They explained it by stating that in the study of Pak and Cho (1998), the increase in viscosity was large, which might have suppressed the turbulence, reducing the heat transfer. Hence, they indicated that not only the volume fraction but also the particle dimensions and material properties are important, and if the fluid is designed properly, a substantial rise in heat transfer coefficient is achievable.

They also plotted the Nusselt number against the Reynolds number (for both of them, effective nanofluid properties were used), as shown in Fig. 5.17. The figure

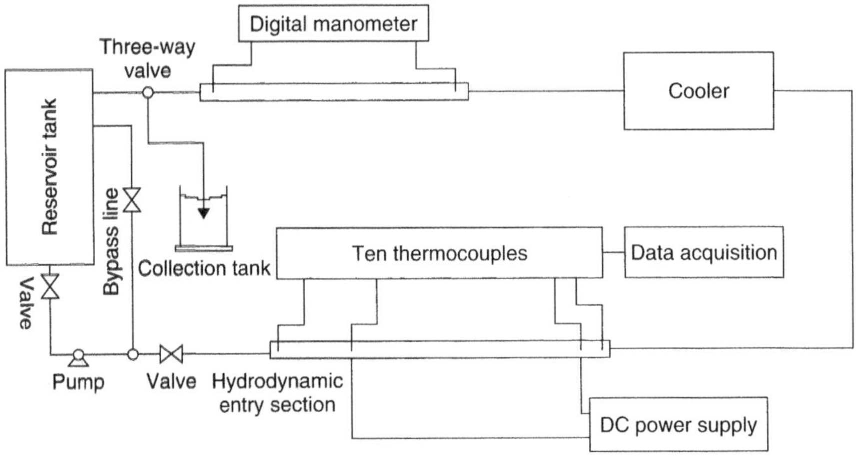

Fig. 5.15 Convection test loop. [From Xuan and Li (2003), with permission from ASME Publishing).]

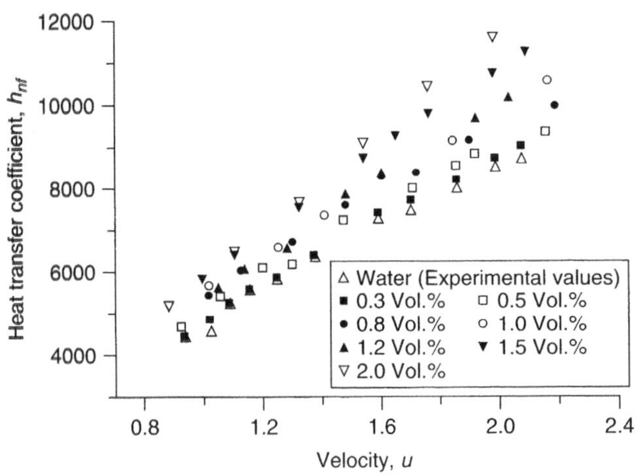

Fig. 5.16 Variation of heat transfer coefficient with flow velocity. [From Xuan and Li (2003), with permission from ASME Publishing.]

indicates clearly that the Dittus–Boelter equation(5.55) with modified nanofluid properties is not enough to describe convection in nanofluids. In other words, a nanofluid cannot be treated as a single fluid just by changing properties to effective properties. In nanofluid convection there are distinct additional effects, such as gravity, Brownian force, drag on the particle, and diffusion. The discrepancy between the Dittus–Boelter equation and nanofluid convection was found to be 39% with 2% Cu particles. Following the theory of thermal dispersion,

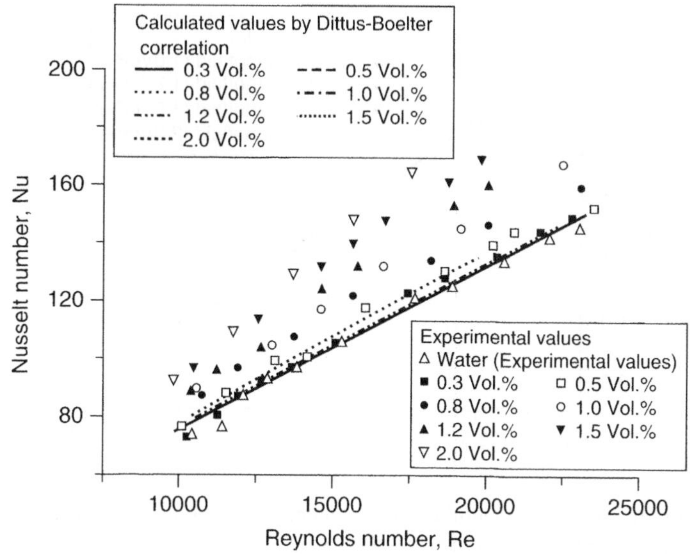

Fig. 5.17 Nusselt numbers of nanofluids with Reynolds numbers and values predicted from the Dittus–Boelter correlation. [From Xuan and Li (2003), with permission from ASME Publishing.]

they suggested a correlation for turbulent nanofluid heat transfer inside a pipe in the form

$$\text{laminar flow Nu} = 0.4328(1 + 11.285\varepsilon_p^{0.754}\text{Pe}_d^{0.218})\text{Re}^{0.333}\text{Pr}^{0.4} \tag{5.98}$$

$$\text{turbulent flow Nu} = 0.0059(1 + 7.6286\varepsilon_p^{0.6886}\text{Pe}_d^{0.001})\text{Re}^{0.9238}\text{Pr}^{0.4} \tag{5.99}$$

where Pe_d is the particle Peclet number $= u_p d_p/\alpha_p$, α_p is the thermal diffusivity of the particle, and ε_p is the particle-volume fraction. They also brought out the pressure-drop study and plotted the friction factor as shown in Fig. 5.18. (to avoid confustion we have changed the symbol they used for the friction factor).

The next immensely important investigation of convection heat transfer in nanofluids was one by Wen and Ding (2004), which is significant in many respects. Most important, it was the first study to observe the entry-length effect. Laminar flows usually have long hydrodynamic and thermal entrance regions. Since the boundary layer is thinner, in these regions the heat transfer coefficient is higher. Wen and Ding (2004) measured the local heat transfer coefficient along the tube during laminar flow (Fig. 5.19). They used different water/γ- Al_2O_3 flowing inside a copper tube 4.5 mm in inner diameter and 970 mm long. For data reduction they estimated viscosity using Einstein's equation (5.96), with the results shown in Fig. 5.19. A substantial rise in heat transfer coefficient is clear, but most interesting is the fact that the increase in heat transfer coefficient is

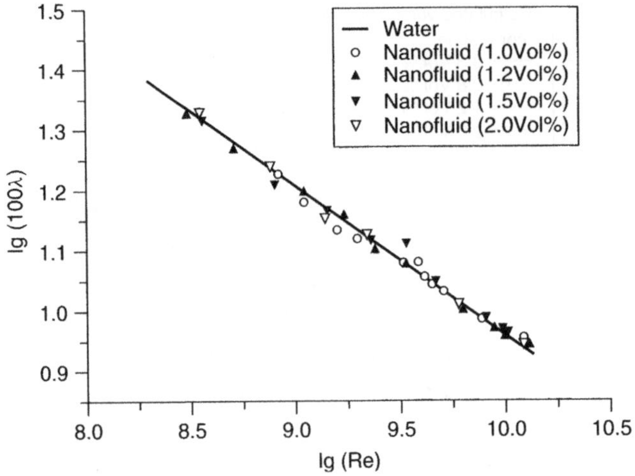

Fig. 5.18 Friction factor for nanofluids in turbulent flow. [From Xuan and Li (2003), with permission from ASME Publishing.]

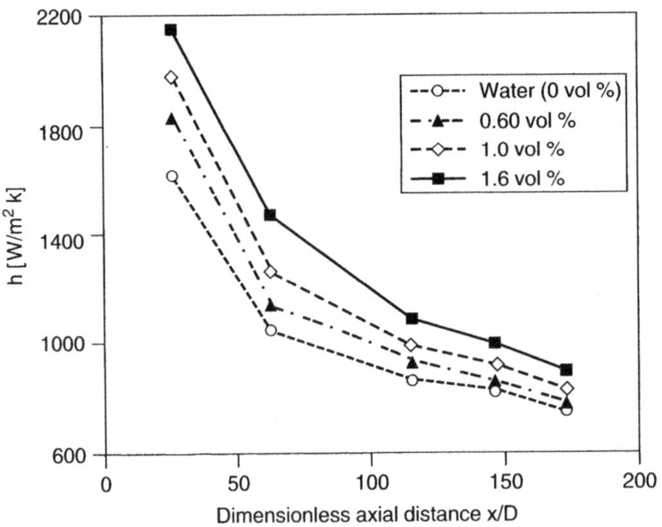

Fig. 5.19 Measured local heat transfer coefficient for convection inside a tube. [From Wen and Ding (2004), with permission from Elsevier.]

greatest at the entry-length region and enhancement increases with particle concentration. This indicates not only a steady entrance region effect but also higher heat transfer enhancement through "smart" options such as boundary layer intruption and creation of an artificial entrance region. For comparison with existing heat transfer correlation, they used Shah's (1975) correlation for thermal entry length with constant heat flow in the form

$$\mathrm{Nu} = \begin{cases} 1.953\left(\mathrm{Re}\cdot\mathrm{Pr}\dfrac{D}{x}\right)^{1/3} & \text{for } \left(\mathrm{Re}\cdot\mathrm{Pr}\frac{D}{x}\right) \geq 33.3 \quad (5.100) \\ 4.364 + 0.0722\left(\mathrm{Re}\cdot\mathrm{Pr}\dfrac{D}{x}\right) & \text{for } \left(\mathrm{RePr}\frac{D}{x}\right) < 33.3 \quad (5.101) \end{cases}$$

It is obvious in the second equation that for a greater length the equation tends to 4.364, which is the standard Nusselt number value for fully developed laminar flow inside tubes with constant heat flux [equation (5.52)].

Wen and Ding (2004) observed that equation (5.101) fails to predict the heat transfer coefficient, indicating that some special effects are present in the convection of nanofluids. Figure 5.20 shows the discrepancy between Shah's (1975) equation and their data measured at a distance of $x/D = 63$. Wen and Ding (2004) also made the important observation that the entry length of nanofluids is longer than that of the pure base fluid. While commenting on the reason for the enhancement and special effects involved, Wen and Ding (2004) indicated a few possibilities. They argued that particle migration and nonuniform distribution of thermal conductivity and viscosity may lead to reduction of the boundary layer thickness increasing the heat transfer. However, this proposition was just a suggestion and was not proven conclusively by their study.

Yang et al. (2005) presented the results of their experiments on a test rig similar to those used by earlier investigators. They used tubes of 4.57 mm inner diameter

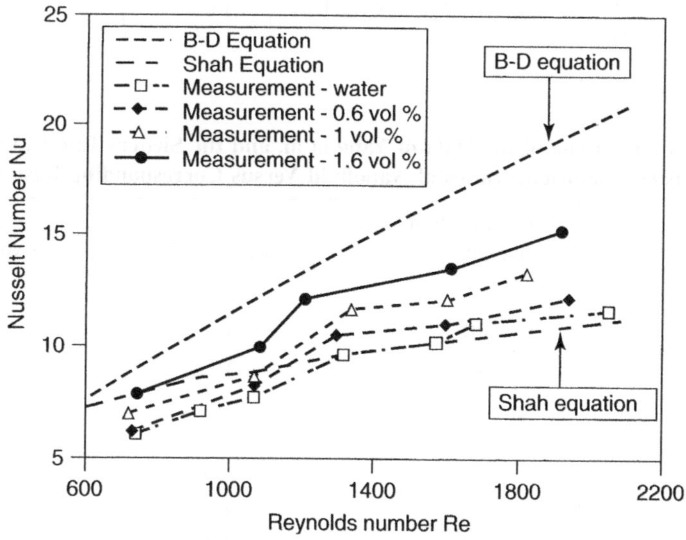

Fig. 5.20 Comparison of the data of Wen and Ding (2004) with Shah's (1975) correlation. [From Wen and Ding (2004), with permission from Elsevier.]

and 457 mm (i.e., 100 diameters) long. One important feature of their test loop was its small volume of holdup fluid, and another was their use of hot water for heating instead of electrical heating. This second feature may be an important one because recently, Kabelac and Kuhnke (2006) indicated that electrical heating may affect particle movements in nanofluids in that the particles are likely to carry electrical charge. Yang et al. (2005) used four different experimental fluids of different combinations of two base fluids (one is an automatic transmission fluid, the other is a mixture of synthetic oils with additives) and graphite particles between 2 and 2.5% concentration. The particles were disk-shaped, 1 to 2 nm in thickness and 20 to 40 nm in diameter.

Yang et al.'s conclusion was that the particle loading, temperature, source of nanoparticles, and base fluid used affected the heat transfer results. The results were plotted using the Sieder–Tate (1936) correlation for laminar developing flow as the datum in the form

$$\Omega = 1.86\mathrm{Re}^{1/3} \qquad \text{where } \Omega = \mathrm{Nu}\cdot\mathrm{Pr}^{-1/3}\left(\frac{L}{D}\right)^{1/3}\left(\frac{\mu_b}{\mu_w}\right)^{-0.14} \tag{5.102}$$

Although overall results are found to agree with this, the scatter is large at the lower Reynolds number. The comparison of the experimental data with equation (5.102) shown in Table 5.2 indicates a much lower level of enhancement. They then predicted values for the constants in the equation. They also compared their results with those of two other correlations: the Oliver (1962) correlation,

$$\mathrm{Nu}\left(\frac{\mu_w}{\mu_b}\right)^{0.4} = 1.75\left(\mathrm{Gz} + 5.6\times10^{-4}\left(\mathrm{GrPr}\frac{L}{D}\right)^{0.7}\right)^{1/3} \tag{5.103}$$

Table 5.2 Comparison of the Data of Yang et al. and the Sieder–Tate Correlation: Heat Transfer Coefficient Ratios of Nanofluid Versus Corresponding Base Fluid

	Heating Fluid Temperature (°C)	Eq. (5.102)	Experiment
$\frac{h(\text{EF 1-1})}{h(\text{BF 1})}$	50	1.19	1.03
	70	1.19	1.03
$\frac{h(\text{EF 1-2})}{h(\text{BF 1})}$	50	1.36	1.22
	70	1.36	1.15
$\frac{h(\text{EF 2-1})}{h(\text{BF 1-1})}$	50	1.02	1.01
	70	1.02	1.01
$\frac{h(\text{EF 2-2})}{h(\text{BF 2})}$	50	1.14	1.08
	70	1.14	1.07

Source: Yang et al. (2005), with permission from Elsevier.

where

$$\text{Gz} = \text{Graetz number} = \frac{\dot{m}C_p}{kL}$$

$$\text{Gr} = \text{Grashof number} = \frac{\rho_f^2 \beta_f g \Delta T D^3}{\mu^2}$$

and the Eubank and Proctor (1951) correlation for laminar flow in a horizontal tube,

$$\text{Nu}\left(\frac{\mu_w}{\mu_p}\right)^{0.14} = 1.75\left[\text{Gz} + 12.6\left(\text{Gr}\cdot\text{Pr}\frac{D}{L}\right)^{0.4}\right]^{1/3} \tag{5.104}$$

The results fall in between. However, Yang et al. failed to indicate that in reality, the slope of the experimental curve is higher than that suggested by both equations, indicating a much higher entry-length effect than these equations suggest. This work by Yang et al. (2005) shows several deviations from other data. This may be due to the fact that the particles are of disk type and the major dimension (diameter) is too large to qualify them as nanoparticles. Hence, there is some doubt whether this work falls in the category of nanofluids at all.

The other work that has reached conclusions similar to those of Xuan and Li (2003) is that of Heris et al. (2006). Here, the test was conducted in a 6-mm-diameter copper tube for water–CuO and water–Al_2O_3 nanofluids. Substantial enhancement was reported with higher enhancement for Al_2O_3-based nanofluid. It was found that the Sieder–Tate correlation for turbulent flow is inadequate to predict the heat transfer enhancement with these nanofluids. The two important features of this work were the observations that heat transfer enhancement increases significantly with particle-volume fraction and the enhancement is greater at higher Peclet numbers. Figure 5.21 and 5.22 show these two effects clearly.

Heris et al.'s explanation regarding this increase was similar to that of Xuan and Li (2003): namely, dispersion, chaotic particle movement, Brownian motion, and so on. These results are significant on many counts. First, here the authors carried out experiments with steam as the heating medium, and hence the problems envisaged by Kabelac and Kuhnke (2006) regarding the electrical effect on the particles should not be present. Second, unlike Xuan and Li (2003), nonmetallic particles were used here and the enhancement was still high enough to be marked, and traditional turbulent flow equations do not seem to be adequate to describe it. In general, it seems that particle source, method of preparation, technique of dispersion, size distribution, pH value, and a large number of other issues are responsible for these divergent trends in experimental data between Pak and Cho (1998) and Yang et al. (2005) on the one hand and Xuan and Li (2003), Wen and Ding (2004), and Heris et al. (2006) on the other.

One other interesting experiment is that of Ding et al. (2006) on the convection of carbon nanotube (CNTs) containing nanofluids. They used multiwalled carbon nanotubes (MWCNTs) and a simple setup (Fig. 5.23) with a

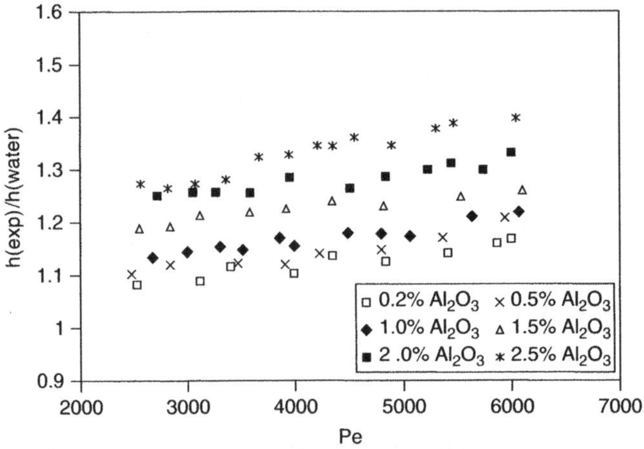

Fig. 5.21 Increase in the heat transfer coefficient of water–Al_2O_3 nanofluids against Peclet number. [From S. Z. Heris, M. N. Esfahany, and S.G. Etemad, Experimental investigation of convective heat transfer of Al_2O_3–water nanofluid in circular tube, *Int. J. Heat Fluid Flow*, in press, with permission from Elsevier.]

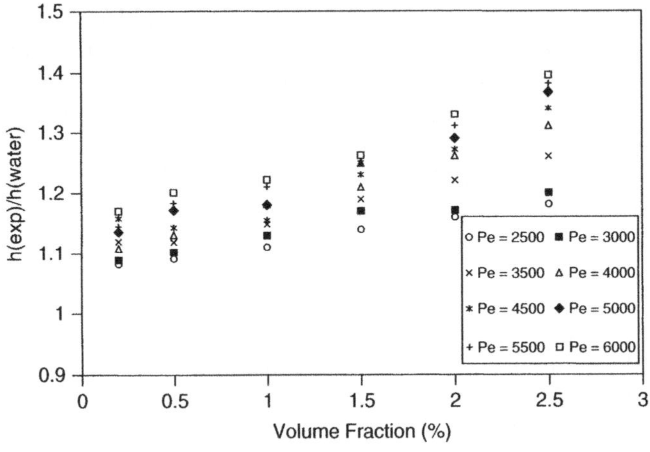

Fig. 5.22 Increase in the heat transfer coefficient of water–Al_2O_3 nanofluids against particle-volume fraction. [From S. Z. Heris, M. N. Esfahany, and S. G. Etemad, Experimental investigation of convective heat transfer of Al_2O_3–water nanofluid in circular tube, *Int. J. Heat Fluid Flow*, in press, with permission from Elsevier.]

4.5-nm-inner-diameter tube heated electrically. This is probably the only study on convection with CNTs containing nanofluids. Since CNTs have a tendency to agglomerate, they used a high-speed (24,000-rpm) rotor to disperse them properly. Figure 5.24(a) shows the CNTs in "as received" and "after dispersion" form [Fig. 5.24(b)]. They first measured the thermal conductivity of the nanofluids,

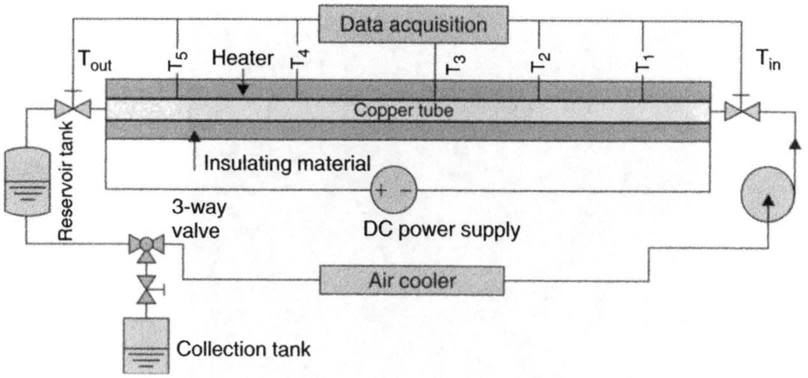

Fig. 5.23 Experimental setup for water–CNT nanofluid convection studies. [From Ding et al. (2006), with permission from Elsevier.]

which showed up to 50% enhancement with 0.7% CNTs. It was also interesting to note that there was a tremendous temperature effect on conductivity (Fig. 5.25) over just a 10% rise in suspension temperature.

The viscosities of these fluids were discussed earlier. Coming to convective heat transfer, they showed dramatic improvement. Since gum arabic was used as the stabilizing agent in this study, the studies were always compared with water having gum arabic alone (0% CNTs). The enhancement was tested against parameters such as particle concentration, axial distance, Reynolds number, and pH value. Figures 5.26 and 5.27 show some of the effects they observed. As much as 350% enhancement of heat transfer coefficient was observed at $Re = 800$. The CNT volume concentration used was small (<0.5%). The entrance-length effect was observed, but unlike Wen and Ding (2004), the enhancement increased along the entry length, reached a maximum, and then decreased. The point of maximum enhancement was found to increase with Reynolds number and particle concentration. The enhancement increases with Reynolds number and were not much affected by the pH value of the suspension. The large amount of enhancement shown in this study cannot be attributed to increased thermal conductivity alone. They found that the enhancement suddenly increases enormously beyond a particular Reynolds number, which they attributed to shear thinning. Further, they suggested that particle rearrangement of high-aspect-ratio (>100) CNTs and reduction of boundary layer thickness by nanotubes are important additional mechanisms.

5.3.4. Natural Convection

Compared to forced convection, the numbers of studies are limited in natural convection. The first work in this area was from Putra et al. (2003). Using water with 131.2-nm Al_2O_3 particles and 87.3-nm CuO particles, they studied natural convection in a horizontal cylindrical cavity filled with nanofluids.

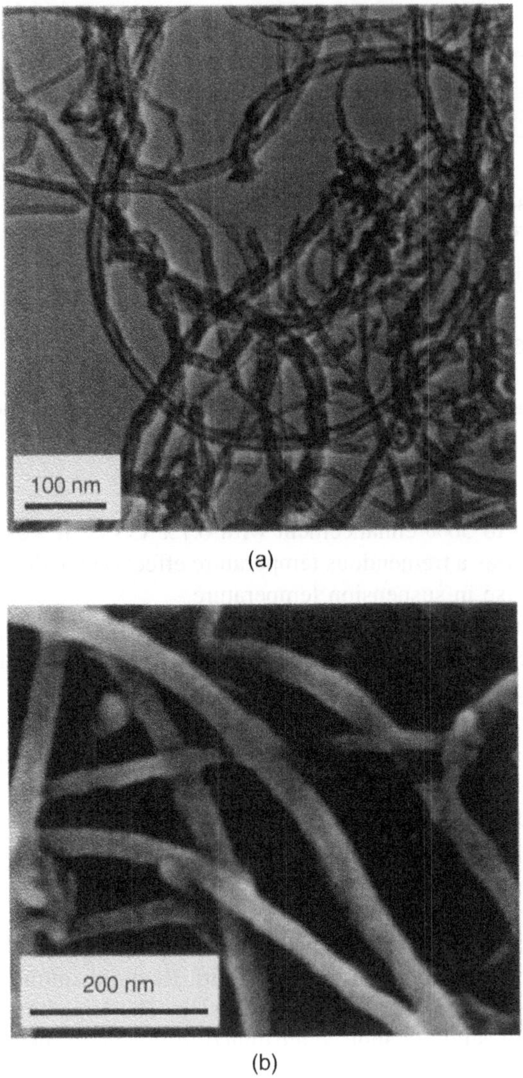

Fig. 5.24 CNTs used by Ding et al. in their experiments: (a) as-received; (b) after dispersion. [From Ding et al. (2006), with permission from Elsevier.]

The experimental setup is shown in Fig. 5.28. One end of the cavity is heated and the other is cooled by circulating water through the piston. The lateral surface is insulated and thermocouples are located at various axial positions. After initial transients the natural convection sets in and the final temperatures at various locations are shown in Fig. 5.29. for a liquid aspect ratio of 1.0 The resulting Nusselt numbers against Rayleigh number for various particle concentrations and fluid column aspects ratios are shown in Figs. 5.30 and 5.31, respectively.

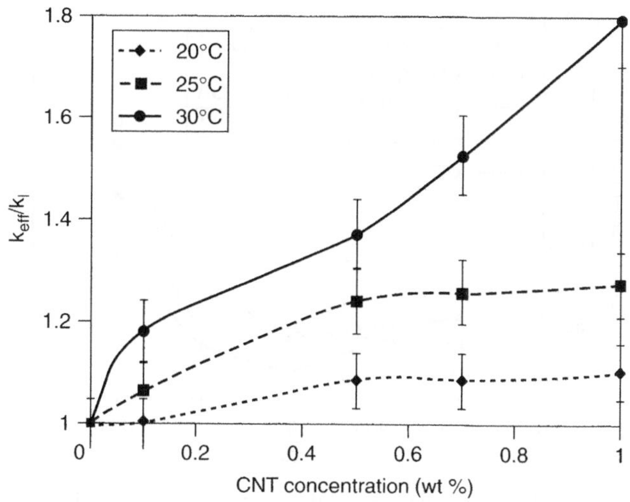

Fig. 5.25 Measured thermal conductivity of CNT nanofluids. [From Ding et al. (2006), with permission from Elsevier.]

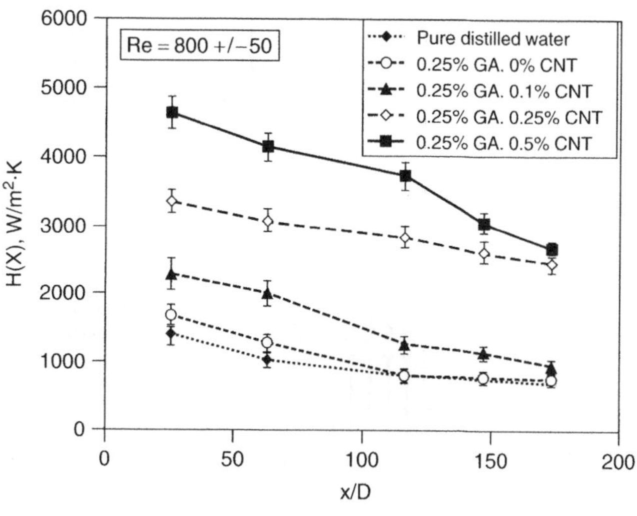

Fig. 5.26 Axial variation of the heat transfer coefficient of water–CNT nanofluids. [From Ding et al. (2006), with permission from Elsevier.]

The results clearly show that the natural convective heat transfer in nanofluids is lower than pure water with an increase in particle concentration. The same observation was made for water–CuO nanofluids as well. The deterioration of CuO nanofluids was greater than that of Al_2O_3 nanofluids, as shown in Fig. 5.32. Putra et al. observed the nature of this deterioration to be different compared to

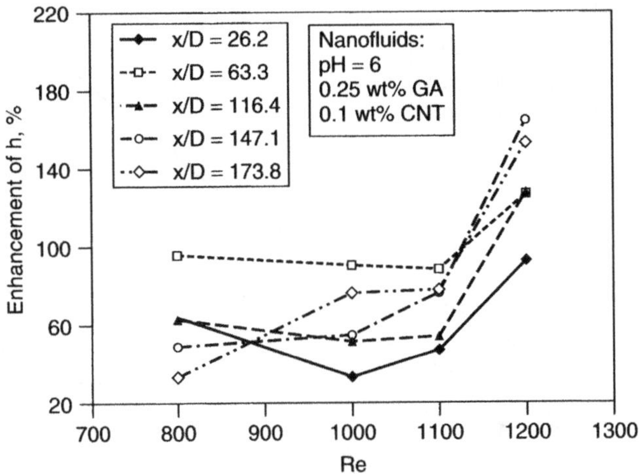

Fig. 5.27 Effect of Reynolds number on the heat transfer coefficient of water–CNT nanofluids. [From Ding et al. (2006), with permission from Elsevier.]

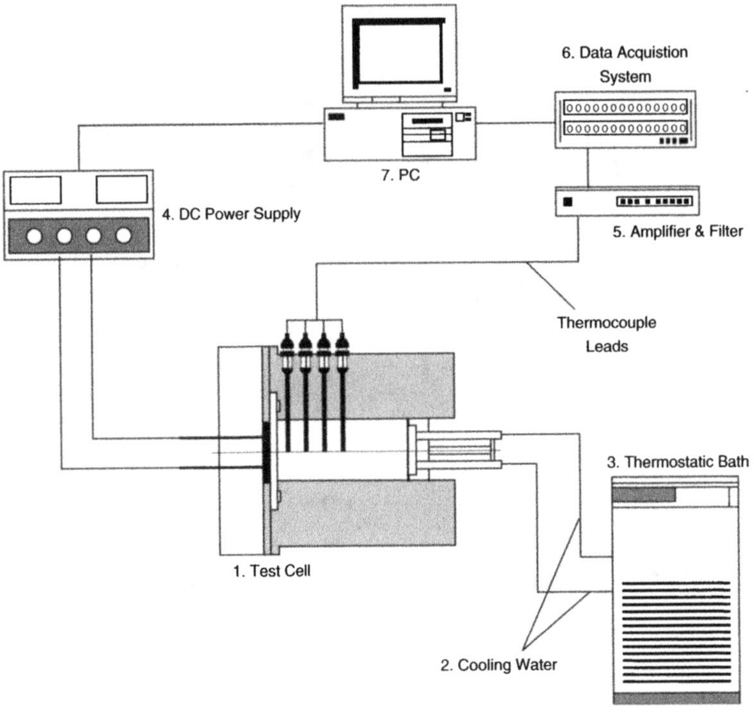

Fig. 5.28 Experimental apparatus for study of natural convection (Putra et al., 2003).

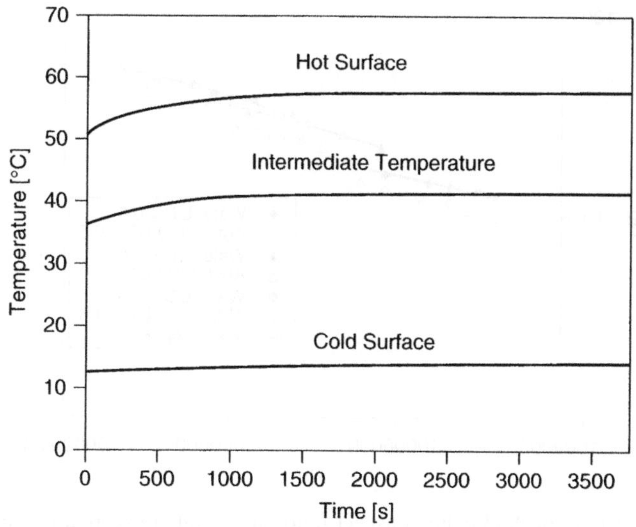

Fig. 5.29 Initial transients of the surface and midpoint ($L/D = 1$, 1% Al_2O_3).

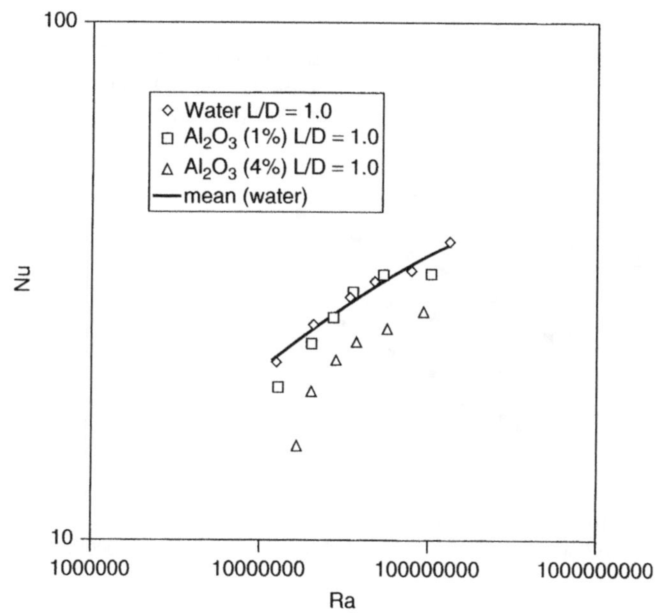

Fig. 5.30 Nu–Ra behavior of Al_2O_3–water nanofluids for a fluid aspect ratio of 1.0.

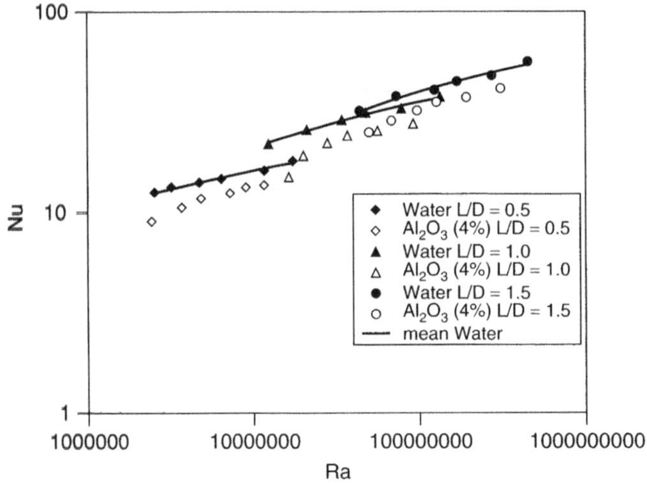

Fig. 5.31 Effect of fluid column aspect ratio on natural convection in nanofluids.

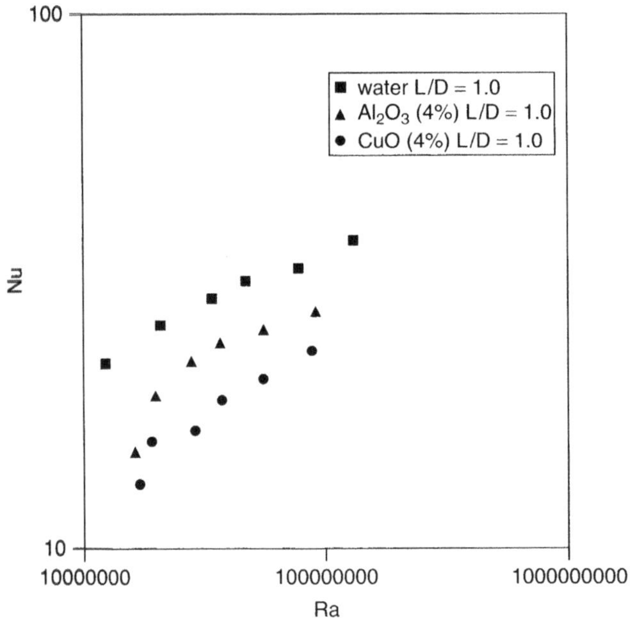

Fig. 5.32 Comparison of natural convection in Cuo and Al_2O_3 nano fluids.

common slurries, and it is not a double diffusive feature (diffusion of heat and mass simultaneously). On the contrary, they attributed it to the slip between the fluid and the particles because the denser CuO particles showed more deterioration.

The other experimental work on natural convection in nanofluids was by Wen and Ding (2005). First they measured the zeta potential to determine the pH value at which the TiO_2 particles will be stable in a water acid solution. Their experimental setup is shown in Fig. 5.33. The results plotted by them reconfirm the deterioration of heat transfer by natural convection in nanofluids, as shown in Fig. 5.34. They attributed this deterioration to convection driven by concentration

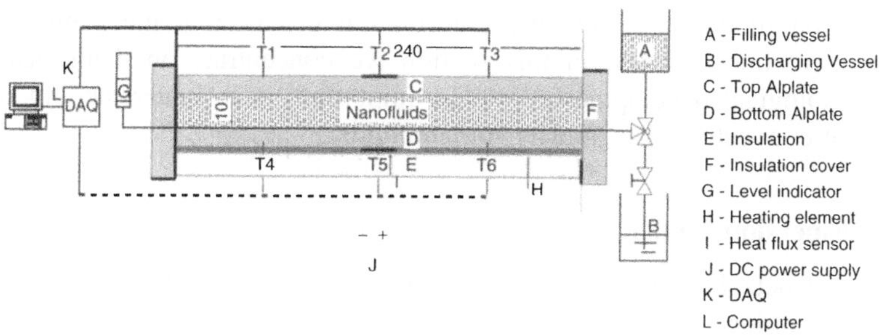

Fig. 5.33 Experimental setup for the study of natural convection. [From Wen and Ding (2005), with permission from Elsevier.]

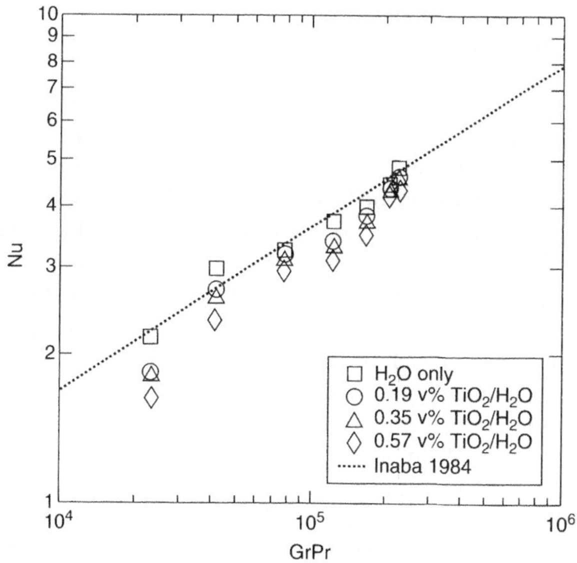

Fig. 5.34 Steady-state heat transfer results of Wen and Ding for natural convection in nanofluids. [From Wen, and Ding (2005), with permission from Elsevier.]

gradient, particle–surface and particle–particle interaction, and modification of dispersion properties.

5.4. ANALYSIS OF CONVECTION IN NANOFLUIDS

The experimental results presented in previous sections show that in most cases, investigators obtained substantial enhancement of forced convection, and the enhancement depends, among other factors, on particle type, particle material, volume fraction, and flow properties. However, for natural convection a decrease in heat transfer was observed. There have been simultaneously tremendous efforts to theorize the convection process in nanofluids. The efforts are concentrated primarily on modeling, estimating contribution of different mechanisms, and measurement uncertainties. In this section we concentrate on theoretical and phenomenological developments that are important to an understanding of the process of convective heat transfer in nanofluids.

5.4.1. Dispersion Model

Xuan and Roetzel (2000) at the University of the Federal Armed Force, Hamburg and Nanjing University of Science and Technology, China were probably the first to put forth a concept for theorizing convection in nanofluids. Their concept, although abstract and lacking definite proof that such an approach is acceptable, gave a direction to modeling nanofluid convection. They proposed a dispersion model that takes care of the additional enhancement over effective fluid treatment. This concept originated from the mass dispersion theory of Taylor (1953) and Aris (1956), which was supported by Dankwert (1953). In this work the authors observed how the concentration profile of a binary fluid gets changed (or dispersed) when it flows through a tube. Later, this concept was used on porous media (Kaviany, 1991) and suspensions (Kaviany, 1994).

The concept of dispersion is that due to the presence of solid particles, the flow of fluid and the transport of heat will not follow the same path as that of the pure fluid. It will be a more torturous path and its effect can be modeled by adding an equivalent diffusion term in the corresponding energy equation. However, the equivalent thermal conductivity for this modeling is not a real conductivity (or a property of medium) but is a flow and particle property because this amount of dispersion depends on such factors as particle–particle interaction and particle–surface interaction, which in turn depend on particle size, particle movement, particle concentration, fluid velocity, and so on. Thus, in essence, we assume that due to all these features there is an additional amount of heat transfer, and this corresponds to an additional fictitious conductivity for modeling purposes. This conductivity is known as the *thermal dispersion coefficient*. Xuan and Roetzel (2000) quoted Kaviany (1994) to present the basic equations related to thermal dispersion. They considered that the particles induce a velocity and temperature perturbation in the nanofluid given by u' and T', respectively. Thus,

$$\mathrm{T} = \langle \mathrm{T} \rangle^f + \mathrm{T}' \qquad \mathbf{u} = \langle \mathbf{u} \rangle^f + \mathbf{u}' \tag{5.105}$$

Here the volume-averaged temperature and velocity vectors are

$$\langle T \rangle^f = \frac{1}{V_f} \int_{V_f} T\, dV \qquad \langle \mathrm{u} \rangle^f = \frac{1}{V_f} \int_{V_f} \mathrm{u}\, dV \tag{5.106}$$

Then the basic convection equation in the vector $\mathbf{u}$ can be written for fluid f as

$$\frac{\partial T}{\partial t} + \nabla \cdot \mathbf{u} T = \nabla \cdot (\alpha_f \nabla T) \tag{5.107}$$

Substituting equation (5.105) in equation (5.107) and simplifying using the procedure of Kaviany gives

$$(\rho C_p)_{\mathrm{nf}} \left[\frac{\partial \langle T \rangle^f}{\partial t} + \langle \mathbf{u} \rangle^f \nabla \langle T \rangle^f \right] = \nabla (k_{\mathrm{nf}} \nabla \langle T \rangle^f) - (\rho C_p)_{\mathrm{nf}} \nabla \langle \mathbf{u}' T' \rangle^f \tag{5.108}$$

Here the subscript nf indicates a nanofluid.

In the dispersion model the additional term in equation (5.108) is due to perturbations in velocity and temperature and the last term is modeled like a conduction flux:

$$(\rho C_p)_{\mathrm{nf}} \nabla \langle \mathbf{u} T \rangle^f = -\mathbf{k}_d \nabla \langle T \rangle^f \tag{5.109}$$

where $\mathbf{k}_d$ is the tensor of dispersive thermal conductivity (also called the dispersion coefficient). This makes equation(5.108) solvable, but the dispersion coefficient in $\mathbf{k}_d$ needs to be known. As an example of this type of analysis, they considered the flow of nanofluids inside a tube that gives the equation

$$\frac{\partial T}{\partial t} + u \frac{\partial T}{\partial x} = \frac{1}{r} \frac{\partial}{\partial r} \left[\left(\alpha_{\mathrm{nf}} + \frac{k_{d,r}}{(\rho C_p)_{\mathrm{nf}}} \right) r \frac{\partial T}{\partial r} \right] + \frac{\partial}{\partial x} \left[\alpha_{\mathrm{nf}} + \frac{k_{d,x}}{(\rho C_p)_{\mathrm{nf}}} \frac{\partial T}{\partial x} \right] \tag{5.110}$$

Here we have dropped the $\langle \cdot \rangle^f$ notation and used T and u for volume-averaged values. $k_{d,r}$ gives the dispersion coefficient in the radial direction, and $k_{d,x}$, that in the axial direction. Now the task remains to determine these dispersion coefficients. This remains a challenge in this concept of modeling. Probably, extensive experiments can help to find out the dispersion coefficient and the nature of its variation with such parameters as particle loading, Reynolds number, geometry, and particle size. One interesting such experimental technique may be the one used by Roetzel et al. (1993) for determination of the dispersion coefficient in a dented tube. The method uses a periodic temperature profile at the entry to the tube which after heat transfer and dispersion experiences an amplitude attenuation and phase shift. Since these two measurable quantities are available, two unknown quantities, such as the heat transfer and thermal dispersion coefficients, can be evaluated from them. However, Xuan and Roetzel (2000) provided some intuitive suggestions for the dispersion coefficient from similar research work as

$$\kappa_d = C(\rho C_p)_{\text{nf}} u d_p R \varepsilon_p \quad \text{or} \quad C^*(\rho C_p)_{\text{nf}} u R \tag{5.111}$$

where R is the tube radius and C and C^* are constants. These expressions are purely intuitive, and the determination of dispersion coefficient and predicting its value for a variety of situations remains a task to be taken up in future.

Xuan and Li (2000) further advanced the concept of using the dispersion coefficient by solving equation (5.110) under the assumption that axial dispersion in negligible, giving

$$\frac{\partial T}{\partial t} + u\frac{\partial T}{\partial x} = \frac{1}{r}\frac{\partial}{\partial r}\left[\left(\alpha_{\text{nf}} + \frac{k_d}{(\rho C_p)_{\text{nf}}}\right) r\frac{\partial T}{\partial r}\right] \tag{5.112}$$

Considering constant inlet and wall temperatures, the boundary conditions are given by

$$T|_{r=R} = T_w \qquad T|_{x=0} = T_0 \tag{5.113}$$

considering laminar fully developed flow, this equation can be solved by separation of variables to give

$$\frac{T - T_w}{T_0 - T_w} = 2\sum_{n=1}^{\alpha} e^{-\beta_m^2 \bar{x}/\overline{\text{Pe}}} \frac{J_0(\beta_m \bar{r})}{J_1(\beta_m)\beta_m} \qquad \text{where} \quad \bar{r} = \frac{r}{R}, \bar{x} = \frac{x}{L} \tag{5.114}$$

$$\text{Nu} = \frac{h(2R)}{k_{\text{nf}}^*}$$

$$\text{Pe}^* = \frac{uL}{\alpha_{\text{eff}}^*} \quad \text{and} \quad \overline{\text{Pe}} = \text{Pe}^*\left(\frac{R}{L}\right)^2$$

Here $k_{\text{eff}}^* = k_{\text{nf}} + k_d$. The β_m's are the positive roots of the equation

$$J_o(\beta_m) = 0$$

From this temperature profile, the Nusselt number can be deduced as

$$\text{Nu} = \frac{\sum_{m=1}^{\alpha} e^{-\beta_m^2 \bar{x}/\overline{\text{Pe}}}}{\sum_{m=1}^{\alpha} e^{-\beta_m^2 \bar{x}/\overline{\text{Pe}}}/\beta_m^2} \tag{5.115}$$

Although this expression is similar to the solution of pure fluid, here the conductivity used for Nu is the sum of effective nanofluid conductivity and dispersive conductivity. If the axial dispersion is not neglected, we need an axial boundary condition. If we assure that the thermal dispersion begins at the entry, for dispersive flow we get a temperature drop at the entry as shown in Fig. 5.35. This is due to the well-known Dankwert (1953) boundary condition, which can be expressed as

$$-k_{\text{eff}}^{*}\frac{\partial T}{\partial x} = uA\rho C_p(T_0 - T) \qquad \text{at } x = 0 \tag{5.116}$$

At the exit of the tube the we can use the usual derivative boundary condition,

$$\frac{\partial T}{\partial x} = 0 \qquad \text{at } x = L \tag{5.117}$$

Under these conditions the Nusselt number is

$$\text{Nu} = \frac{\sum\limits_{m=1}^{\alpha} X(\overline{x})/[X(0) - X'(0)/\text{Pe}^*]}{\sum\limits_{m=1}^{\alpha} X(\overline{x})/[X(0) - X'(0)/\text{Pe}^*]/\beta_m^2} \tag{5.118}$$

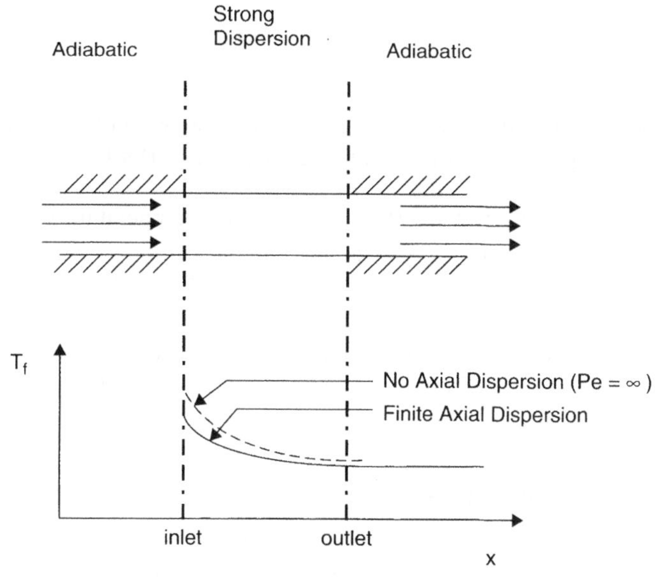

Fig. 5.35 Temperature drop at the entry due to the Dankwert boundary condition.

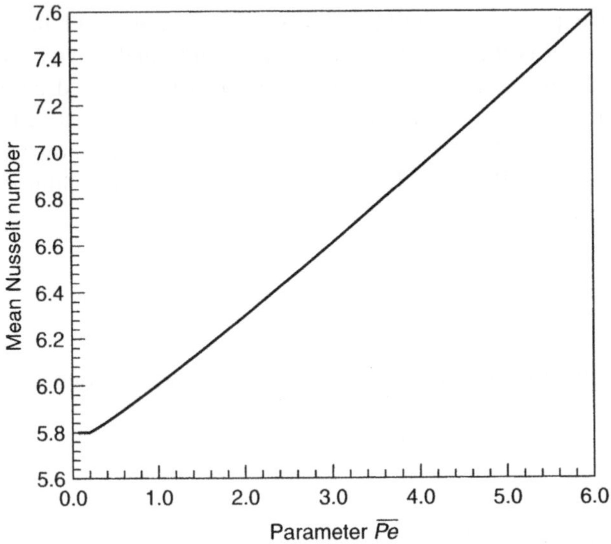

Fig. 5.36 Nusselt number according to equation (5.118). [From Xuan and Roetzel (2000), with permission from Elsevier.]

where

$$X(\overline{x}) = m_2 e^{m_2 + m_1 \overline{x}} - m_1 e^{m_1 + m_2 \overline{x}}$$

$$X'(\overline{x}) = m_1 m_2 (e^{m_2 + m_1 \overline{x}} - e^{m_1 + m_2 \overline{x}})$$

$$m_{1,2} = \frac{\mathrm{Pe}^* \pm \sqrt{\mathrm{Pe}^{*2} + 4\beta_m (L/R)^2}}{2}$$

This solution was plotted by Xuan and Li (2000) as shown in Fig. 5.36 using the expression for dispersion coefficient of Beckman et al. (1990):

$$\frac{k_d}{\rho C_p} = \begin{cases} 10.1\mathrm{Ru}\sqrt{f/2} + 5.03\mathrm{Ru} & \text{for a large temperature gradient} \\ \dfrac{10.1\mathrm{Ru}\sqrt{f/2} + 5.03\mathrm{Ru}}{1 + l/\sqrt{2f}} & \text{for adiabatic cases} \end{cases} \tag{5.119}$$

where f is the friction factor. They indicated that the dispersion coefficient is the effective embodiment of effects such as Brownian diffusion, sedimentation, dispersion, and so on.

5.4.2. Particle Migration Effect

While the dispersion model tells about an overall modeling strategy, it is important to identify the mechanisms that are probably responsible behind the dispersive behaviors. It is a fact a that the true contribution of these mechanisms can only be revealed by highly sophisticated experiments which have not yet been conducted. Analytical treatments taking these effects into consideration and comparison of the results with available data can give some indication of the importance of these mechanisms. Among these mechanisms, particle migration seems to be the most logical and is likely to play an important role in the convection of nanofluids.

Ding and Wen (2005) investigated the particle migration effect in nanofluids. They used a well-known *mass balance approach*, the constitutive equation for which was laid by Phillips et al. (1992). The basic concept of this model is that the particles migrate here under the action of a shear force from a region of higher shear to a region of lower shear under conditions of higher viscosity to lower viscosity and due to Brownian diffusion from a region of higher particle concentration to one of lower particle concentration. Now, considering the mass balance over the control volume in Fig. 5.37, one obtains

$$J + r\frac{dJ}{dr} = 0 \tag{5.120}$$

where J is the total particle flux in the r (radial) direction. In equation (5.120) the particle phase is taken to be continuous. This particle migration flux consists of these components arising out of three mechanisms:

$$J = J_{\mu} + J_b + J_c \tag{5.121}$$

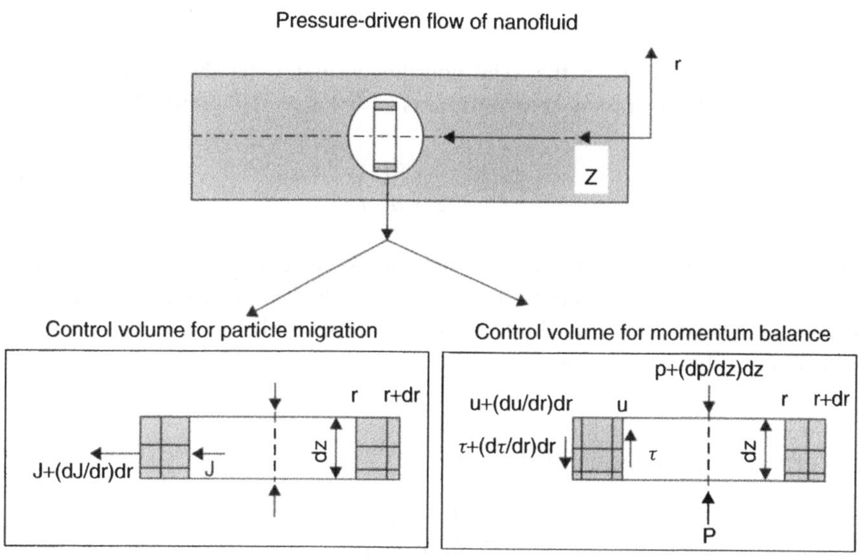

Fig. 5.37 Control volume considered by Ding and Wen mass flux equation. [From Ding and Wen (2005), with permission from Elsevier.]

where

$$J_{\mu} = \text{flux due to viscosity gradient} = -K_{\mu}\dot{\gamma}\varepsilon_p^2\left(\frac{d_p^2}{\mu}\right)\frac{d\mu}{d\varepsilon_p}\nabla\varepsilon_p$$

$$J_b = \text{flux due to nonuniform shear} = -K_c d_p^2(\varepsilon_p^2\nabla\dot{\gamma} + \varepsilon_p\dot{\gamma}\nabla\varepsilon_p)$$

$$J_c = \text{flux due to Brownian motion} = -D_b\nabla\varepsilon_p$$

where K_{μ} and K_c are constants, $\dot{\gamma}$ is the shear rate, μ the viscosity, and d_p the particle diameter. D_b is the Brownian diffusion coefficient, given by

$$D_b = \frac{k_b T}{3\pi\mu d_p} \tag{5.122}$$

where k_b is the Boltzmann constant. Integration of equation (5.121) and imposition of a symmetric boundary condition ($J = 0$ at $r = 0$) yields the following equation for the one-dimensional case:

$$K_{\mu}\dot{\gamma}\varepsilon_p^2\frac{d_p^2}{\mu}\frac{d\mu}{dr} + K_c d_p^2\varepsilon_p^2\frac{d\dot{\gamma}}{dr} + K_c d_p^2\varepsilon_p\dot{\gamma}\frac{d\varepsilon_p}{dr} + D_b\frac{d\varepsilon_p}{dr} = 0 \tag{5.123}$$

Also in Fig. 5.37, the momentum balance in the control volume is shown, which yields the equation

$$\frac{1}{r}\left(\frac{dr\tau}{dr}\right) = -\frac{dp}{dz} \tag{5.124}$$

Where p is the pressure, z the axial coordinate, and τ the shear stress. Due to symmetry, the shear stress is zero at the axis ($\tau = 0$ at $r = 0$), giving the solution:

$$\tau = -\frac{r}{2}\left(\frac{dp}{dz}\right) \tag{5.125}$$

To get an explicit solution to this, one must know the correlation for shear stress, which depends on the rheology of the fluid. Since most investigations suggests Newtonian behavior for nanofluids, a linear correlation between shear stress and shear rate can be assumed as

$$\tau = -\mu\dot{\gamma} \tag{5.126}$$

This reduces the momentum balance to

$$\dot{\gamma} = \frac{1}{2\mu}\left(\frac{dp}{dz}\right)r \tag{5.127}$$

where the shear rate is given by $\dot{\gamma} = du/dr$.

A model has to be used for viscosity. In this case, Ding and Wen (2005) used Bachelor's formula (5.95). Now the equations can be nondimensionalized as

$$\bar{\dot{\gamma}} = -\frac{\bar{r}}{\bar{\mu}} = \frac{d\bar{u}}{d\bar{r}} \tag{5.128}$$

$$\frac{1}{\bar{\mu}}\frac{d\bar{\mu}}{d\bar{r}} + \left(\frac{K_c}{K_\mu}\right)\frac{1}{\bar{\dot{\gamma}}}\frac{d\bar{\dot{\gamma}}}{d\bar{r}} + \left(\frac{K_c}{K_\mu}\right)\frac{1}{\varepsilon_p}\frac{d\varepsilon_p}{d\bar{r}} = \frac{1}{K_\mu \mathrm{Pe}}\frac{1}{\varepsilon_p^2 \bar{r}}\frac{d\varepsilon_p}{d\bar{r}} \tag{5.129}$$

where

$$\bar{\mu} = \frac{\mu}{\mu_f}$$

$$\bar{\dot{\gamma}} = \dot{\gamma}\frac{2\mu_f}{(dp/dz)R}$$

$$\bar{r} = \frac{r}{R}$$

$$\bar{u} = \frac{2\mu_f}{(dp/dz)R^2}u$$

$$\mathrm{Pe} = \frac{3\pi d_p^3(-dp/dz)R}{2k_B T}$$

Solving the equations above, particle concentration profiles across the fluid can be obtained. The key parameter in this equation is the particle Peclet number, Pe. For 90- to 500-nm particles, the variation of Pe with particle diameter, d_p, is as given in Fig. 5.38. The distribution of particles in the radial direction for

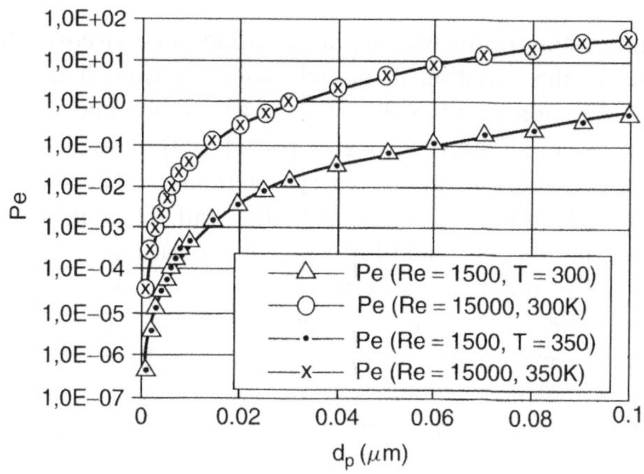

Fig. 5.38 Effect of temperature and particle size on the particle Peclet number. [From Ding and Wen (2005), with permission from Elsevier.]

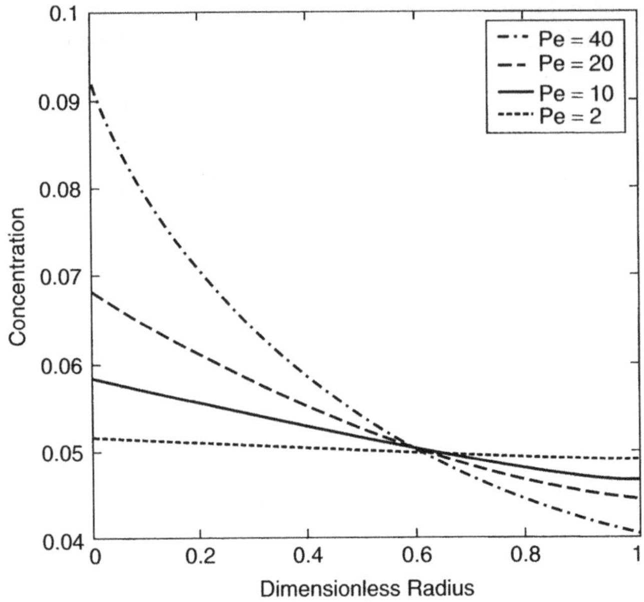

Fig. 5.39 Influence of the Peclet number on particle distribution. [From Ding and Wen (2005), with permission from Elsevier.]

different Peclet numbers is given in Fig. 5.39. It is clear that the higher the Peclet number, the more variation is there in particle concentration. This result also indicates that there is a possibility of agglomeration at the core region due to high particle concentration, which is unlikely to be dispersed by the shear, due to the low magnitude of shear at the core region. However, these results may get changed when shear thinning behavior is considered due to non-Newtonian behavior.

The other interesting result was the large variation of viscosity from the core to the wall due to the variation in particle concentration. It has to be kept in mind that this variation of viscosity (Fig. 5.40) is not due to shear thinning (which is not considered here) but due to particle migration. This also explains why nanofluids are likely to give higher heat transfer with a relatively lower pressure penalty, due to lower viscosity near the wall. They also showed that the particle distribution is more nonuniform for higher particle concentrations. This also indicates that in the analysis of thermal transport in nanofluids, one must be concerned that the near-wall region may have a lower particle concentration, leading to lower thermal conductivity near the wall and lower heat transfer (also lower shear stress). Hence, Ding and Wen (2005) predicted that there may be an optimum particle size for a compromise between heat transfer and pressure drop.

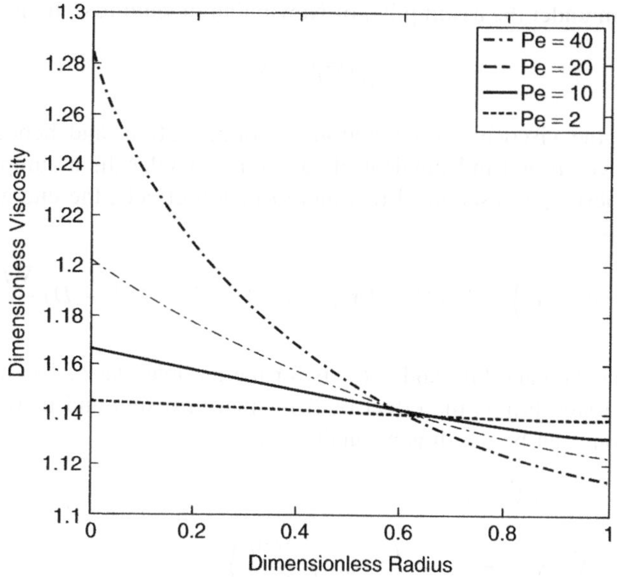

Fig. 5.40 Influence of the Peclet number on viscosity distribution. [From Ding and Wen, (2005), with permission from Elsevier.]

The consideration of additional mechanisms was taken one step further by Buongiorno (2006). He first discussed all possible mechanisms of fluid particle slip during convection of nanofluids: (1) inertia, (2) Brownian diffusion, (3) thermophoresis, (4) diffusionphoresis, (5) Magnus effect, (6) fluid drainage, and (7) gravity. By comparing all the time scales of these processes he concluded that for laminar flow (also in the viscous sublayer of the turbulent flow) thermophoresis and Brownian diffusion are important mechanisms, while in the turbulent region the nanoparticles are carried by turbulent eddies without slip and the diffusion mechanisms above are negligible there. Based on these assumptions, the continuity equation for nanofluids and nanoparticles were derived in the form

$$\nabla \cdot \vec{v} = 0 \tag{5.130}$$

$$\frac{\partial \varepsilon_p}{\partial t} + \vec{v} \cdot \nabla \varepsilon_p = \nabla \cdot \left(D_B \ \nabla \varepsilon_p + D_T \frac{\nabla T}{T} \right) \tag{5.131}$$

The momentum equation was proposed by Bird et al. (1960) as

$$\rho \left[\frac{\partial \vec{v}}{\partial t} + \vec{v} \cdot \nabla \vec{v} \right] = -\nabla P - \nabla \cdot \tau \tag{5.132}$$

The stress term under the assumption of Newtonian incompressible flow becomes

$$\tau = -\mu \lfloor \nabla\vec{v} + (\nabla\vec{v})^t \rfloor \tag{5.133}$$

In nanofluids the viscosity is a function of concentration, and hence the three equations above are not independent of each other. Under the assumption of the presence of thermophoresis and Brownian diffusion effects, the energy equation takes the form

$$\rho C\left(\frac{\partial T}{\partial t} + \vec{v}\cdot\nabla T\right) = \nabla\cdot(k\ \nabla T) + \rho_p C_p\left(D_B \nabla\varepsilon_p\ \nabla T + D_T \frac{\nabla T\cdot\nabla T}{T}\right) \tag{5.134}$$

Here ρC is for the nanofluid and $\rho_p C_p$ is for the particle phase. The last term on the right-hand side brings the effect of Brownian diffusion and thermophoresis. These equations can be nondimensionalized as

$$\nabla\cdot\vec{V} = 0 \tag{5.135}$$

$$\frac{\partial\overline{\varepsilon_p}}{\partial\xi} + \vec{V}\cdot\nabla\varepsilon_p = \frac{1}{R_e S_c}\left(\nabla^2\overline{\varepsilon_p} + \frac{\nabla^2\theta}{N_{BT}}\right) \tag{5.136}$$

$$\frac{\partial\vec{v}}{\partial\xi} + \vec{V}\nabla\vec{V} = -\nabla\psi + \frac{\nabla^2\vec{V}}{\text{Re}} \tag{5.137}$$

$$\frac{\partial\theta}{\partial\xi} + \vec{V}\nabla\theta = \frac{1}{\text{Re}\cdot\text{Pr}}\left(\nabla^2\theta + \frac{\nabla\overline{\varepsilon_p}\cdot\nabla\theta}{\text{Le}} + \frac{\nabla\theta\cdot\nabla\theta}{\text{Le}N_{BT}}\right) \tag{5.138}$$

where

$$\vec{V} = \frac{\vec{v}}{\overline{v}} \qquad \overline{\varepsilon_p} = \frac{\varepsilon_p}{\varepsilon_{pb}} \qquad \theta = \frac{T - T_b}{\Delta T}$$

$$\psi = \frac{P}{\rho\overline{v}^2} \qquad R = \frac{r}{D} \qquad \xi = \frac{t}{D/\overline{v}}$$

where $\overline{v}$, ε_{pb}, ΔT, and D are the reference values for these quantities and the nondimensional numbers

$$\text{Re} = \frac{\rho\overline{v}D}{\mu} = \text{Reynolds number}$$

$$\text{Sc} = \frac{\mu}{\mu D_B}\text{Schmidt number}$$

$$N_{BT} = \frac{\varepsilon_{pb} D_B T_b}{D_T \Delta T} = \frac{\text{Brownian diffusivity}}{\text{Thermophoresis}}$$

$$\text{Pr} = \text{Prandtl number}$$

$$\text{Le} = \text{Lewis number} = \frac{k}{\rho_p C_p D_B \varepsilon_{pb}}$$

Assuming that the axial transport terms are small compared to the radial terms, the turbulent transport equations were derived as

$$\frac{d}{dy}\left[(D_B + D_p)\frac{d\varepsilon_p}{dy} + \frac{D_T}{T}\frac{dT}{dy}\right] = 0 \tag{5.139}$$

$$\frac{d}{dy}\left[\mu + \rho D_M \frac{dv}{dy}\right] = 0 \tag{5.140}$$

$$\frac{d}{dy}\left[(k + \rho c D_H)\frac{dT}{dy}\right] = 0 \tag{5.141}$$

Here y is the radial coordinate as shown in Fig. 5.41 and D_p, D_m, and D_H are diffusivities of particle eddy, momentum, and heat in the turbulent sublayer. They assumed that $D_p \sim D$ and $\varepsilon_p \sim \varepsilon_{pb}$, due to mixing with eddies. Eliminating the temperature gradient within the equations above for a laminar sublayer, the equations can be solved to get the particle concentration distribution as

$$\varepsilon = \varepsilon_b e^{1/N_{\mathrm{BT}}(1-y/\delta_c)} \tag{5.142}$$

The plot of this result is shown in Fig. 5.42, which shows a distribution quantitatively different from that of Ding and Wen (2005), due mainly to the inclusion of thermophoresis and neglecting the migration under shear and viscosity gradient.

The final heat transfer equation Boungiorno obtained was after comparison with the Prandtl analogy correlation and Gniclinski correlation [equation (5.60)]:

$$\mathrm{Nu}_b = \frac{f/8(\mathrm{Re}_b - 1000)\,\mathrm{Pr}_b}{1 + \delta_v^+ \sqrt{f/8}(\mathrm{Pr}_v^{2/3} - 1)} \tag{5.143}$$

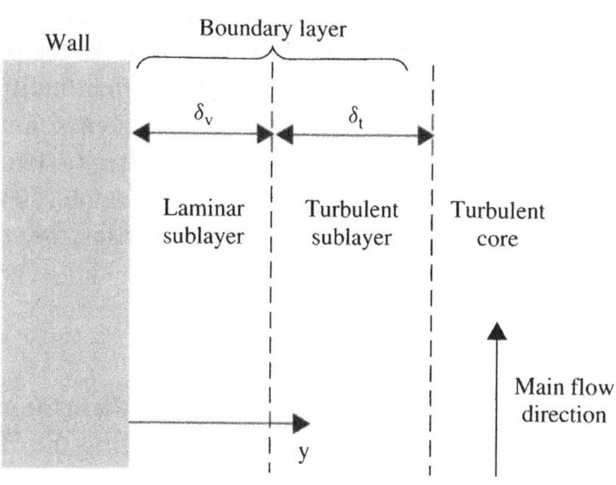

Fig. 5.41 Flow structure near the wall. [From Buongiorno (2006), with permission from ASME Publishing.]

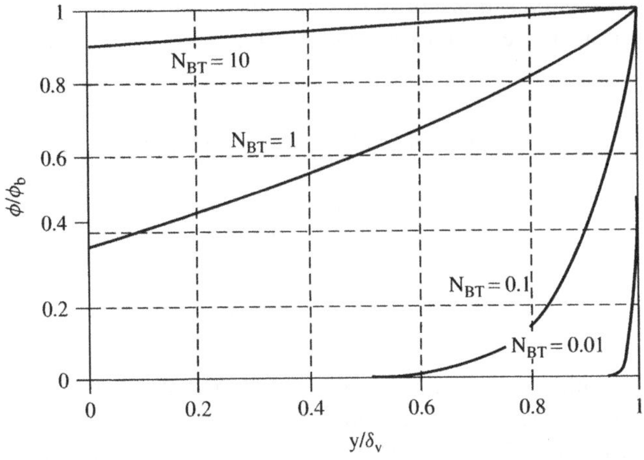

Fig. 5.42 Particle-volume-fraction variation. [From Boungiorno (2006), with permission from ASME Publishing.]

where δ_v^+ is an empirical constant and Pr_b is the Prandtl number evaluated at mean viscous sublayer temperature. The results were plotted against the correlation obtained from the data of Xuan and Li (2003) and Pak and Cho (1998) along with the well-known Dittus–Boelter equation. The results are shown in Fig. 5.43 for Al_2O_3 water nanofluids. The results show that the present model agrees with Pak and Cho (1998) but underpredicts the Xuan and Li (2003) correlation at $\varepsilon_p > 0$. They attributed this to the temperature effects on thermal conductivity and variation of viscosity near the wall. This does not explain the entire story of nanofluid convection, although it gives an important insight.

5.4.3. Natural Convection and Stability in Nanofluids

Natural convection is set in a medium by convective instability. For example, let's take a fluid kept between two parallel plates (Fig. 5.44). As the buoyancy-driven instability sets in and natural convection starts in the form of multiple cells, which is shown in the figure. This is called *Rayleigh–Benard convection*. Chandrasekar (1961) analyzed the linear stability criterion for Rayleigh–Benard convection. Such instabilities in nanofluids were dealt with by Kim et al. (2004). For suspensions where the Soret effect (discussed earlier) is significant, the stability criterion is given as

$$Ra_c = \frac{1708}{1 + \psi} \qquad \text{for} \quad \psi > 0 \tag{5.144}$$

where $\psi = S\alpha_s/K_s\alpha$ (S is the Soret coefficient, α the thermal expansion coefficient, and K_s the thermal diffusivity of the solute). Using the Brinkman model for viscosity yields

$$\frac{\mu_{nf}}{\mu_f} = \frac{1}{(1 - \varepsilon_p)^{2.5}} \tag{5.145}$$

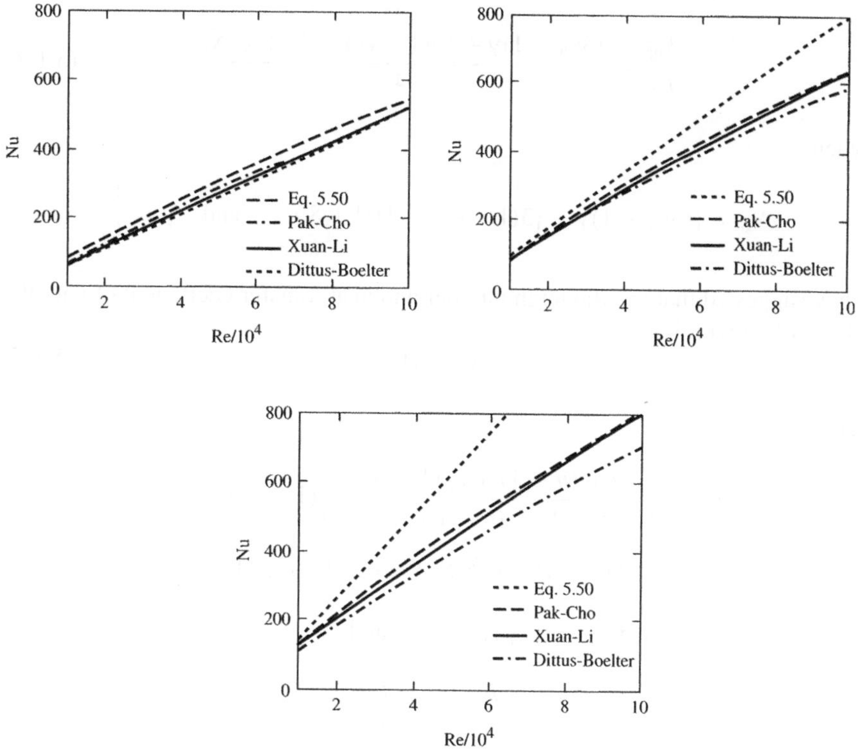

Fig. 5.43 Heat transfer in alumina–water nanofluids. [From Boungiorno (2006), with permission from ASME Publishing.]

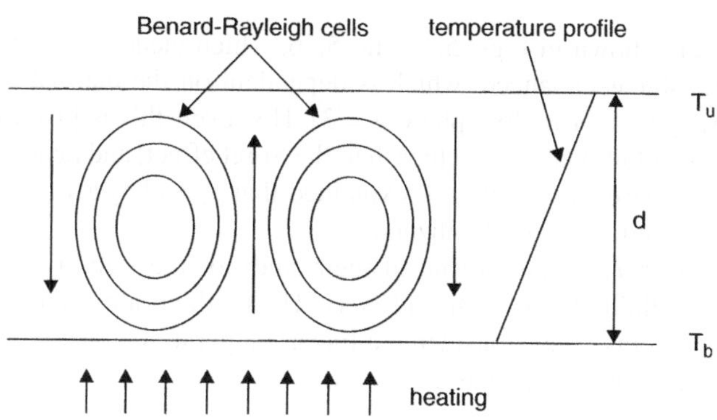

Fig. 5.44 Fluid motion in Rayleigh–Benard convection. The bottom plate is heated while the top is cooled beyond a certain value of temperature difference.

Using the Bruggemann conductivity model gives

$$\frac{k_{\rm nf}}{k_f} = \frac{(3\varepsilon_p - 1)\gamma + \{3(1-\varepsilon_b) - 1\} + \sqrt{\Delta_B}}{4} \tag{5.146}$$

where

$$\Delta_B = [(3\varepsilon_p - 1)\gamma + \{3(1-\varepsilon_p) - 1\}^2] + 8\gamma \qquad \text{and} \quad \gamma = \frac{k_p}{k_f}$$

They suggested that the Rayleigh number and heat transfer coefficient of nanofluids can be given by

$$\mathrm{Ra} = f_1 \mathrm{Ra}_f \tag{5.147}$$

where

$$f_1 = \frac{\gamma + (n-1) + \varepsilon_p(1-\gamma)}{\gamma + (n-1) - (n-1)\varepsilon_p(1-\gamma)}(1-\varepsilon_p)$$
$$\times\, [((1-\varepsilon_p) + \varepsilon_p\delta_1)((1-\varepsilon_p) + \varepsilon_p\delta_2)(1-\varepsilon_p)]^{2.5}$$

where $\delta_1 = \rho_p/\rho_f$ and $\delta_2 = (\rho c_p)_{\rm nf}/(\rho c_p)_{\rm f}$ and

$$\frac{h_{\rm nf}}{h_f} = f^m \frac{k_{\rm nf}}{k_f}$$

where m is the exponent of the heat transfer equation

$$\mathrm{Nu}_{nf} = A\mathrm{Ra}_{nf}^m \tag{5.148}$$

Their results are shown in Figs. 5.45 and 5.46, which clearly indicate that in all cases heat transfer is enhanced, which is dependent on the thermal conductivity ratio γ and shape factor n (for sphere $n = 3$). However, this is just a predication and assumes no additional effect other than the Soret effect, and hence needs to be confirmed by experiment. They also explained that δ_1 and δ_2 act as destabilizing factors while γ and x act as stabilizers.

Kim et al. (2007) further advanced their study of the stability of nanofluids with the binary fluids H_2O–LiBr and NH_3–H_2O, which are used in absorption refrigeration systems. They considered a linear temperature gradient giving linear concentration gradient in the basic state: as

$$C_o = C_i[1 - B_T\Delta(d - z)] \tag{5.149}$$

where C is the concentration, β_T the temperature gradient, d the fluid layer thickness, and z the vertical coordinate from the bottom plate. They considered both the Soret effect (particle diffusion under a temperature gradient) and the

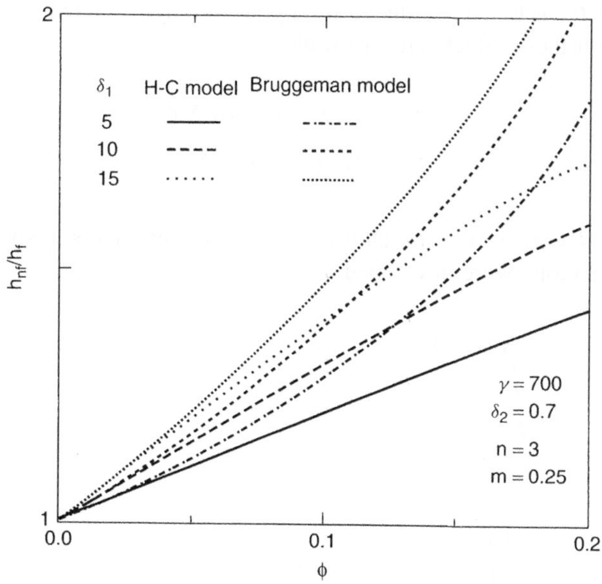

Fig. 5.45 Ratio of the heat transfer coefficient versus ϕ for various δ_1. [From Kim et al. (2004), with permission from the American Institute of Physics.]

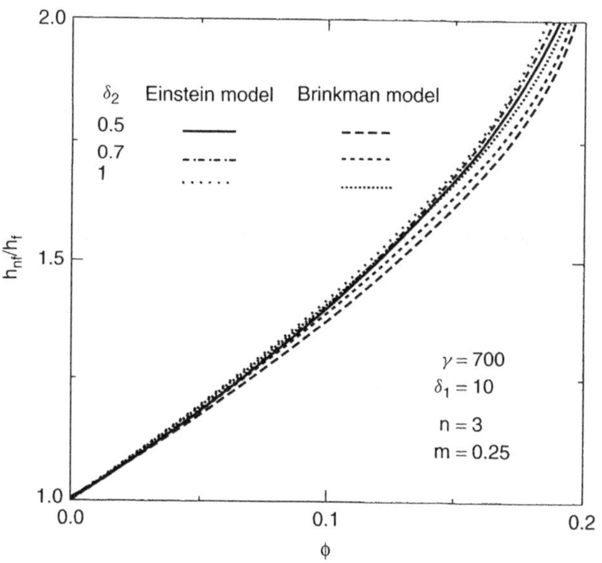

Fig. 5.46 Ratio of the heat transfer coefficient versus ϕ for various δ_2. [From Kim et al. (2004), with permission from the American Institute of Physics.]

Dufour effect (heat transfer induced by a concentration gradient). The heat and mass flow with these effects are given by

$$-J_h = k\nabla T + \rho c_p \alpha_{Df} \nabla C \tag{5.150}$$

$$-J_m = D\nabla C + D_{\text{Sr}} \nabla T \tag{5.151}$$

Here α_{Df} and D_{Sr} represent the Dufour and Soret coefficients, respectively. The governing equations were developed as

$$\nabla \vec{U} = 0 \tag{5.152}$$

$$\rho_R \frac{D}{Dt}\vec{U} = -\nabla P + \mu \nabla^2 \vec{U} + \rho \vec{g} \tag{5.153}$$

$$\frac{DT}{Dt} = \alpha \nabla^2 T + \alpha_{\text{Df}} \nabla^2 C \tag{5.154}$$

$$\frac{DC}{Dt} = D\nabla^2 C + D_{\text{Sr}} \nabla^2 T \tag{5.155}$$

$$\rho = \rho_R[1 - \beta_T(T - T_R) + \beta_S(C - C_R)] \tag{5.156}$$

The equations above are for continuity, momentum, energy, and concentration, and *D/Dt* is the total differential:

$$\frac{D}{Dt} = \frac{\partial}{\partial t} + \vec{U} \cdot \nabla \tag{5.157}$$

From these equations they derived the liner stability equation,

$$(D^2 - a^2)^3 w^* = \overline{\text{Ra}} a^2 w^* \quad \text{where} \quad \overline{\text{Ra}} = \text{Ra}(1 + F_s + F_{\text{Sr}} + F_{\text{Df}})(1 - K)^{-1} \tag{5.158}$$

where

$$\text{Ra} = \frac{g\beta_T \, \Delta T d^3}{\alpha \gamma}$$

$$F_s = \frac{\beta_s C_i \alpha}{D}$$

$$F_{\text{sr}} = \frac{\beta_s D_{\text{Sr}}}{\beta_T D}$$

$$F_{\text{Df}} = \frac{\beta_T C_i \alpha_{\text{Df}}}{D}$$

$$K = \frac{\alpha_{\text{Df}} D_{\text{Sr}}}{\alpha D}$$

With the parameters above they calculated the stability parameters for different systems of nanofluids (copper and silver). The separation factor, which is a prime stability parameter, is given by the following equation for no Dufour effect:

$$\psi = \frac{(1-\phi_w) + \phi_w \delta_4}{(1-\phi_w) + \phi_w \delta_3}\psi_{\mathrm{bf}} \tag{5.159}$$

where ψ_{bf} is the separation factor for the base fluid, ϕ_{w} the weight fraction of particles, and

$$\delta_3 = \frac{D\text{ nanoparticles}}{D_s} \qquad \delta_4 = \frac{D_{\mathrm{Sr}},\ \text{nanoparticles}}{D_{\mathrm{Sr}},\ \text{solute in binary fluid}} \tag{5.160}$$

One of their stability diagrams is given in Fig. 5.47. They concluded that the Dufour and Soret effects make nanofluids unstable, and that for heat transfer the Soret effect is more important. They also concluded that denser initial concentration makes a nanofluid more unstable.

A word of caution is important for both the work by Kim and his group and the forced convection work of Boungiorno. They both proposed some mechanisms to be important which need local precise measurement to be confirmed. To the best of our knowledge, this has not been done yet, hence they remain hypotheses rather than theories.

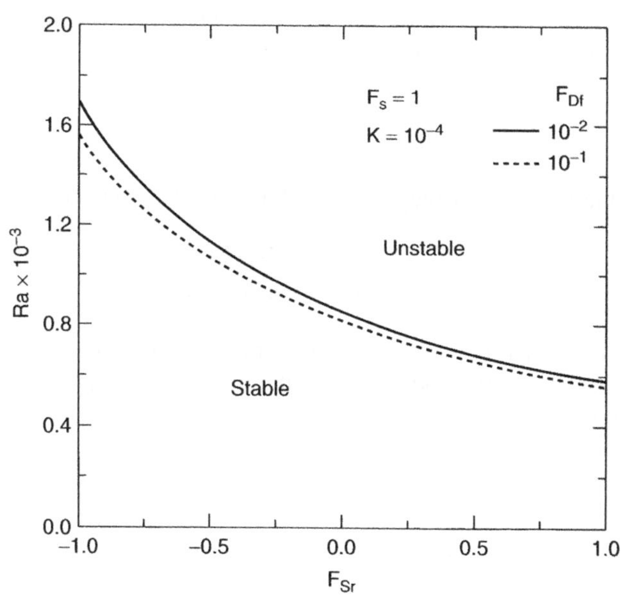

Fig. 5.47 Ra versus F_{Sr} for various values of F_{Df}. [From Kim et al. (2007), with permission from Elsevier.]

5.4.4. Design Aspects of Convection: Physical Properties and Optimization of Thermohydraulics

For convection with nanofluids, a number of issues are associated, as discussed earlier. However, these effects and strategies for tackling them do not end questions regarding the analysis of nanofluids—it only opens up possibilities for many more questions. One such question is that of the role of uncertainties in physical properties on convective heat transfer in nanofluids, as discussed by Mausour et al. (2007).

To assess the performance of a nanofluid while replacing a base fluid during a cooling exercise, one needs to examine the variation in pumping power and mass flow rate for a given heat transfer or variation of bulk fluid temperature or pumping power for a given heat transfer rate and mass flow rate. In engineering terms this is a *rating* of the cooling arrangement, or more simply, it is an assessment of a hydraulic penalty (power consumption due to pressure drops) for a given tube geometry and heat flux. Mansour et al. (2007) showed that the nature of variation of these parameters (power ratio or mass flow rate) depends critically on which models of effective properties are used. As it is not yet very certain which models for viscosity, specific heat, and thermal conductivity are actually applicable to nanofluids, they took two sets of equations, naming them GdS and BMGN. In the GdS model they used the Pak and Cho (1998) model for specific heat:

$$C\hat{p} = (1 - \varepsilon_p)(\rho C_p)_f \quad + \quad \varepsilon_p(\rho C_p)_p \tag{5.161}$$

where a subscript f indicates a base fluid, and p, a particle. They used Brinkmans (1952) model of viscosity for the GdS model:

$$\mu_{\text{eff}} = \mu_f \frac{1}{(1 - \varepsilon_p)^{2.5}} \tag{5.162}$$

Finally, for the GdS model they use the Hamilton–Crosser model for spherical particles:

$$\frac{k_{\text{eff}}}{k_0} = \frac{k_p - 2k_f - 2\varepsilon_p(k_f - k_p)}{k_p - 2k_f - \varepsilon_p(k_f - k_p)} \tag{5.163}$$

For the BMGN model they used the Wang et al. (1999) model for viscosity, the Xuan and Roetzel (2000) model for specific heat, and Yu and Choi (2003) for thermal conductivity:

$$\frac{\mu_{\text{eff}}}{\mu_f} = 123\varepsilon_p^2 + 7.3\varepsilon_p + 1 \tag{5.164}$$

$$(\rho C_p)_{\text{eff}} = (1 - \varepsilon_p)(\rho C_p)_f + \varepsilon_p(\rho C_p)_p \tag{5.165}$$

$$\frac{k_{\text{eff}}}{k_f} = \frac{k_p + 2k_f + 2(k_p - k_f)(1 + \beta)^3\varepsilon_p}{k_p + 2k_f - 2(k_p - k_f)(1 + \beta)^3\varepsilon_p} \tag{5.166}$$

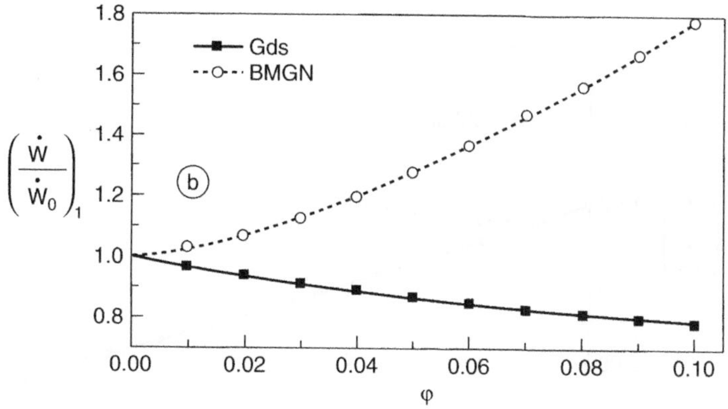

Fig. 5.48 Effect of particle loading on pumping power for fixed heat and mass flow rates. [From Mansour et al. (2007), with permission from Elsevier.]

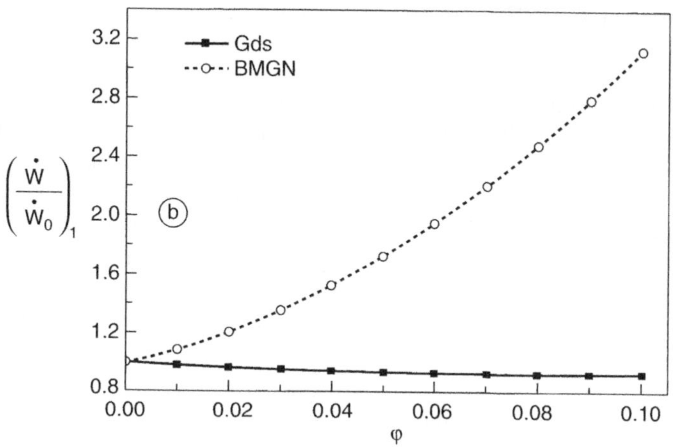

Fig. 5.49 Effect of particle loading on pumping power for a fixed heat rate and bulk temperature ratio. [From Mansour et al. (2007), with permission from Elsevier.]

Their results showed that for a fixed mass flow rate the pumping power ratio (Fig. 5.48), and for a fixed change in bulk temperature the pumping power ratio (Fig. 5.49), behave differently for the GdS and BMGN models. In fact, for a fixed mass flow rate the two models show completely different trends: one showing an increase in pumping power with volume fraction while the other shows a decrease.

The design of cooling equipment is an estimation of the equipment dimensions for a given performance. Essentially, this means that for a given mass flow rate, heating rate, and bulk temperature rise, the dimensions are to be evaluated. This is known as *sizing* in engineering terminology. Figure 5.50 shows such a sizing exercise, which clearly indicates that the two models of property evaluation differ

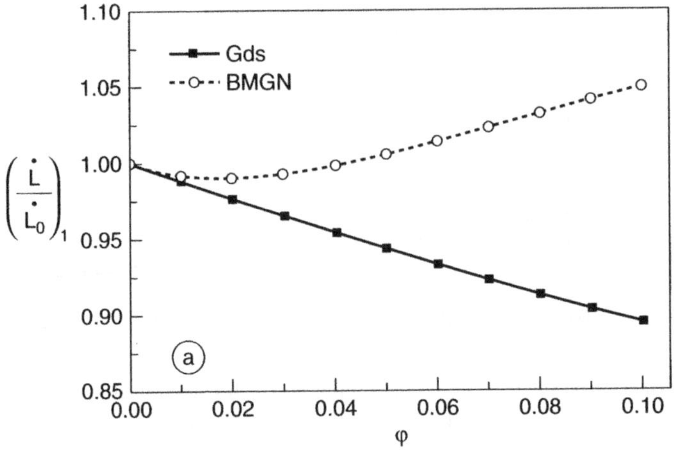

Fig. 5.50 Length ratio for a given heat rate, bulk temperature rise, and mass flow rate. [From Mansour et al. (2007), with permission from Elsevier.]

significantly in design or sizing estimation. The results above clearly indicate that before concluding the thermal and hydraulic performance of convection with nanofluids, we must have an accurate model for property evaluation.

Another important work is that of Gosselin and da Silva (2004), who focused on optimizing particle loading for laminar and turbulent forced and natural convections in nanofluids. They defined the optimum as the highest value of relative heat transfer Ω_{FC} (heat transfer in nanofluids divided by heat transfer in base fluids) for a fixed pumping power. For laminar and turbulent forced flow it is given by

$$\Omega_{\text{FC, laminar}} = \frac{\tilde{k}_{\text{eff}}^{2/3} \tilde{\rho}_{\text{eff}}^{2/5} \tilde{C}_{p,\text{eff}}^{1/3}}{\tilde{\mu}_{\text{eff}}^{4/15}} \tag{5.167}$$

$$\Omega_{\text{FC, turbulent}} = \frac{\tilde{k}_{\text{eff}}^{2/3} \tilde{\rho}_{\text{eff}}^{4/7} \tilde{C}_{p,\text{eff}}^{1/3}}{\tilde{\mu}_{\text{eff}}^{11/21}} \tag{5.168}$$

where $\tilde{k}_{\text{eff}} = k_{\text{eff}}/k_f$, $\tilde{\rho}_{\text{eff}} = \rho_{\text{eff}}/\rho_f$, and $\tilde{\mu}_{\text{eff}} = \mu_{\text{eff}}/\mu_f$. For laminar natural convection it is

$$\Omega_{\text{NC,laminar}} = \frac{\tilde{\beta}_{\text{eff}}^{1/4} \tilde{\delta}_{\text{eff}}^{1/2} \tilde{C}_{p\text{eff}}^{1/4} \tilde{k}_{\text{eff}}^{3/4}}{\tilde{\mu}_{\text{eff}}^{1/4}} \tag{5.169}$$

Figure 5.51 shows how this relative heat transfer varies with volume fraction for laminar forced convection. Similarly, for forced convection Fig. 5.52 shows the optimum value of the particle fraction and relative heat transfer rate at different particle rates (particle shape factor). Such figures can also be drawn for natural

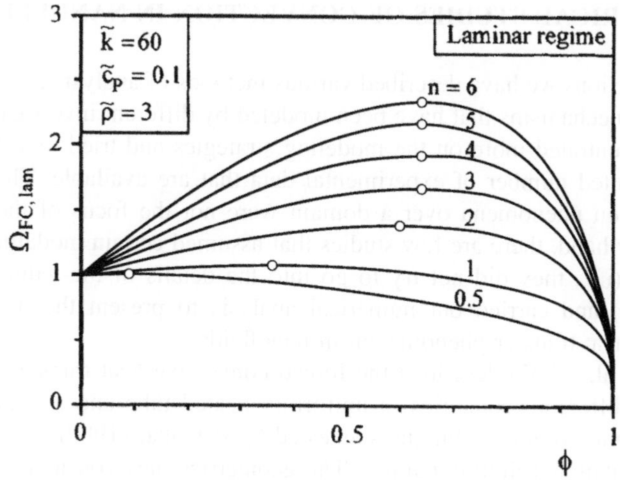

Fig. 5.51 Relative heat transfer against particle volume fraction for different particle shapes. [From Gosselin and da Silva (2004), with permission from the American Institute of Physics.]

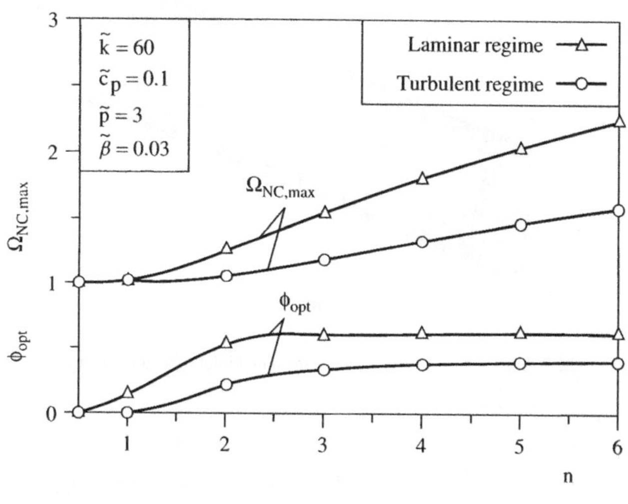

Fig. 5.52 Optimum values of particle fraction and relative heat rate for forced convection. [From Gosselin and da Silva (2004), with permission from the American Institute of Physics.]

convection. This work clearly demonstrates that an optimization in particle loading is possible. At the same time, it should be noted that the entire exercise is limited by the accuracy of the correlations used in the optimization process, a conclusion similar to that of Mansour et al. (2007).

5.5. NUMERICAL STUDIES OF CONVECTION IN NANOFLUIDS

In earlier sections we have described various methods of analysis and the various effects and mechanisms that have been modeled by different investigators. These studies concentrated more on the modeling strategies and tried to validate them with the limited number of experimental data that are available. The details of the convection phenomena over a domain were not the focus of these studies. On the other hand, there are few studies that assumed certain modeling concepts to be valid (i.e., they did not try to go into the details of the validity of these assumptions) and carried out numerical analysis to present the details of the convective heat transfer phenomenon in nanofluids.

Maiga et al. (2005) described the forced convective heat transfer in nanofluids in two different geometries (a uniformly heated tube and a radial channel) using the finite volume technique suggested by Patankar (1980), which is popular in computational fluid dynamics. The geometrics they considered are shown in Fig. 5.53. Their formulation was that of a single-fluid approach, which we

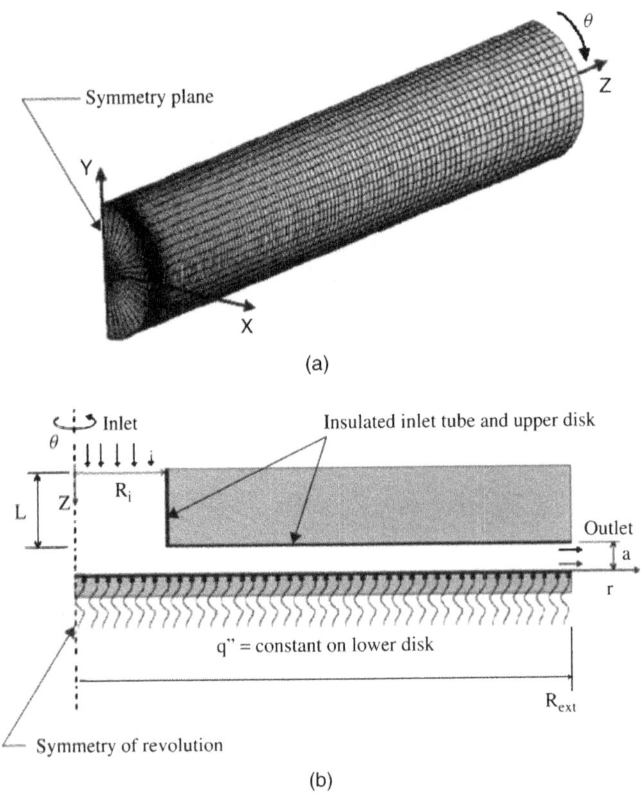

Fig. 5.53 Geometries studied by Maiga et al.: (a) uniformly heated tube; (b) radial channel between heated disks. [From Maiga et al. (2005), with permission from Elsevier.]

discussed at the beginning of this chapter, with fluid properties replaced by effective nanofluid properties. Hence, the basic conservation equations for mass momentum and energy are the usual convection equations:

$$\nabla \cdot (\rho \vec{V}) = 0 \tag{5.170}$$

$$\nabla \cdot (\rho \vec{V} \vec{V}) = -\nabla P + \mu \nabla^2 \vec{V} \tag{5.171}$$

$$\nabla \cdot (\rho \vec{V} C_p T) = \nabla \cdot (k \nabla T) \tag{5.172}$$

As boundary conditions they considered both constant wall temperature and constant wall heat flux at the tube wall. Symmetry is assumed about a vertical plane through the axis. For the radial channel, constant heat flux was assumed at the impinging wall and insulated upper wall.

For computation, the most important decision here is to choose the effective property of nanofluids. They decided to use the following equations for this purpose:

$$\rho_{\text{nf}} = (1 - \varepsilon_{\text{p}})\rho_{\text{bf}} + \varepsilon_p \rho_p \tag{5.173}$$

$$(C_p)_{\text{nf}} = (1 - \varepsilon_p)(C_p)_{\text{bf}} + \varepsilon_p (C_p)_p \tag{5.174}$$

$$\mu_{\text{nf}} = \mu_{\text{bf}}(123\varepsilon_p^2 + 7.3\varepsilon_p + 1) \quad \text{for water–Al}_2\text{O}_3 \tag{5.175}$$

$$= \mu_{\text{bf}}(306\varepsilon_p^2 - 0.19\varepsilon_p + 1) \quad \text{for ethylene glycol–Al}_2\text{O}_3 \tag{5.176}$$

$$k_{\text{nf}} = k_{\text{bf}}(4.97\varepsilon_p^2 + 2.72\varepsilon_p + 1) \quad \text{for water–Al}_2\text{O}_3 \tag{5.177}$$

$$= k_{\text{bf}}(28.905\varepsilon_p^2 + 2.8273\varepsilon_p + 1) \quad \text{for ethylene glycol–Al}_2\text{O}_3 \tag{5.178}$$

Here the subscript bf stands for base fluid, and nf, for nanofluid. These equations were chosen by fitting curves through regression analysis of experimental data available for nanofluids (particularly for viscosity and thermal conductivity). However, we note here that these correlations are very much questionable. Also, treatment of the nanofluid as a single fluid is difficult to accept because experimental data by a large number of investigators clearly indicated that nanofluid convective heat transfer behavior cannot be attributed to property variation alone. Finally, they validated the computational scheme by standardizing a single-fluid problem, which is again, not quite logical.

They found that at lower Reynolds numbers for both water–Al_2O_3 and ethylene glycol–Al_2O_3 nanofluids, enhancement of the Nusselt number with particle-volume fraction is low, whereas at a higher volume fraction the Nusselt number increases rapidly. One such result is shown in Fig. 5.54. Based on their simulation, they suggested the following correlations:

$$\overline{\text{Nu}} = 0.086\text{Re}^{0.55}\text{Pr}^{0.5} \quad \text{for contact wall heat flux} \tag{5.179}$$

$$\text{Nu} = 0.28\text{Re}^{0.35}\text{Pr}^{0.36} \quad \text{for contact wall temperature} \tag{5.180}$$

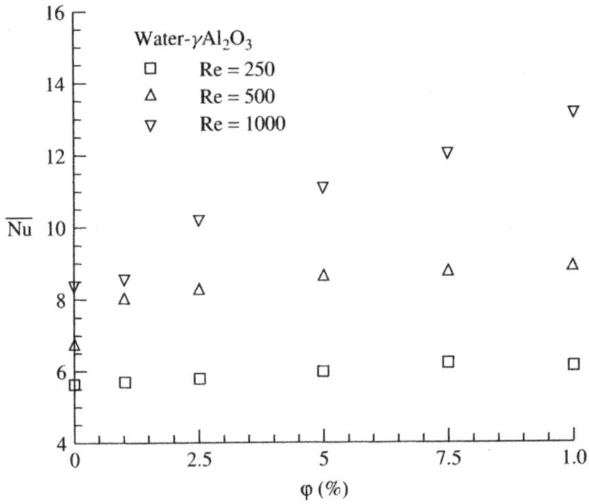

Fig. 5.54 Influence of Re and φ (ε =$_p$) on the enhancement of Nusselt numbers for convection in nanofluids. [From Maiga et al. (2005), with permission from Elsevier.]

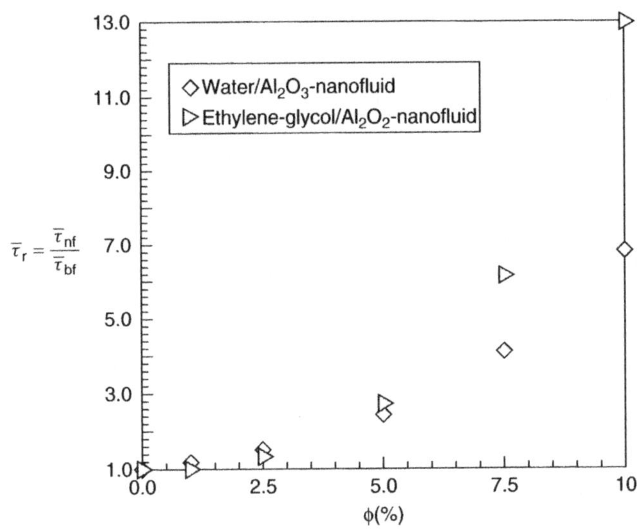

Fig. 5.55 Increase in the shear stress ratio for nanofluids. [From Maiga et al. (2005), with permission from Elsevier.]

However, they found that the shear stress at the wall is also increased significantly, as shown in Fig. 5.55. For the radial flow situation, they found that both the gap between disks and the Reynolds number have very little effect on heat transfer enhancement. The wall temperature for this type of flow is shown in Fig. 5.56. It must be kept in mind that the results above are purely numerical,

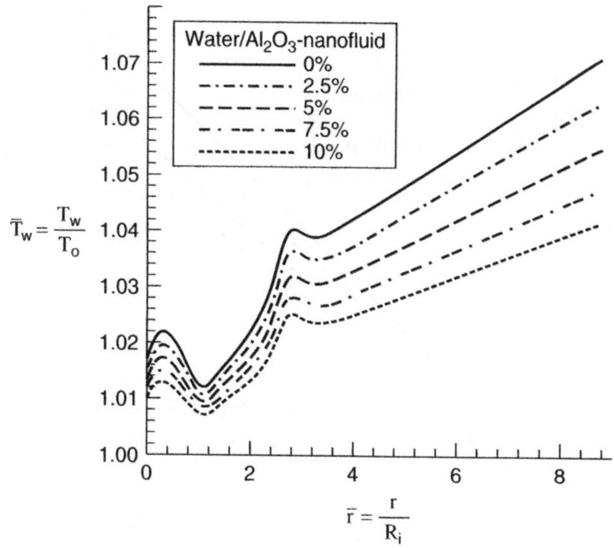

Fig. 5.56 Wall temperature in radial laminar flow. [From Maiga et al. (2005), with permission from Elsevier.]

and no experimental evidence was provided in the paper. In view of the fact that a single-fluid model was used, the results need to be validated at least qualitatively by a carefully designed experiment.

The same results with much more elaboration of the numerical results were published by Maiga et al. (2005) and Roy et al. (2004). Palm et al. (2006) extended the radial channel flow problem with consideration of temperature-dependent properties. They evaluated temperature-dependent properties by fitting curves to the experimental data of Putra et al. (2003) for Al_2O_3–water nanofluids in the form

$$
\begin{aligned}
\mu_{\text{nf}} &= 2.9 \times 10^{-7} T^2 - 2 \times 10^{-4} T + 3.4 \times 10^{-2} \text{ at } \varepsilon_p = 1\% \\
&= 3.4 \times 10^{-7} T^2 - 2.3 \times 10^{-4} T + 3.9 \times 10^{-2} \text{ at } \varepsilon_p = 4\% \qquad (5.181)
\end{aligned}
$$

$$
\begin{aligned}
k_{\text{nf}} &= 0.003352T - 0.3708(\text{W/m}\cdot\text{K}) \text{ at } \varepsilon_p = 1\% \\
&= 0.004961T - 0.8078(\text{W/m}\cdot\text{K}) \text{ at } \varepsilon_p = 4\% \qquad (5.182)
\end{aligned}
$$

They observed that use of a variable property model predicts higher thermal and hydraulic performance. For example, the local wall temperature was found to be reduced when a variable property model is used, as shown in Fig. 5.57. Also, the average heat transfer coefficient is increased by the use of variable properties, and the wall shear stress is decreased with their use. This is encouraging because both thermal and hydraulic performances are enhanced, as shown in Figs. 5.58 and 5.59.

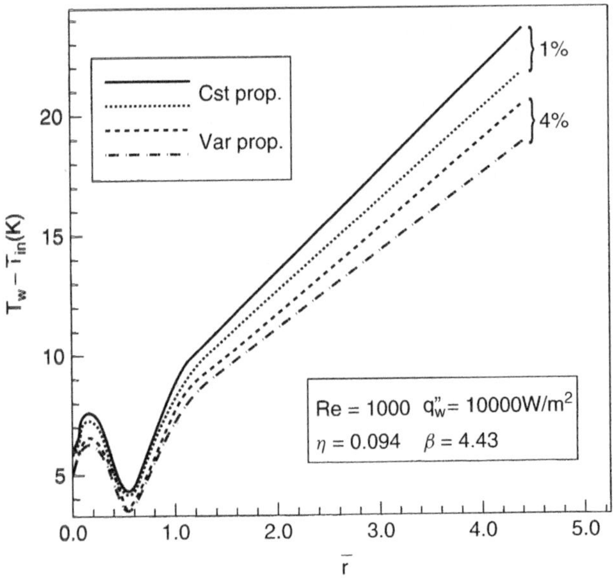

Fig. 5.57 Local wall temperature at different particle volumes for a radial flow channel. [From Palm et al. (2006), with permission from Elsevier.]

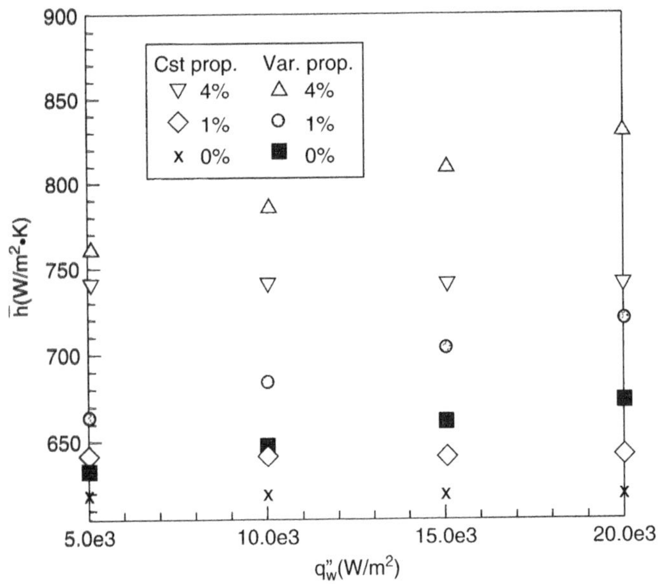

Fig. 5.58 Effect of variable properties on the average heat transfer coefficient of a radial channel. [From Palm et al. (2006), with permission from Elsevier.]

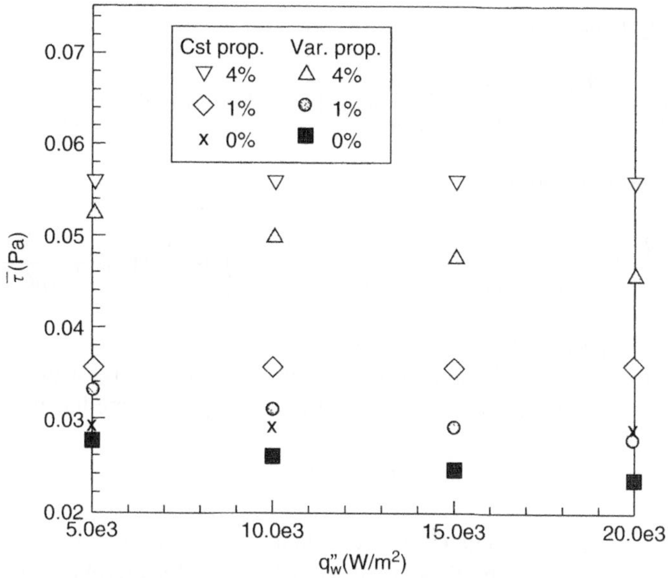

Fig. 5.59 Effect of variable properties on the average wall shear stress of a radial channel. [From Palm et al. (2006), with permission from Elsevier.]

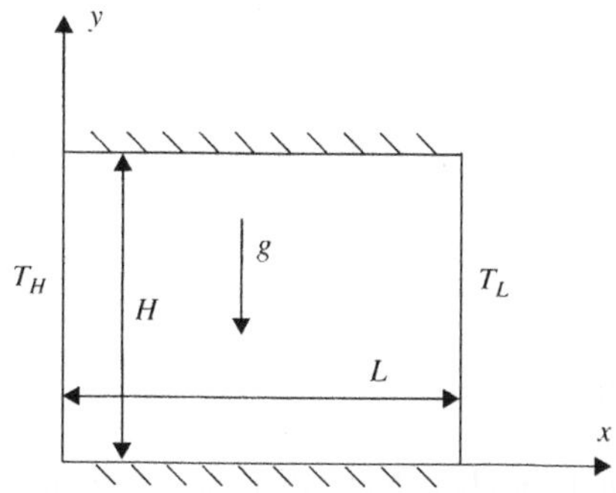

Fig. 5.60 Differentially heated cavity for natural convection. [From Khanafer et al. (2003), with permission from Elsevier.]

Numerical work on the natural convection of nanofluids was carried out by Khanafer et al. (2003). They carried out their work on a differentially heated cavity with hot and cold vertical walls and an adiabatic horizontal wall (Fig. 5.60). They used the well-known stream function vorticity formulation in which the primary variables are replaced by stream function ψ and vorticity ω, mainly to avoid the pressure term in the momentum equation, which requires complicated algorithms in incompressible flow. This gives the kinematics equation,

$$\frac{\partial^2 \psi}{\partial x^2} + \frac{\partial^2 \psi}{\partial y^2} = -\omega \tag{5.183}$$

the vorticity equation,

$$\frac{\partial \omega}{\partial t} + u\frac{\partial \omega}{\partial x} = v\frac{\partial \omega}{\partial y} = \frac{\mu_{\text{eff}}}{\rho_{\text{nf},0}}[\phi\rho_{s,0}\beta_s + (1-\phi)\rho_{f,0}\beta_f]g\frac{\partial T}{\partial x} \tag{5.184}$$

and the Energy equation,

$$\frac{\partial T}{\partial t} + u\frac{\partial T}{\partial x} + v\frac{\partial T}{\partial y} = \frac{\partial}{\partial x}\left[\left(\alpha_{\text{nf}} + \frac{k_d}{(\rho C_p)_{\text{nf}}}\right)\frac{\partial T}{\partial x}\right] + \frac{\partial}{\partial y}\left[\left(\alpha_{\text{nf}} + \frac{k_d}{(\rho C_p)_{\text{nf}}}\right)\frac{\partial T}{\partial y}\right] \tag{5.185}$$

where $\phi = \varepsilon_p$ (particle-volume fraction).

In this formulation, along with the nanofluid conductivity, a dispersive conductivity k_d is assumed. Following the theory of porous media, this term was modeled as

$$k_d = C(\rho C_p)_{\text{nf}}|\overline{V}|\phi\, d_p \tag{5.186}$$

where $|\overline{V}| = \sqrt{u^2 + v^2}$ and C is a constant. They used the Brinkman (1952) model for viscosity and the Wasp et al. (1977) model for thermal conductivity of nanofluids. They used the finite difference technique with the ADI algorithm and a power law scheme to solve the transient equations and validated them by comparison with the solutions of FIDAP software as well as the experimental value of pure fluids. Subsequently, they carried out studies on natural convection in a differentially heated cavity with nanofluids. Considerable differences in velocity and temperature were obtained between pure fluid and nanofluid as shown in Fig. 5.61. They used different modeling concepts with different thermal expansion coefficients and concluded that in general, there is a substantial increase in heat transfer in natural convection of nanofluids, as shown in Fig. 5.62.

They also presented a correlation for an average Nusselt number in the form

$$\overline{\text{Nu}} = (0.5163)(0.4436 + \varepsilon_p^{1.0809})\text{Gr}^{0.3123} \tag{5.187}$$

where Gr is the Grashof number. Note that the predictions of this work are in direct contradiction with the experimental observations of Putra et al. (2003), which needs to be explained.

Similar work was carried out by Jou and Tzeng (2006) inside a differentially heated cavity. They also used the stream function vorticity formulation in a way identical to that used in a previous study by Khanafer et al. (2003). In addition to the Grashof number effects, then described the effect of the cavity aspect ratio (width/height) on thermal behavior. Figure 5.63 shows the effects of the aspect ratio on the isotherms for natural convection in nanofluids with 20% particle concentration. However, it is difficult to use such results in practice because with a 20% volume fraction it is extremely difficult to make stable nanofluids. Also, at such volume fractions, Newtonian behavior of the fluid is doubtful.

A recent numerical study by Behzadmehr et al. (2007) paints a vivid picture of the effects of modeling strategy on turbulent flow simulation. They considered

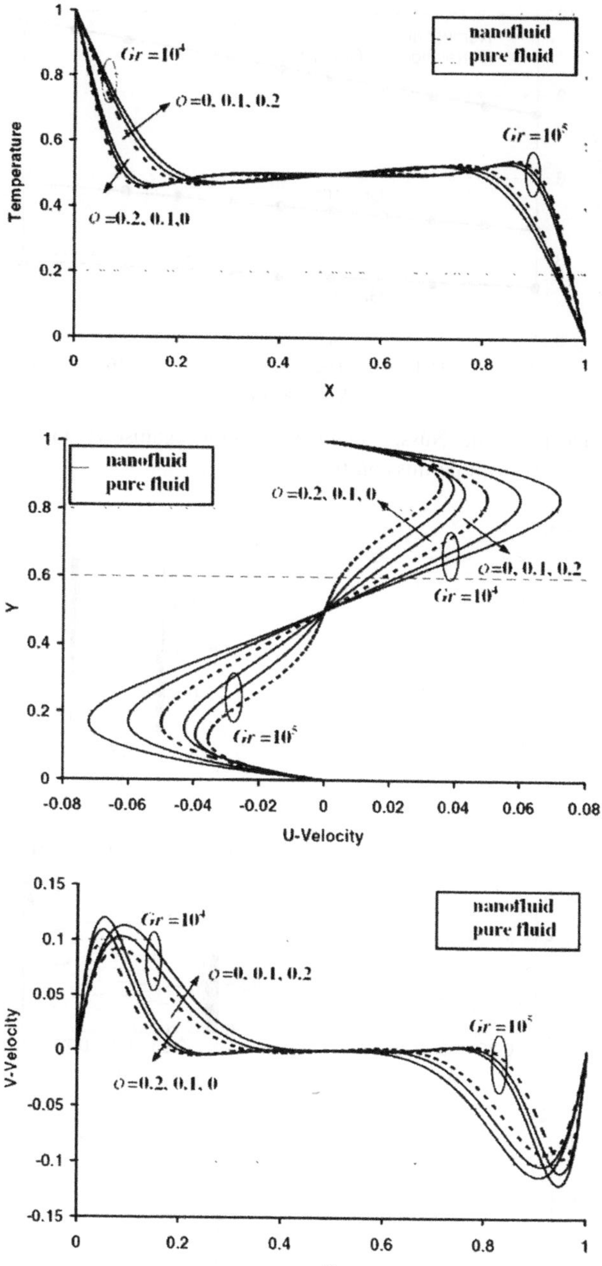

Fig. 5.61 Comparison of temperature and velocity profiles between nanofluid and pure fluid. [From Khanafer et al. (2003), with permission from Elsevier.]

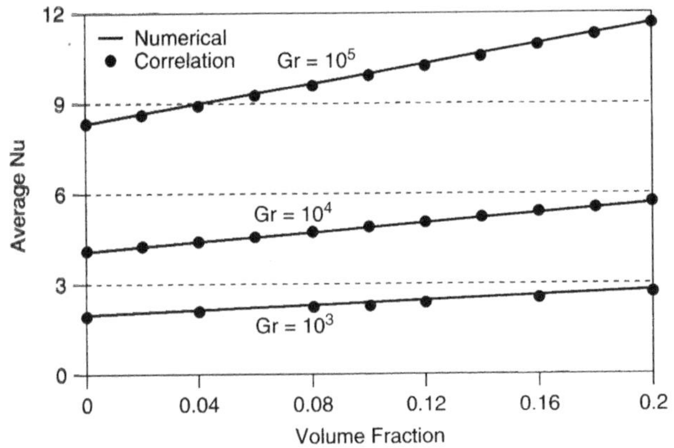

Fig. 5.62 Average nanofluid Nusselt number variation against volume fraction. [From Khanafer et al. (2003), with permission from Elsevier.]

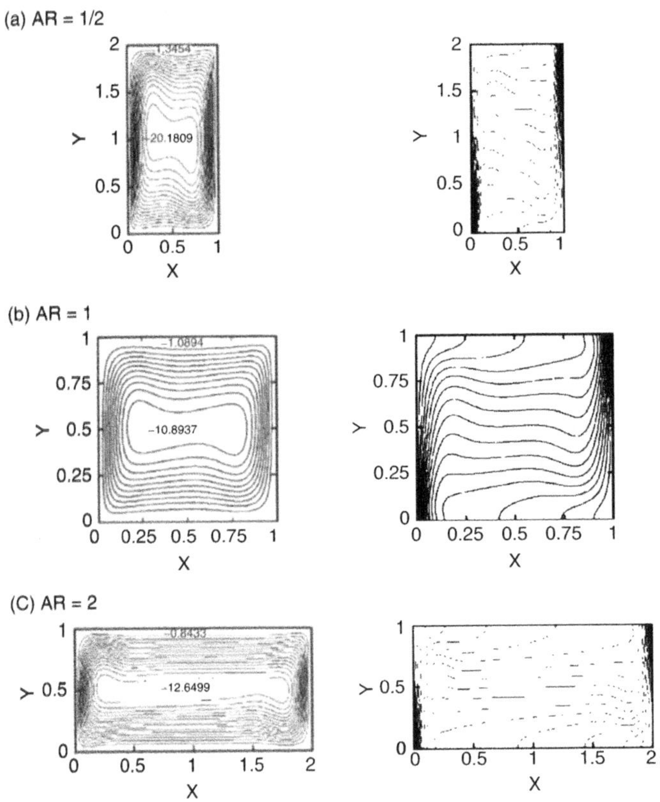

Fig. 5.63 Stream line and isotherm at various aspect ratios ($Gr = 10^5$, $Pr = 6.2$, volume fraction 20%) [From Jou and Tzeng (2006), with permission from Elsevier.]

two modeling concepts: the mixture and single-phase models. In the mixture model the fluid is considered to be a single fluid with two phases, and the coupling between them is strong. But each phase has its own velocity vectors, and within a given volume fraction there is a certain volume fraction of each phase. The governing equations are written for the mixture as

$$\nabla(\rho_m V_m) = 0 \tag{5.188}$$

$$\nabla \cdot (\rho_m V_m V_m) = -\nabla p_m + \nabla \cdot (\tau - \tau_l) + \rho_m g + \nabla \cdot \left(\sum_{k=1}^{n} \phi_k \rho_k V_{\mathrm{dr},k} V_{\mathrm{dr},k} \right) \tag{5.189}$$

$$\nabla \cdot (\phi_p \rho_p V_m) = -\nabla \cdot (\phi_p \rho_p V_{\mathrm{dr},p}) \tag{5.190}$$

$$\nabla \cdot [\phi_k V_k(\rho_k h_k + p)] = \nabla \cdot (k_{\mathrm{eff}} \nabla T - C_p \rho_m v T) \tag{5.191}$$

where m stands for mixture and k for the kth phase. Here V_{dr} is the drift velocity of phase k given by

$$V_{\mathrm{dr},k} = V_k - V_m \tag{5.192}$$

Mixture density and viscosity are given by

$$\rho_m = \sum_{k=1}^{n} \phi_k \rho_k \tag{5.193}$$

$$\mu_m = \sum_{k=1}^{n} \phi_k \mu_k \tag{5.194}$$

and the shear relation by

$$\tau = \mu_m \nabla V \tag{5.195}$$

$$\tau_l = -\sum_{k=1}^{n} \phi_k \rho_k \overline{V_k' V_k'} \tag{5.196}$$

Here V' and T' are fluctuating components of V and T in turbulent flow. The drift velocity is calculated from the relative velocity:

$$\mathrm{V}_{pf} = \mathrm{V}_p - \mathrm{V}_f \tag{5.197}$$

V_p is the particle velocity and V_f the primary phase velocity:

$$V_{\mathrm{dr},p} = V_{pf} - \sum_{k=1}^{n} \frac{\phi_k \rho_k}{\rho_m} V_{fk} \tag{5.198}$$

V_{pf} is calculated as

$$V_{pf} = \frac{\rho_p d_p^2(\rho_p - \rho_m)}{18\mu_f f_{\text{drag}}\rho_m} a \tag{5.199}$$

where

$$f_{\text{drag}} = 1 + 0.15\text{Re}_p^{0.687}, \qquad \text{Re}_p \leq 1000$$

$$= 0.0183\text{Re}_p, \qquad \text{Re}_p > 1000$$

$$a = g - (V_m \cdot \nabla)V_m$$

In addition the equations for turbulent kinetic energy (k) and its dissipation rate (ε) are also set:

$$\nabla \cdot (\rho_m V_m k) = \nabla \cdot \left(\frac{\mu_{t,m}}{\sigma_k}\nabla k\right) + G_{k,m} - \rho_m \varepsilon \tag{5.200}$$

$$\nabla \cdot (\rho_m V_m \varepsilon) = \nabla \cdot \left(\frac{\mu_{t,m}}{\sigma_\varepsilon}\nabla \varepsilon\right) + \frac{\varepsilon}{k}(C_1 G_{k,m} - C_2\rho_m\varepsilon) \tag{5.201}$$

where the turbulent viscosity is given by

$$\mu_{t,m} = \rho_m C_\mu \frac{k^2}{\varepsilon} \tag{5.202}$$

and $G_{k,m} = \mu_{t,m}(\nabla V_m + (\nabla V_m)T)$, $C_1 = 1.44$, $C_2 = 1.92$, $C_\mu = 0.09$, $\sigma_k = 1$, and $\sigma_\varepsilon = 1.3$. The boundary conditions can be set for tube flow considering the cylindrical coordinate, as at the entrance ($z = 0$) $Vz = V_0$, $V_\theta = V_r = 0$, $T = T_0$, *and* $I = I_0$, where I is the turbulence intensity and I_0 is its value at the tube entrance. Under an assumption of isotropic turbulence, the turbulent kinetic energy at the entrance is given by

$$k_0 = 1.5(I_0 V_0)^2 \tag{5.203}$$

At the tube outlet ($Z = L$) the diffusive flux in the axial direction should vanish. At the fluid wall interface ($r = D/Z$),

$$V_r = V_\theta = Vz = 0, \qquad k = \varepsilon = 0, \qquad q_w = -k_{\text{eff}}\frac{\partial T}{\partial r} \tag{5.204}$$

With the equations and boundary conditions above, the numerical method can be carried out for the mixture model. For the single-phase model, the equations are even simpler. It is simply the continuity, momentum, and energy equation of a single fluid with its properties replaced by the effective nanofluid properties. Both models were solved by the finite volume technique (Patankar, 1980) with a second-order upwind scheme used for both diffusive and convective forms. The pressure field was evolved through the SIMPLE algorithm to ensure pressure–velocity coupling.

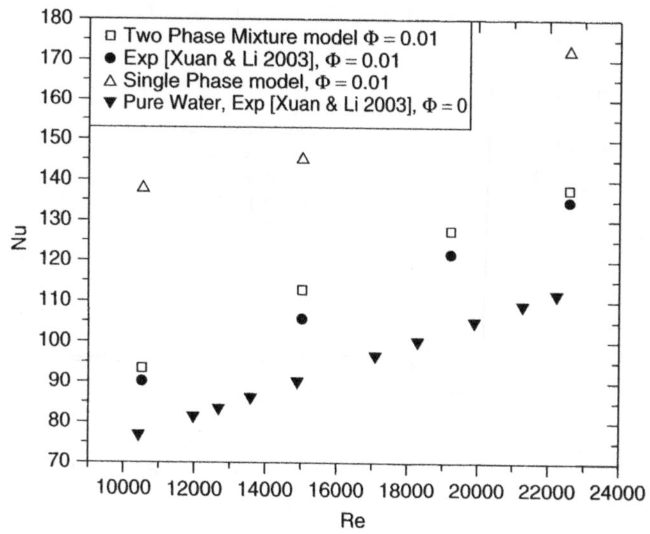

Fig. 5.64 Comparison of measured and calculated Nusselt numbers for nanofluid flow. [From Behzadmehr et al. (2007), with permission from Elsevier.]

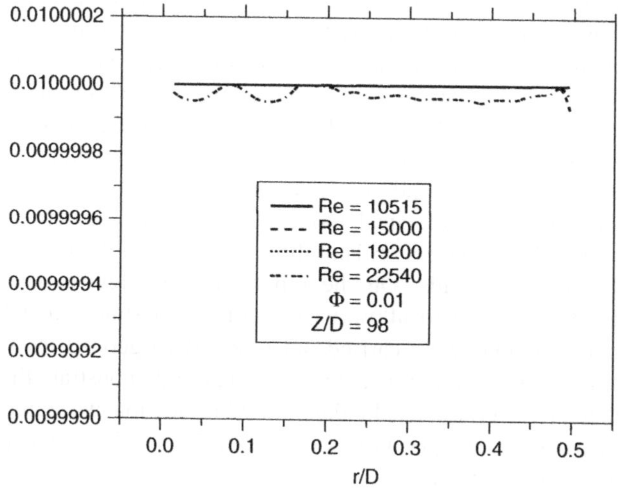

Fig. 5.65 Radial distribution of fully developed particle volume fraction. [From Behzadmehr et al. (2007), with permission from Elsevier.]

The results presented by Behzadmehr et al. (2007) clearly indicated the success of the mixture model over the single-phase model in predicting the Nusselt number data produced by Xuan and Li (2003) for water–Cu nanofluids (Fig. 5.64). They clearly indicated that for high values of Re ϕ^{-1}, the assumption of uniform particle distribution is not valid, which can be seen from Fig. 5.65. The work

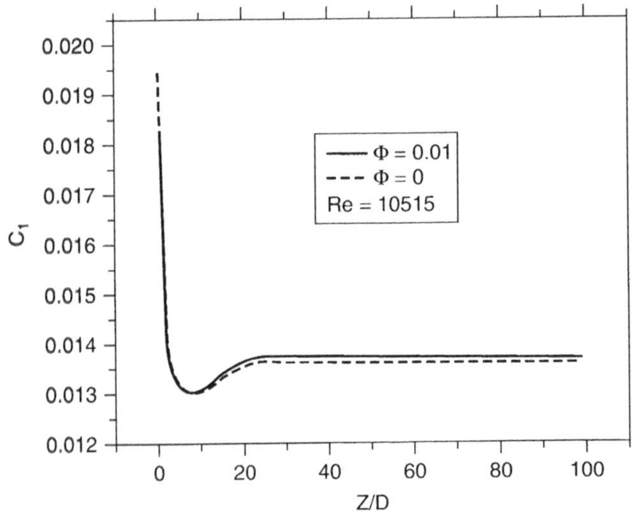

Fig. 5.66 Frictional behavior of pure and nanofluids. [From Behzadmehr et al. (2007), with permission from Elsevier.]

also agreed with the observation of Xuan and Li (2003) that the nanoparticles do not have a significant effect on the frictional behavior of the fluid (Fig. 5.66).

Thus, the numerical works discussed above considered an entire range of theoretical concepts for simulation. This includes the two-phase and single-phase models, the dispersion model, and the Eulerian and Lagrangian approaches. One of the major issues in these simulations is found to be the evaluation of thermophysical properties in general and viscosity and thermal conductivity in particular for nanofluids. The use of classical models such as the Einstein or Brinkman for viscosity and the Maxwell or Hamilton–Crosser for thermal conductivity is questionable for nanofluids. On the other hand too few experimental data on nanofluids are available to allow us to settle on a single model. However, most of the computations were carried out in standard geometries such as pipe flow or flow inside differentially heated enclosures to demonstrate the nanoeffect for well-benchmarked problems. In the following section we demonstrate one such computational process that brings out the utility of the convective features observed in standard geometries.

5.6. CONVECTIVE SIMULATION FOR CHIP COOLING APPLICATION

The convective features of nanofluids discussed in this chapter can be shown to be effective for a practical problem such as cooling an electronic chip. This example demonstrates how the heat sink of a normal computer chip can bring about enhanced cooling, of a nanofluid, particularly in the entrance region.

A 3-GHz Pentium IV computer chip generates 70 to 120 W of heat over a 35 mm × 35 mm footprint, requiring dissipation of about 9×10^4 W/m^2. A copper heat sink (35 mm × 35 mm × 5 mm) mounted over the chip helps in dissipating heat from the chip. The heat sink consists of 10 holes (2 mm in diameter) drilled along the length. A laminar developing flow is maintained in the passages, which eventually takes heat away from the sink. A schematic representation of the problem is shown in Fig. 5.67.

We want to evaluate the heat transfer capabilities of nanofluid using the heat sink. Apart from thermal conductivity enhancement several factors may affect heat transfer enhancement using nanofluids. The phenomena of Brownian diffusion, particle migration, and dispersion may exist in the flow of a nanofluid. Thus, a modified single-phase model, taking into the account some of the foregoing factors, is considered. We also observe that heat transfer enhancement is more prominent in the entrance region when using nanofluids. A numerical scheme that takes this fact into account is modeled. The nanofluid is assumed to behave as a single-phase fluid with changed thermal properties. To account for the disorderly movement of particles in enhancing heat transfer, thermal conductivity was further augmented using a thermal dispersion model. The numerical scheme is then extended to the chip cooling problem by changing appropriate boundary conditions. A commercial CFD software package, FLUENT 6.1, is used which adopts a control volume–based approach to solving the governing continuity momentum and energy equation. The laminar viscous model was used because the Reynolds number was low. A careful grid independence study showed that grid points with a total of 480,000 nodes were sufficient to predict temperature distribution in the heat sink. For the flow in the tube, uniform velocity boundary conditions were imposed at the inlet, and a pressure outlet boundary (equal to the atmosphereic pressure) was used at the exit. For the treatment of pressure, the SIMPLE algorithm was adopted. For higher accuracy second-order unwinding for convection terms and the central difference for diffusive terms were utilized. Also, it was ensured that in all cases, a converged solution was obtained.

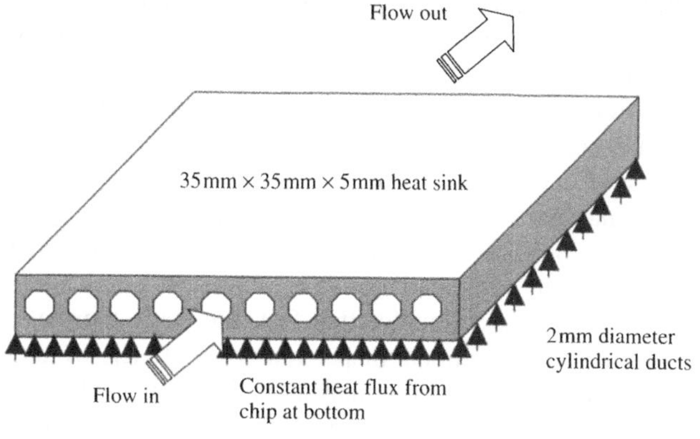

Fig. 5.67 Schematic showing heat sink and boundary conditions.

In this numerical investigation the nanofluid is assumed to behave as a single-phase fluid with a local thermal equilibrium between the base fluid and the nanoparticles suspended in it. Thus, all the governing equations of mass, momentum, and energy can be applied to nanofluids, as in the case of pure fluids, by changing the appropriate thermal and physical properties.

The effective density of a nanofluid containing suspended particles at a reference temperature is given by

$$\rho_{\mathrm{nf}} = (1 - \Phi)\rho_{bf} + \Phi\rho_p \tag{5.205}$$

This is in line with the property determination of two-phase mixtures. Similarly, the specific heat value of a nanofluid is given as

$$C_{p_{\mathrm{nf}}} = (1 - \Phi)C_{p_{\mathrm{bf}}} + \Phi C_{p_p} \tag{5.206}$$

Einstein's formula (Drew and Passman, 1999) for evaluating the effective viscosity of a fluid containing a dilute suspension of small rigid spherical particles is given as

$$\mu_{\mathrm{nf}} = \mu_{\mathrm{bf}}(1 + 2.5\Phi) \tag{5.207}$$

The most important property determining the heat transfer characteristic is the effective thermal conductivity of nanofluids. In the present study a microconvection model by Patel et al. (2005) taking into account the entire set of factors above is considered for evaluation of thermal conductivity of the nanofluid. The percentage enhancement in thermal conductivity is given as

$$\%\text{Enhancement} = \left(\frac{k_{\mathrm{eff}}}{k_{\mathrm{bf}}} - 1\right) \times 100 = \frac{k_p}{k_{\mathrm{bf}}}\left(1 + c\frac{u_p d_p}{\alpha_{\mathrm{bf}}}\right)\frac{d_m}{d_p}\frac{\phi}{1-\phi} \times 100 \tag{5.208}$$

where d_m is the molecular size of the liquid, u_p the Brownian motion velocity, α_{bf} the thermal diffusivity of the liquid, d_p the diameter of the particle, and c is a constant determined empirically. $c = 25{,}000$ was found to give very accurate predictions for a wide range of experiments. The thermal conductivity determined from the model above is denoted by k_{static}.

From observations of the experimental data of Wen and Ding (2004), it can be noted that the enhancement in heat transfer is significantly higher at the entrance region and decreases with axial distance. On further investigation it can be seen that maximum enhancement of thermal conductivity is about 10% from both the microconvection model and experiments (Wen and Ding, 2004) whereas the enhancement in the convective heat transfer is much higher that this in the entrance region. To account for this fact, the static thermal conductivity (k_{static}) is augmented with additional dispersion conductivity, which is similar to the approach adopted in heat transfer analysis in a porous medium (refer to Hsu and Cheng, 1990; Xuan and Roetzel, 2000; Khanafer et al. (2003)). This means that the slip velocity between the particle and the fluid is not zero, even though the particle sizes are very low. Particle migration that could result in nonuniform particle concentration has also been reported in the literature (Drew and Passman, (1999)). From the numerical exercises conducted in the present analysis it was observed that dispersion in the radial direction was more prominent when dealing with entrance region heat transfer. Then the effective thermal conductivity of the nanofluid will take the form

$$k_{\text{nf}} = k_{\text{static}} + k_d \tag{5.209}$$

$$k_d = C(\rho g p)_{\text{nf}}|v|\Phi d_p \tag{5.210}$$

where C is an unknown constant that is to be determined by matching with experimental data. It has to be noted here that the radial component of the velocity is taken for evaluation of the dispersion thermal conductivity, as it is prominent in the entrance region. In the present investigation this constant, C, was estimated to be $7.5\ x\ 10^6$.

Before carrying out the numerical scheme developed for the chip cooling problem, nanofluid flow through a tube subjected to a constant heat flux was simulated to evaluate the constant C by matching the experimental results of Wen and Ding (2004). A two-dimensional axisymmetric tube with a diameter of 4.5 mm and a length of 970 mm is considered subjected to a wall heat flux of 20,000 W/m^2. Particle concentration was also varied. The nominal diameter of the particles was taken to be 35 nm. The Reynolds number varied over a range in the laminar regime. The local heat transfer coefficient and Nusselt number is given as

$$h(x) = \frac{q}{T_w(x) - T_f(x)} \tag{5.211}$$

where T_w and T_f are the wall and fluid temperatures respectively. The wall temperature is determined from the numerical simulation and the fluid temperature profile is obtained through the energy balance:

$$T_f(x) = T_{\text{in}} + \frac{qSx}{\rho C_p u A} \tag{5.212}$$

where A and S are the cross-sectional area and perimeter of the test tube and u is the average fluid velocity. Since the thermal transport properties are functions of temperature, the properties were evaluated at the bulk fluid temperature during the complete analysis. The properties of the fluid were varied with the help of the user-defined function (UDF) facility in FLUENT.

Variation, in the ratio of the heat transfer coefficient of the nanofluid to that of the base fluid is plotted along the length of the tube. Figure 5.68 shows this variation for a Reynolds number of 1050. The inclusion of a thermal conductivity dispersion term is found more appropriate while analyzing the entrance region. Further validation at a higher Reynolds number of 1600 is made as shown in Fig. 5.69. From the combined observation of Figs. 5.68 and 5.69, it can be

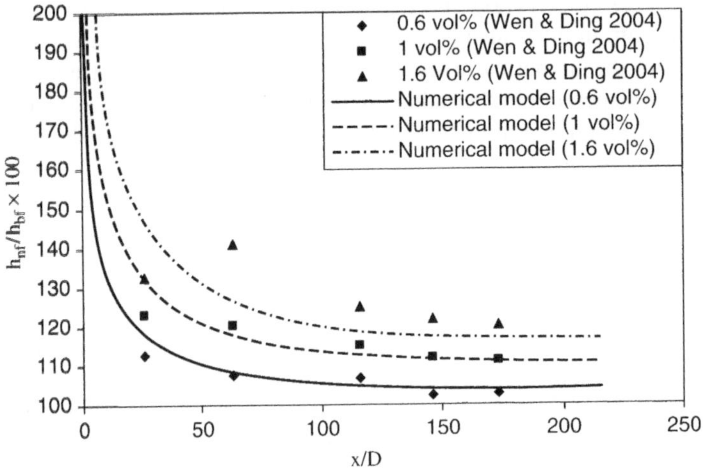

Fig. 5.68 Heat transfer enhancement ratio versus. nondimensional axial distance (Re = 1050).

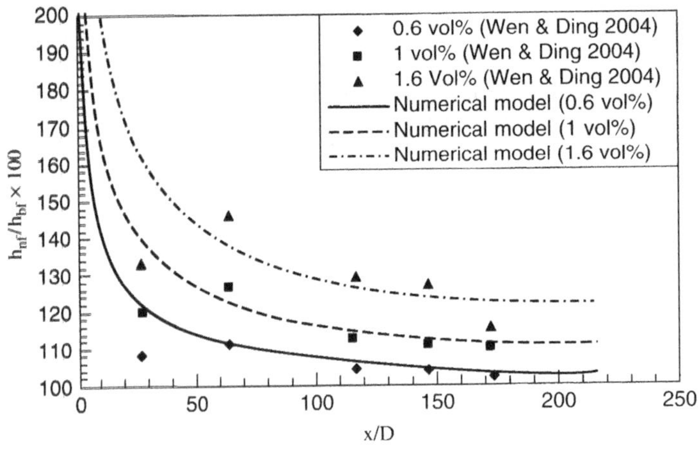

Fig. 5.69 Heat transfer enhanced ratio versus nondimensional axial distance (Re = 1600).

inferred that the numerical model is working well within the laminar range and is able to predict the heat transfer behavior in the entrance region. Figure 5.68 also shows the effect of the particle volume fraction on heat transfer enhancement. It can be noted that the enhancement in heat transfer due to particle concentration is more significant at the entrance region, which is well captured by the present model. In all the numerical simulations, the constant C in the dispersion thermal conductivity term was 7.5×10^6. After evaluating the constant C, the numerical model is extended in applying to the chip cooling problem. The chip was simulated by giving constant heat flux at the bottom of the heat sink. In the problem defined, a constant heat flux of 98,000 W/m^2 was subjected to the bottom of the heat sink and the conduction and convection in the cooling block and the channels were treated as a conjugate problem. All other external surfaces were considered to be adiabatic. The fluid flowing through the channels take away the heat coming from the chip. Different mass flow rates were considered in the simulation with the inlet temperature of the fluid to the chip kept constant (300 K) in all cases. The heat transfer performance of the nanofluid is evaluated by measuring the centerline surface temperature at the bottom of the heat sink along the flow direction. Here T_s denotes the centerline surface temperature and x, the coordinate in the flow direction.

Figure 5.70 shows the nondimensional surface temperature variation along the flow direction for a total mass flow rate of 2.8×10^{-3} kg/s divided equally among the 10 channels. It can be observed that the nanofluid reduced the surface temperature considerably (which in turn leads to cooling of the chip). It can also be observed from the figure that as the particle concentration is increased, the surface temperature is reduced. Figure 5.71 shows the axial variation of the surface temperature nondimensionalized with the maximum surface temperature of a flow, with a mass flow rate of to 2.8×10^{-3} kg/s. It is clear from the figure that the temperature of the surface increases in the flow direction. A thorough examination also indicates that heat transfer is greater for a nanofluid than for water at the same flow rate in the entrance region.

The thermal resistance variation of the heat sink for a particular power input(120 W) is plotted against the flow rate in Fig. 5.72. The thermal resistance of the heat sink is defined as resistance$_{th} = (T_{s,max} - T_{in})/Q$. It is seen that the nanofluid reduces the thermal resistance of a given heat sink for a particular flow

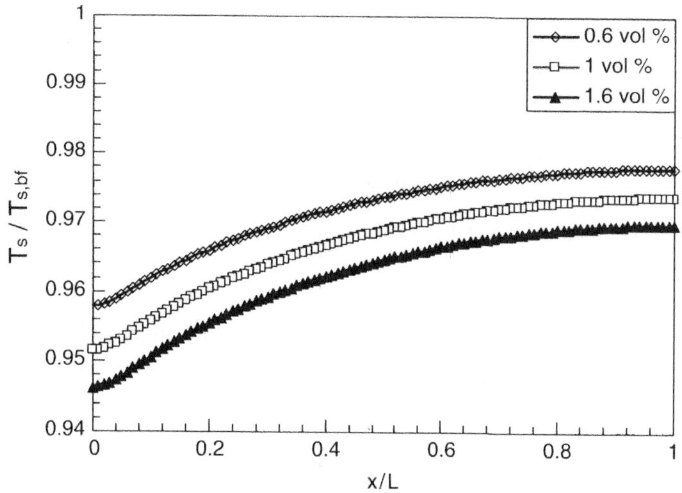

Fig. 5.70 Axial variation of nondimensional centerline surface temperature for a mass flow rate of 2.8×10^{-3} kg/s.

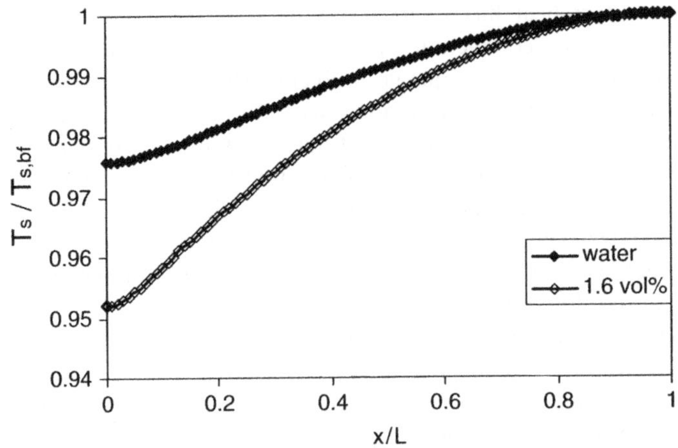

Fig. 5.71 Axial variation of the nondimensional centerline surface temperature (with maximum surface temperature) for a mass flow rate of 2.8×10^{-3} kg/s.

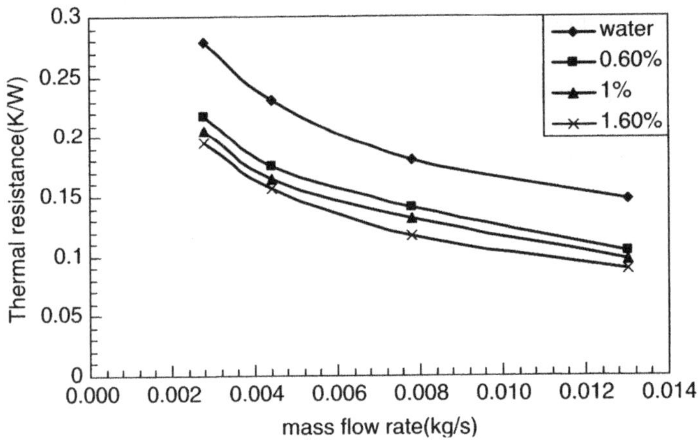

Fig. 5.72 Thermal resistance of the heat sink for a power input of 120 W.

rate. Thermal resistance is also reduced with an increase in particle concentration, as is evident from the same figure.

Thus, it is clear that the effect of a nanofluid is more predominant in the entrance region. Increasing the particle concentration always increased the heat transfer properties of the nanofluid. It was also observed that the inclusion of nanoparticles in the base fluid reduced the thermal resistance of the heat sink. From these observations it can thus be concluded that nanofluids promise to be prominent candidates for cooling of electronic equipment.

REFERENCES

Aris, R. (1956). On the dispersion of a solute in a fluid flowing through a tube, *Proc. Roy. Soc.* (London) A235, 67–77.

Batchelor, G. K. (1977). The effect of Brownian motion on the bulk stress in a suspension of spherical particles, *J. Fluid Mech.*, 83(1): 97–117.

Beckman, L. V., V. J. Law, R. V. Bailey, and D. U. von Rosenberg (1990). Axial dispersion for turbulent flow with a large radial heat flux, *AIChE J.*, 36: 598–604.

Behzadmehr, A., M. Saffar-Avval, and N. Galanis (2007). Prediction of turbulent forced convection of a nanofluid in a tube with uniform heat flux using a two phase approach, *Int. J. Heat Fluid Flow*, 28(2), 211–219.

Bird, R. B., W. E. Stewart, and E. N. Lightfoot (1960). *Transport Phenomena*, Wiley, New York.

Brinkman, H. C. (1952). The viscosity of concentrated suspensions and solution, *J. Chem. Phys.*, 20: 571–581.

Buongiorno, J. (2006). Convective transport in nanofluids, *J. Heat Transfer*, 128: 240.

Chandrasekhar, S. (1961). *Hydrodynamic and Hydromagnetic Stability*, Oxford University Press, London.

Churchill, S. W., and M. Bernstein (1977). A correlating equation for forced convection in gases and liquids to a circular cylinder in cross flow, *J. of Heat Transfer*, 99: 300–306.

Churchill, S. W., and H. H. S. Chu (1975). Correlating equations for laminar and turbulent free convection from a horizontal cylinder, *Int. J. Heat Mass Transfer*, 18: 1049–1053.

Dankwert, P. V. (1953). Continuous flow systems, distribution of resistance times, *Chem. Eng. Sci.*, 2: 1–10.

Das, S. K., N. Putra, and W. Roetzel (2003). Pool boiling characteristics of nano-fluids, *Int. J. Heat and Mass Trans* 46(5): 851–862.

Ding, Y., and D. Wen (2005). Particle migration in a flow of nanoparticle suspensions, *Powder Technol*. 149(2–3): 84–92.

Ding, Y., H. Alias, D. Wen, and A. R. Williams (2006). Heat transfer of aqueous suspensions of carbon nanotubes (CNT nanofluids), *Int. J. Heat and Mass Transfer*, 49: 240–250.

Dittus, F. W., and L. M. K. Boelter (1930). Heat transfer in automobile radiators of the tabular type, *Univ. Calif. Berkeley, Publ. Eng.*, 2: 433.

Drew, D. A., and S. L. Passman (1999). *Theory of Multicomponent Fluids*, Springer-Verlag, Berlin.

Eubank, C. C., and W. S. Proctor (1951). Effect of natural convection on heat transfer with laminar flow in tubes, M.Sc. thesis in chemical engineering, Massachusetts Institute of Technology, Cambridge, MA.

Gnielinski, V. (1976). Equations for heat and mass transfer in turbulent pipe and channel flow, *Int. Chem. Eng.*, 16: 359–368.

Gosselin, L., and A. K. da Silva (2004). Combined heat transfer and power dissipation optimization of nanofluid flows, *Appl. Phys. Lett.*, 85(18): 4160–4162.

Hausen, H. (1943). VDIZ, *Beih. Verfahrenstech.*, 4: 91–98.

Heris, S. Z., S. G. Etemad, and M. N. Esfahany (2006). Experimental investigation of oxide nanofluids laminar flow convective heat transfer, *Int. Commun. Heat Mass Transfer*, 33(4): 529–535.

Hsu, C. T., and P. Cheng (1990). Thermal dispersion in a porous medium, *Int. J. Heat Mass Transfer*, 33: 1587–1597.

Jou, R., and S. Tzeng (2006). Numerical research of nature convective heat transfer enhancement filled with nanofluids in rectangular enclosures, *Int. Common. Heat Mass Transfer*, 3(6): 727–736.

Kabelac, S., and J. F. Kuhnke (2006). Heat transfer mechanisms in nanofluids: experiments and theory keynote lecture presented at the International Heat Transfer Conferences, Sydney, Australia, Aug. 13–18, 2006.

Kaviany, M. (1991). *Principles of Heat Transfer in Porous Media*, Springer-Verlag, New York.

Kaviany, M. (1994). *Convective Heat Transfer*, Springer-Verlag, New York.

Khanafer, K., K. Vafai, and M. Lightstone (2003). Buoyancy-driven heat transfer enhancement in a two-dimensional enclosure utilizing nanofluids, *Int. J. Heat Mass Transfer*, 46: 3639–3653.

Kim, K., and Y. Lee (2001). Hydrodynamic and heat transfer characteristic of glass bead–water flow in a vertical tube, *Desalination*. 233–243.

Kim, J., Y. T. Kang, and C. K. Choi (2004). Analysis of convective instability and heat transfer characteristics of nanofluids. *Phys. Fluids*, 16(7): 2395–2401.

Kim, J., Y. T. Kang, and C. K. Choi (2007). Soret and Dufour effects on convective instabilities in binary nanofluids for absorption application, *Int. J. Refrig*. 30(2), 323–328.

Lee, S., S. U. S. Choi, S. Li, and J. A. Eastman (1999). Measuring thermal conductivity of fluids containing oxide nanoparticles, *J. Heat Transfer*, 121: 280–289.

Li, A., and G. Ahmadi (1992). Dispersion and deposition of spherical particles from point sources in a turbulent channel flow, *Aerosol Sci. Technol.*, 16: 209–226.

Maiga, S. E., S. J. Palm, C. T. Nguyen, G. Roy, and N. Galanis (2005). Heat transfer enhancement by using nanofluids in forced convection flows, *Int. J. Heat Fluid Flow*, 26, 530–546.

Mansour, R. B. N. Galanis, and C. T. Nguyen (2007). Effect of uncertainties in physical properties on forced convection heat transfer with nanofluids, *Appl. Therm. Eng.*, 27(1): 240–249.

McAdams, W. H. (1954). *Heat Transmission*, 3rd ed., McGraw-Hill, New York.

Michaelides, E. E. (1986). Heat transfer in particulate flows, *Int. J. Heat Mass Transfer*, 29: 265–273.

Oliver, D. R. (1962). Effect of natural convection on viscous-flow heat transfer in horizontal tubes, *Chem. Eng. Sci.*, 17: 335–350.

Pak, B., and Y. I. Cho (1998). Hydrodynamic and heat transfer study of dispersed fluids with submicron metallic oxide particle, *Exp. Heat Transfer*, 11: 151–170.

Palm, S. J. G. Roy, and C. T. Nguyen (2006). At transfer enhancement with the use of nanofluids in radial flow cooling systems considering temperature-dependent properties, *App. Therm. Eng.*, 26(17–18): 2209–2218.

Patankar, S. V. (1980). *Numerical Heat Transfer and Fluid Flow*, Hemisphere Publishing Corporation, NY.

Patel, H. E. T. Sundararajan, T. Pradeep, A. Dasgupta, N. Dasgupta, and S. K. Das (2005). A micro-convection model for thermal conductivity of nanofluid, *Pramana J. Phys.*, 65(5): 863–869.

Petukhov, B. S. (1970). Heat transfer and friction in turbulent pipe flow with variable physical properties, in *Advances in Heat Transfer*, J. P. Harnett, and T. F. Irvine, Eds., Academic Press, New York, pp. 504–564.

Phillips, R. J., R. C. Armstrong, R. A. Brown, A. L. Graham, and J. R. Abbott (1992). A constitutive equation for concentrated suspensions that accounts for shear-induced particle migration, *Phys. Fluids, A Fluid Dyn.*, 4: 30–40.

Putra, N., W. Roetzel, and S. K. Das (2003). Natural convection of nano-fluids, *Heat and Mass Trans*, 39(8–9), 775–784.

Roetzel, W., X. Luo, and Y. Xuan (1993). Measurement of heat transfer coefficient and axial dispersion coefficient using temperature oscillation, *Exp. and Fluid Sci.* 7: 345–353.

Roy, G., C. T. Nguyen, and P. Lajoie (2004). Numerical investigation of laminar flow and heat transfer in a radial flow cooling system with the use of nanofluids, *Superlattices and Microstruct.*, 35(3–6): 497–511.

Saffman, P. G. (1965). The lift on a small sphere in a slow shear flow., *J. Fluid Mech.*, 22(2): 385–400.

Shah, R. K. (1975). Thermal entry length solutions for the circular tube and parallel plates, Paper HMT -11 -75. *Proc. 3rd National Heat Mass Transfer Conference* Indian Institute of Technology, Bombay, India, Vol. 1.

Sieder, E. N., and G. E. Tate (1936). heat transfer and pressure drop of liquids in tubes, *Ind. Eng. Chem.*, 28: 1429–1435.

Talbot, L., R. K. Cheng, R. W. Schefer, and D. R. Willis (1980). Thermophoresis of particles in heated boundary layer., *J. Fluid Mech*. 101: 737–758.

Taylor, G. I. (1953). Dispersion of soluble matter in solvent flowing through a tube, *Proc. R. Soc.* (London), A219, 186–203.

Tzeng, S. C., C. W. Lin, and K. D. Huang (2005). Heat transfer enhancement of nanofluids in rotary blade coupling of four-wheel-drive vehicles, *Acta Mech.* (printed in Austria), 179: 11–23.

Wang, X., X. Xu, and S. U. S. Choi (1999). Thermal conductivity of nanoparticles-fluid mixture, *J. Thermophys. Heat Transfer*, 13(4), 474–480.

Wasp, E. J., J. P. Kenny, and R. L. Gandhi (1997). *Solid–Liquid Flow Slurry Pipeline Transportation, Series on Bulk Materials Handling*, Trans Tech Publications Clausthal, Germany.

Wen, D., and Y. Ding (2004). Experimental investigation into convective heat transfer of nanofluids at the entrance region under laminar flow conditions, *Int. J. Heat Mass Transfer*, 47: 5181–5188.

Wen, D., and Y. Ding (2005). Formulation of nanofluids for natural convective heat transfer applications, *Int. J. Heat Fluid Flow*, 26: 855–864.

Whitaker, S. (1972). Forced convection heat transfer pipes calculations for flow in pipes, past flat plates, single cylinders, and for flow in packed beds and tube bundles, *AIChE J.*, 18: 361–371.

Yu, W., and S. U. S. Choi (2003). The role of interfacial layers in the enhanced thermal conductivity of nanofluids: a renovated Maxwell model, *J. Nanoparticle Res.*, 5, 167–171.

Xuan, Y., and Q. Li (2000). Heat transfer enhancement in nanofluids, *Int. J. Heat and Fluid Flow*, 21, 58–64.

Xuan, Y., and Q. Li (2003). Investigation on convective heat transfer and flow features of nanofluids, *J. Heat Transfer.*, 125(1): 151–155.

Xuan, Y., and W. Roetzel (2000). Conceptions for heat transfer correlation of nano-fluids, *Int. J. Heat Mass Transfer.*, 43: 3701–3707.

Yamamoto, K., and Y. Ishihara (1988). Thermophoresis of spherical particle in a rarefied gas of a transition regime, *Phys. Fluids*, 31: 3618–3624.

Yang, Y., Z. G. Zhang, E. A. Grulke, W. B. Anderson, and G. Wu (2005). Heat transfer properties of nanoparticle-in-fluid dispersions (nanofluids) in laminar flow, *Int. J. Heat Mass Transfer*, 48(6): 1107–16.

6 Boiling of Nanofluids

A primary interest with regard to nanofluids is on suspensions without phase change. However, a great deal of thermal and chemical process equipment involves heat transfer with phase change: liquid to vapor, or vice versa. In fact, there is hardly any chemical, power, or refrigeration process in which all the heat transfer processes take place with a single-phase liquid. Equipment with multiphase heat transfer requires special consideration, and runs the risk of overheating, leading to failure if proper care is not taken in design. The role of nanofluids in such equipment should prove interesting to investigate.

Heat transfer with phase change between the liquid and vapor states can take place in the form of boiling or condensation. Both these types of heat transfer are quite complex, involving different mechanisms, such as evaporation–condensation, transient conduction, local liquid agitation, and natural or forced convection. The physics of boiling and condensation processes is still only partially understood. A large amount of research is under way, as reviewed in articles such as those by Dhir (2000) and Rose et al. (1999). It is neither possible nor desirable to discuss two-phase heat transfer in detail before taking up the boiling of nanofluids. In the first section of the chapter we very briefly overview two-phase heat transfer and present correlations useful in the description of boiling heat transfer in nanofluids. These fundamentals will be important for readers from disciplines in which boiling is not a core subject. For details on phase-change heat transfer, readers are referred to texts such as that by Stephan (1992).

6.1. FUNDAMENTALS OF BOILING

Boiling is the process of changing liquid into vapor at a constant temperature known as the *saturation temperature*. This change can take place in a number of ways. Depending on the way in which phase change takes place, boiling is divided into two categories: pool boiling and flow boiling. This classification is somewhat analogous to classifying single-phase convection into external and internal convection.

Nanofluids: Science and Technology, By Sarit K. Das, Stephen U. S. Choi, Wenhua Yu, and T. Pradeep

6.1.1. Pool Boiling

Initially, pool boiling of a liquid takes place at a stagnant state, and the heated surface from which boiling starts is submerged in it. It should be mentioned here that observations of pool boiling also occur, at least qualitatively, when the liquid flow is imposed on the heating surface instead of a stagnant liquid. The pool boiling behavior of liquids can be better understood by plotting logarithms of the boiling heat flux against the difference in temperature between the wall of the heating surface and the bulk liquid. This temperature difference, known as the *wall superheat* is given by

$$\Delta T_w = T_w - T_{sat} \tag{6.1}$$

where T_{sat} is the saturation temperature of the liquid under a given pressure. The heat flux versus wall superheat curve for boiling on a submerged horizontal surface is known as the *Nukiyama* (1934) curve. The curve in Fig. 6.1 shows different behaviors in different regimes. The first part of the curve, at the lowest heat flux, merges with the natural convection curve (A to B) since the mechanism of heat transfer from the heated surface is by natural convection. This gives the following heat transfer coefficients:

$$h \propto \begin{cases} \Delta T^{1/4} & \text{for laminar flow} \\ \Delta T^{1/3} & \text{for turbulent flow} \end{cases} \tag{6.2}$$

The heat flux, q, is related to the heat transfer coefficient, h, as

$$q = h\Delta T_w \tag{6.3}$$

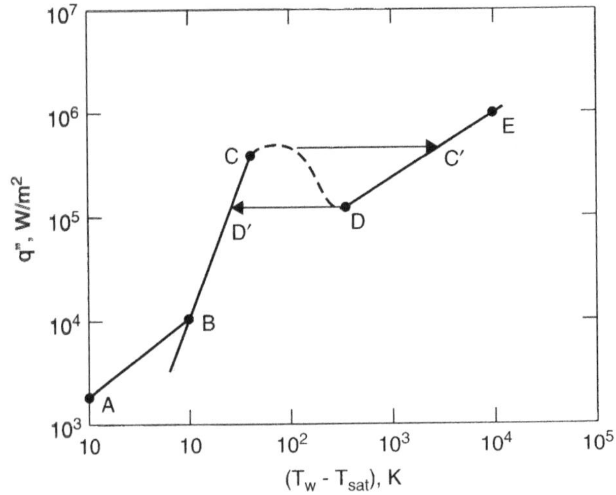

Fig. 6.1 Nukiyama curve for pool boiling.

resulting in the h–q correlation

$$h \propto \begin{cases} q^{1/5} & \text{for laminar flow} \\ q^{1/4} & \text{for turbulent flow} \end{cases} \tag{6.4}$$

However, evaporation from the top surface of the liquid takes place at this stage, and hence it is often called *quiet boiling*. The surface remains slightly (≈0.03 K for water at 1 bar) above the saturation temperature.

With the increase in wall superheat, bubbles start forming. They grow in size, detach from the surface, and move upward due to buoyancy. At lower heat flux the bubbles collapse within the liquid, but as the heat flux grows, the bubbles come to the surface of the liquid and burst. This creates vigorous agitation of the liquid. This curve (B to C) is called the *nucleate boiling curve*. This is the most desirable form of boiling, and most studies of pool boiling are concentrated in this area. The heat transfer correlation here is given by

$$h \quad \propto \Delta T_w^3 \tag{6.5}$$

This sudden change, with the increase in heat flux (the exponent of ΔT_{w} changes from 0.33 to 3.0) can be observed as a change in slope of the Nukiyama curve.

After nucleate boiling the heat flux reaches a maximum (point C), where a small change in heat flux causes a large change in the wall superheat. This point represents a phenomenon known as *burnout* because the sudden change in wall superheat may cause the heater surface to burn out. However, the heater does not always burn at this point, so the name has nothing to do with physical burn out. This point is also referred to by other names, such as *critical heat flux, Departure from Nucleate Boiling* (DNB), and *boiling crisis*.

In the next section the heat flux decreases with an increase in wall superheat (C to D). This is because the rate of evaporation becomes so high that it cannot be removed and the surface becomes covered increasingly by a vapor film (Fig. 6.2). This film appears and disappears, and hence this part (C to D) is called unstable *film boiling*.

The lowest point of the heat flux reached here (point D) is called the *Leidenfrost point*. At this point stable film occurs for the first time. In the subsequent section, during film boiling (D to E), the heat flux again increases due to radiation inside the vapor film, and finally, the heater surface burns out physically somewhere (point E).

It should be kept in mind that this entire curve is available only if the wall temperature of the heater is kept constant (e.g., by flowing a high-boiling-point saturated vapor inside the heater tube). For cases where the wall heat flux is constant (e.g., electrical heaters) a hysteresis will be seen with C moving to C' when the heat flux is increased and D moving to D' when the heat flux is decreased.

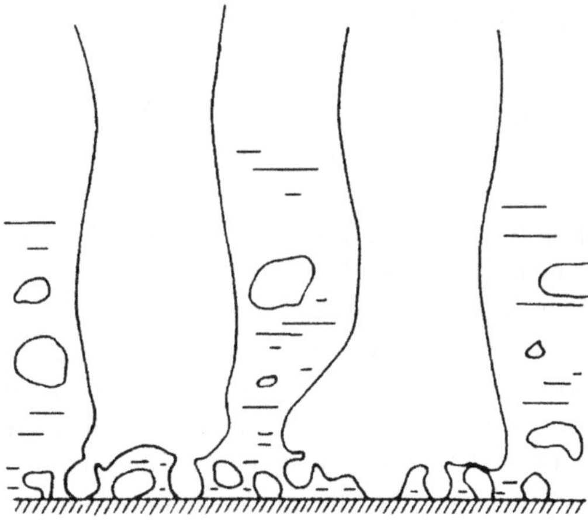

Fig. 6.2 Unstable film on the heater surface.

Nucleate Boiling: Nucleation, Bubble Growth, and Departure The most important part of pool boiling, often used in process heat transfer equipment, is the nucleate boiling. In this regime the bubble life cycle plays a major role in heat transfer. The processes follow in the order nucleation, bubble growth, bubble departure from the surface and collapse or further growth (ultimately, bursting) of the bubble (depending on whether the bulk liquid is subcooled or superheated).

Nucleation For a spherical bubble, at the surface of the bubble the pressure difference between the vapor and the surrounding liquid is balanced by the force of the surface tension (Fig. 6.3).

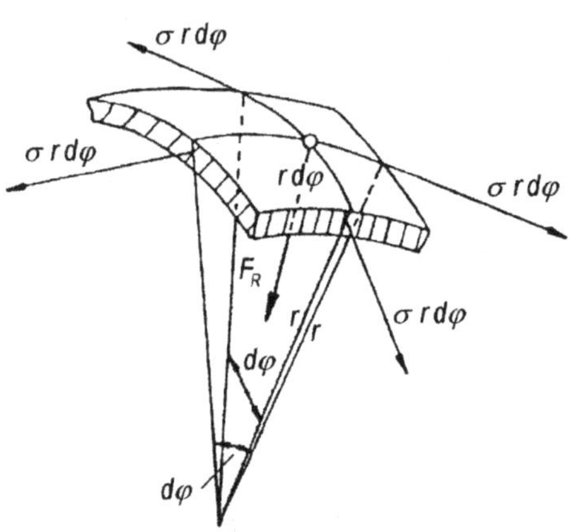

Fig. 6.3 Force balance at the liquid–vapor interface.

For equilibrium this force balance results:

$$P_G = P_L + \frac{2\sigma}{r} \tag{6.6}$$

where P_G is the vapor pressure, P_L the liquid pressure, r the bubble radius, and σ the surface tension.

Now since the bubble is in equilibrium with the liquid at the interface, the liquid and vapor temperatures are equal ($T_G = T_L$). This temperature is indicated in Fig. 6.4. But according to equation (6.6), the liquid pressure is less than the pressure of the vapor, which means that the liquid must remain in a superheated state (Fig. 6.4). If at the liquid pressure the saturation temperature is T_{sat}, the liquid superheat is given by

$$\Delta T = T_L - T_{sat} \tag{6.7}$$

To determine the amount of liquid superheat required, we can use the Clausius–Clapeyron equation of thermodynamics at the interface:

$$\frac{dP}{dT} = \frac{h_{fg}}{v_{fg} T_{sat}} \tag{6.8}$$

where h_{fg} is the latent heat and v_{fg} is the change in specific volume during phase change. For a differential amount this reduces to

$$\frac{P_G - P_L}{T_L - T_{sat}} = \frac{h_{fg}}{v_{fg} T_{sat}} \tag{6.9}$$

Combining this equation with (6.6) gives

$$T_L - T_{sat} = \frac{2\sigma v_{fg} T_{sat}}{r h_{fg}} \tag{6.10}$$

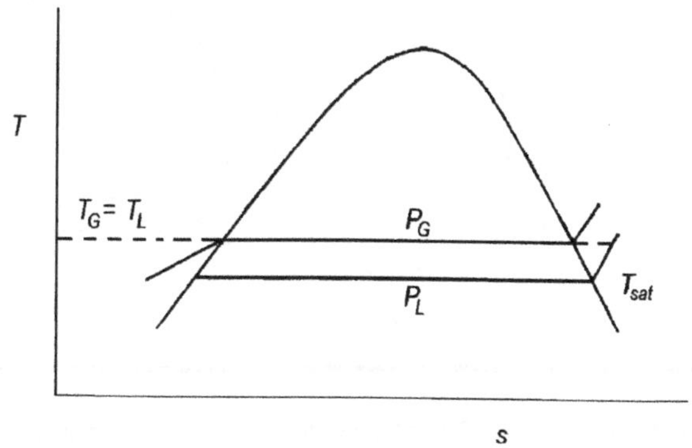

Fig. 6.4 Liquid superheat.

This shows that the liquid superheat required for a bubble to exist varies inversely with bubble diameter and directly as surface tension. Thus, surface tension plays a major role in boiling. For example, the boiling behavior of water is quite different from that of refrigerants that have much less surface tension and produce more smaller bubbles. However, this amount of superheat is not sufficient for inception of the nucleation process on a solid surface. For example, for water at atmospheric pressure the liquid superheat predicted by equation (6.10) is about 5°F, while in reality, on a solid surface bubbles appear only at a superheat of 14°F or more.

The is discrepancy was explained by Hsu and Graham (1976) using a nucleation theory. Hsu assumed that after each bubble departure, some vapor remains trapped within the cavity on the surface, which acts as a nucleation site. Even though cold liquid rushes toward the cavity, it cannot condense this residual vapor, and the residual vapor acts as a starting process in the next bubble growth. The bubble growth and departure process is shown in Fig. 6.5. This process of nucleation can be assessed quantitatively by considering a transient conduction in the liquid after a bubble departs, given by

$$\frac{T - T_{\text{sat}}}{T_w - T_{\text{sar}}} = 1 - \frac{n}{\delta} - \frac{2}{\pi}\sum_{n=1}^{\infty}\frac{1}{n}\sin n\pi\frac{n}{\delta}e^{-n^2\pi^2\alpha\tau/\delta^2} \tag{6.11}$$

where δ is the thickness of the superheated boundary layer on the solid wall. The liquid superheat required for nucleation on a conical cavity (Fig. 6.6) is given by

$$T - T_{\text{sat}} = \frac{2\sigma v_{fg} T_{\text{sat}}}{r h_{fg}} \tag{6.12}$$

Now it is obvious that nucleation will take place only when the liquid superheat given by equation (6.12) is reached through transient conduction in the liquid

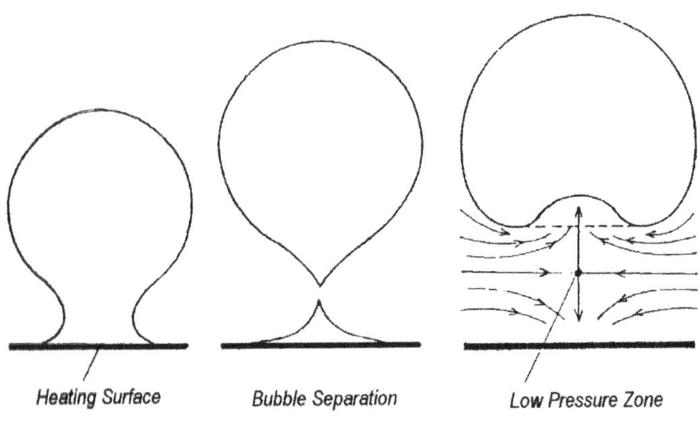

Fig. 6.5 Bubble growth and departure.

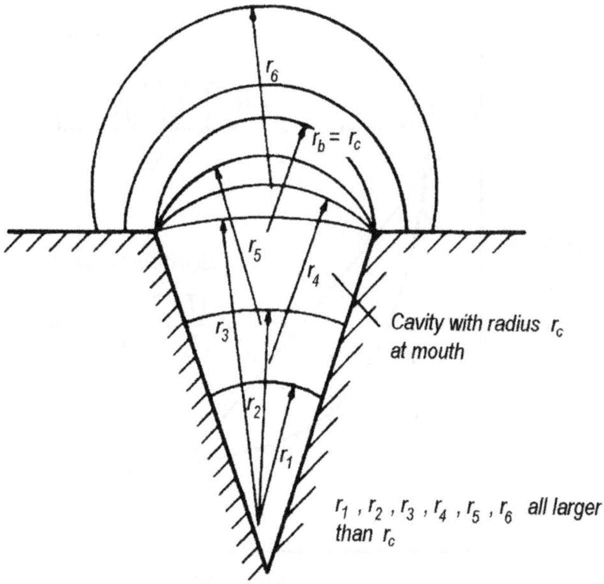

Fig. 6.6 Bubble formation on a conical cavity.

given by (6.11). Hence, nucleation can be obtained by the superposition of curves for these two equations (Fig. 6.7). It can be said that the intersection of equation (6.12) and (6.11) gives the bubble size, corresponding waiting period, and liquid superheat required.

Bubble Growth Bubble growth takes place in a number of stages. The initial phase, known as *isothermal bubble growth* is described by Rayleigh's (1917) equation:

$$\rho_L \left[r\frac{d^2r}{d\tau} + \frac{3}{2}\left(\frac{dr}{d\tau}\right)^2 \right] = \Delta P - \frac{2\sigma}{r} \tag{6.13}$$

The solution of this equation gives several types of bubble growth equations:

1. In the initial state the inertial forces dominate when acceleration and surface tension effects are insignificant, giving

$$r = \left[\frac{2/3(P_\infty^* - P_\infty)}{\rho_L} \right]^{1/2} \tau \tag{6.14}$$

2. At a later stage, heat diffusion becomes important and the energy term that appears in the growth correlation is quite complex.

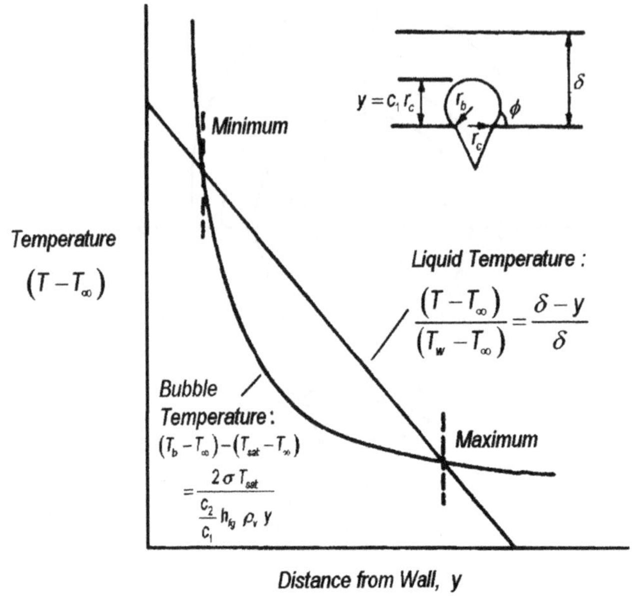

Fig. 6.7 Condition for nucleation on a surface.

Bubble Departure The departure of bubbles is an important phenomenon that influences boiling heat transfer. The first important parameter for a bubble is its departure diameter. A balance of buoyancy and surface tension can give this diameter. Physically, this signifies that a bubble on a horizontal surface departs when the buoyancy force exceeds the surface tension that has been keeping the bubble attached to the surface (Fig. 6.7). This gives the departure diameter known as the *Fritz* (1935) *formula:*

$$d_b = 2.449\theta\sqrt{\frac{\sigma}{g(\rho_L-\rho_G)}} \tag{6.15}$$

where θ is the contact angle (in radians) of the bubble on the surface.

At higher wall superheat values, other effects, such as inertia and convection, come into play and the departure diameter can be given by

$$d_b = 0.25\sqrt{\frac{\sigma}{g(\rho_L-\rho_G)}}\left[1+\left(\frac{Ja}{\mathrm{Pr}_L}\right)^2\frac{1}{\mathrm{Ar}_L}\right]^{1/2} \tag{6.16}$$

where Archimedes' number,

$$\mathrm{Ar} = \frac{g}{v_L^2}\left(\frac{\sigma}{\rho_L g}\right)^{3/2} \quad \text{and} \quad 5\times10^{-7} \leq \frac{\mathrm{Ja}}{\mathrm{Pr}_L}\frac{1}{\mathrm{Ar}} \leq 10^{-1}$$

and the Jacob number, $\mathrm{Ja} = \rho_L C_{pL} \Delta T / \rho_V h_{fg}$, where L and V stand for the liquid and vapor states. In the correlations above it can be seen that a length scale occurs called the *Laplace constant:*

$$L' = \sqrt{\frac{\sigma}{g(\rho_L - \rho_G)}} \tag{6.17}$$

This length scale appears in many boiling correlations where obvious length scales are absent.

The next important departure parameter is the *departure frequency*, the reciprocal of the time period between two nucleations:

$$f_b = \frac{1}{t_d + t_w} \tag{6.18}$$

where t_d is the departure time and t_w is the waiting period between two nucleations. Max Jakob (1949) was the first person to measure departure frequency photographically, suggesting the correlation

$$f d_b = C \tag{6.19}$$

Later, more improved correlations were evolved as

$$f d_b = V_b \tag{6.20}$$

where V_b is the bubble rise velocity and $t_w = 0$. For a finite waiting period,

$$f d_b \frac{t_w + t_d}{t_d} = V_b \tag{6.21}$$

V_b can be estimated as $= 1.18(\sigma g \Delta \rho / \rho_L{}^2)^{1/4}$. Usually, the frequency for the hydrodynamic (initial) and thermodynamic (later) period are different. Thus, departure frequency is a strongly regime- and property-dependent parameter.

Heat Transfer Mechanism in Nucleate Boiling Before going to heat transfer correlations it is important to investigate the mechanism of heat transfer during nucleate boiling. The analysis of bubble growth, nucleation, and bubble dynamics can act as important tools in constructing heat transfer models that agree with the experimentally observed heat flux versus wall superheat correlations. Subsequently, the models described below were built up on theoretical as well as experimental observations.

Microlayer Evaporationt Theory One of the most successful theories was presented by Moore and Mesler (1961). It postulates that there remains a thin liquid microlayer under the bubble which evaporates rapidly [Fig. 6.8(a)]. Measurement shows that on the boiling surface, temperature sometimes falls 10 to 17° C within a few milliseconds, which can only be explained by microlayer evaporation. Later interferometric studies proved the existence of microlayers.

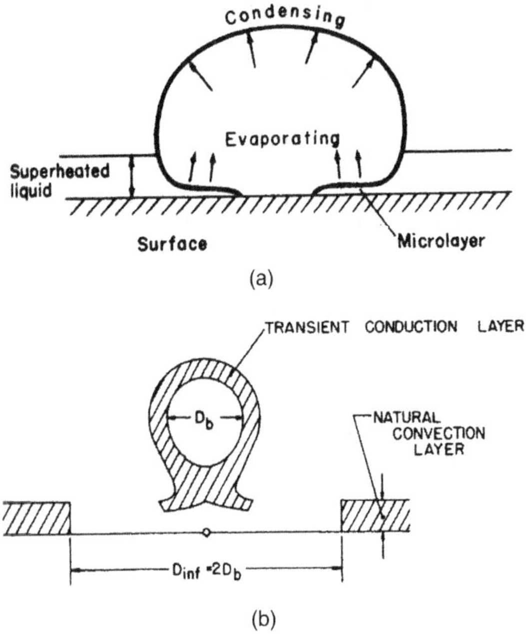

Fig. 6.8 Heat transfer models in boiling: (a) microlayer evaporation model; (b) transient conduction-dominated model.

Transient Conduction-Based Model This model was proposed by Mikic and Rohsenow (1969). The model considers a hot liquid layer on the wall formed by transient conduction. It is assumed that once the bubble departs, the hot boundary layer gets ruptured and subsequently reformed by transient conduction before the next bubble starts growing, as shown in Fig. 6.8. They proposed that the area which gets ruptured has a diameter twice that of a bubble. The major drawback of this model is that it does not take microlayer evaporation into consideration.

Composite Models The foregoing models bring out the truth about heat transfer in boiling but partially. They overestimate one contribution and underestimate the other. In recent times, there have been efforts to take a more balanced view. One such model is that of Benjamin and Balakrishnan (1996). They assumed a total boiling heat flux consisting of:

- Microlayer evaporation (q_{ME}) during bubble growth
- Transient conduction during thermal boundary layer reformation (q_{c})
- Natural convection in the area not influenced by bubbles (q_{nc})

The total heat flux is given by

$$q_{\mathrm{tot}} = \frac{q_{\mathrm{ME}} t_b + q_c t_w}{t_b + t_w} + q_{\mathrm{nc}} \tag{6.22}$$

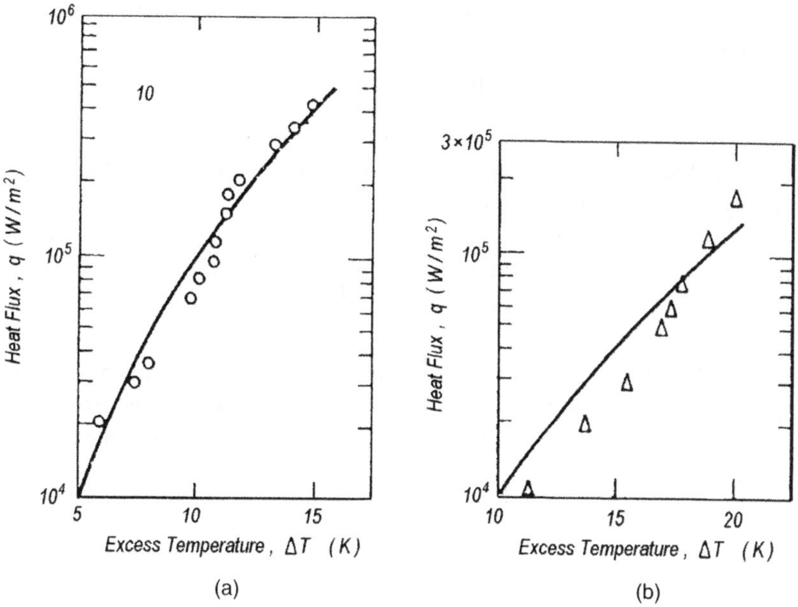

Fig. 6.9 Agreement of Benjamin and Balakrishnan (1996) model with experiment for (a) water and (b) acetone.

where t_b is the bubble growth period and t_w is the waiting period. The heat flux predicted by this model agrees excellently with experiment, as shown in Fig. 6.9. Das and Roetzel (2002) developed this model further by including the effect of sliding bubbles in the case of horizontal tubes.

Heat Transfer Correlations in Pool Boiling Until the 1950s, boiling heat transfer correlations were developed purely empirically. Max Jacob and Warren Rohsenow were the pioneers in this area who started looking at the physical picture of the boiling process and suggested correlations consistent with the physical phenomena. Correlations that are used extensively are based on experiments but take into consideration physical features of the boiling process as well. These correlations can be divided into the following groups.

Rohsenow Correlations Rohsenow identified boiling as primarily a convective process with liquid and vapor displacement. This naturally suggests a convective type of correlation given by

$$\mathrm{Nu} = C \cdot \mathrm{Re}^m \mathrm{Pr}^n \tag{6.23}$$

However, two questions arise: What is the Reynolds number, and what length scale is to be used to define Nusselt and Reynolds numbers? From previous sections it is obvious that the bubble departure diameter is an important parameter.

Hence, the boiling Reynolds number can be taken as the bubble inertia/liquid viscous force ratio:

$$\mathrm{Re}_b = \frac{\rho_G V_b d_b}{\mu_L} \tag{6.24}$$

Now the bubble velocity V_b can be taken as the rate of vapor volume generated divided by the area over which it is generated:

$$V_b = \frac{\text{volume}}{\text{area of surface}} = nf\pi\frac{d_b^3}{6} = \frac{q}{h_{fg}\rho_G} \tag{6.25}$$

where, n is the number of nucleation sites per unit area, f the frequency of bubble formation, and q the heat flux Substituting this correlation and the Fritz formula [equation (6.15) to the definition of Re_b [equation (6.24)] and adjusting θ and the constant into β, we get

$$\mathrm{Re}_b = \frac{q}{\mu_L h_{fg}}\beta\sqrt{\frac{2\sigma}{g(\rho_L - \rho_G)}} \tag{6.26}$$

The Nusselt number should also be based on the bubble departure diameter:

$$\mathrm{Nu}_b = \frac{hd_b}{k_L} = \frac{h}{k_L}\beta\sqrt{\frac{2\sigma}{g(\rho_L - \rho_G)}} \tag{6.27}$$

Rohsenow proposed that

$$\mathrm{Nu}_b = C(\mathrm{Re}_b)^{2/3}\mathrm{Pr}^{-0.7} \tag{6.28}$$

Later he found that it is more convenient to reduce it to the form

$$\frac{\mathrm{Re}_b \mathrm{Pr}_L}{\mathrm{Nu}_b} = C\mathrm{Re}_b^m \mathrm{Pr}_L^n \tag{6.29}$$

This can be simplified by using the definitions of h, Nu_b, Re_b, and Pr_L as

$$\frac{C_{\mathrm{PL}}(T_w - T_{\mathrm{sat}})}{h_{fg}} = C_{sf}\left[\frac{q}{\mu_L h_{fg}}\sqrt{\frac{\sigma}{g(\rho_L - \rho_G)}}\right]^{0.33}\left(\frac{C_{\mathrm{PL}}\mu_L}{k_L}\right)^n \tag{6.30}$$

where $n = 1.0$ for water and 1.7 for other fluids. Since β is absorbed in C_{sf}, it depends on the angle of contact of the bubble and has a different value for each combination liquid and surface. Typical values of C_{sf} are given in Table 6.1. This correlation is the famous *Rohsenow correlation*, which is used extensively. The performance of this correlation against experimental data is shown in Fig. 6.10.

Table 6.1 Values of C_{sf} for Rohsenow's Correlation

Fluid–Heating Surface Combination	C_{sf}	Fluid–Heating Surface Combination	C_{sf}
Water–copper	0.031	Carbon tetrachloride–copper	0.013
Water–platinum	0.013	Carbon tetrachloride–emery-polished copper	0.0070
Water–brass	0.0060	Benzene–chromium	0.010
Water–emery-polished copper	0.0128	*n*-Butyl alcohol–copper	0.00305
Water–ground and polished stainless steel	0.0080	Ethyl alcohol–chromium	0.027
Water–chemically etched stainless steel	0.0133	Isopropyl alcohol–copper	0.00225
Water–mechanically polished stainless steel	0.0132	*n*-Pentane–chromium	0.015
Water–emery-polished and paraffin-treated copper	0.0147	*n*-Pentane–emery-polished copper	0.0154
Water–scored copper	0.0068	*n*-Pentane–emery-polished nickel	0.0127
Water–teflon-pitted stainless steel	0.0058	*n*-Pentane–lapped copper	0.0049
		n-Pentane–emery-rubbed copper	0.0074
		35% K_2CO_3–copper	0.0054
		50% K_2CO_3–copper	0.0027

Empirical Correlations Apart from the correlations above, a large number of empirical correlations exist which arose from a variety of experiments. Some of these correlations are dimensionally consistent (i.e., any system of units can be used for them) but others are dimensional equations where only prescribed units can be used. The most famous of the empirical pool boiling correlations is probably the Stephan and Preusser (1979) correlation, given by

$$\mathrm{Nu} = 0.1\left(\frac{q_0 d_b}{k_L T_s}\right)^{0.67}\left(\frac{\rho_G}{\rho_L}\right)^{0.156}\left(\frac{h_{fg} d_b^2}{\alpha_L^2}\right)^{0.371}\left(\frac{\alpha_L^2 \rho_L}{\sigma d_b}\right)^{0.35}\left(\frac{\mu_L C_{PL}}{k_L}\right)^{-0.16} \tag{6.31}$$

where d_b is the bubble departure diameter.

Another blanket correlation, suggested by Mostinski (1963), is given by

$$h = 0.106 P_{\mathrm{cr}}^{0.69} q^{0.7} f(P^*) \tag{6.32}$$

where

$$f(P^*) = 1.8P^{*0.17} + 4P^{*1.2} + 10P^{*10} \tag{6.33}$$

Here we get h in $\mathrm{W/m^2 \cdot K}$ and have to use P_c in bar. Another correlation was suggested by Cooper (1984):

$$h = A P^{*(0.12-0.2\log_{10}\varepsilon)}(-\log_{10} P^*)^{-0.55} M^{-0.5} q^{0.67} \tag{6.34}$$

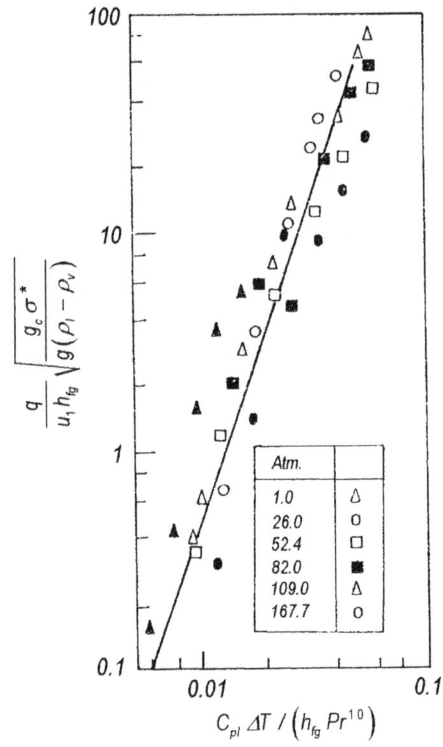

Fig. 6.10 Rohsensow correlation versus experimental data.

where ε is the surface roughness in micrometers and M is the molecular weight. A is 55 for copper plate or stainless steel cylinders and 93.5 for copper cylinders, h is in $W/m^2 \cdot K$, and q is in W/m^2.

Critical Heat Flux in Pool Boiling Critical heat flux (CHF) is an important phenomenon that should be estimated, and the process equipment should be designed such that the CHF is never reached. This is because at CHF the wall temperature suddenly rises, which may cause mechanical failure of the surface due to thermal stress or even lead to burnout of the heater. The phenomenon of critical heat flux can be very well understood from the theory of stability. Stability theory shows that the Taylor wave of vapor and Kelvin–Helmholtz instability of the vapor column can be correlated (Whalley, 1996) to give critical heat flux correlation in the form (refer to Fig. 6.11)

$$q_{\text{crit}} = 0.149 h_{fg} \rho_G^{1/2} [\sigma(\rho_L - \rho_G) g]^{1/4} \tag{6.35}$$

Kutateladze (1959) and Zuber (1958) suggested identical correlations:

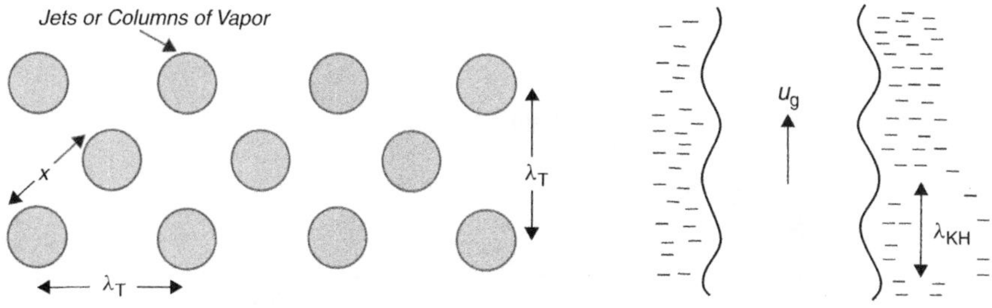

Fig. 6.11 Relation of instability to critical heat flux.

$$q_{\text{crit}} = Kh_{fg}\rho_G^{1/2}[\sigma(\rho_L - \rho_G)g]^{1/4} \tag{6.36}$$

where K is a constant that is to be determined from experiment and which lies between 0.13 and 0.16. A mean value of $K = 0.145$ works well if there is no entry of vapor from the side and if the size of the test section is large. This size effect is shown in Fig. 6.12.

For horizontal tubes instead of plates, the critical heat flux is given by

$$q_{\text{cr}} = Ch_{fg}\rho_G^{1/2}[\sigma(\rho_L - \rho_G)g]^{1/4} \tag{6.37}$$

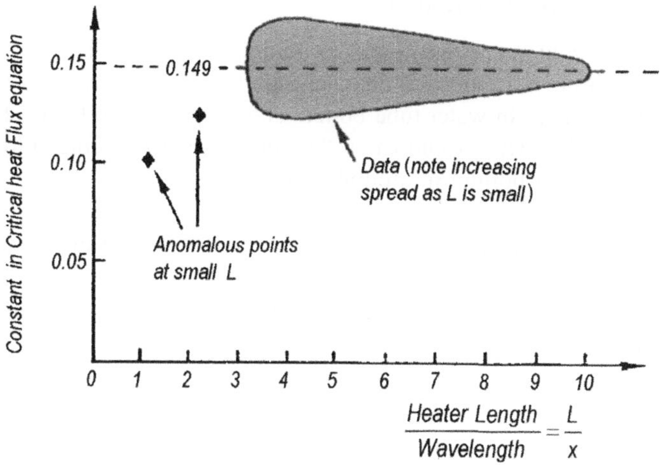

Fig. 6.12 Constant for critical heat flux.

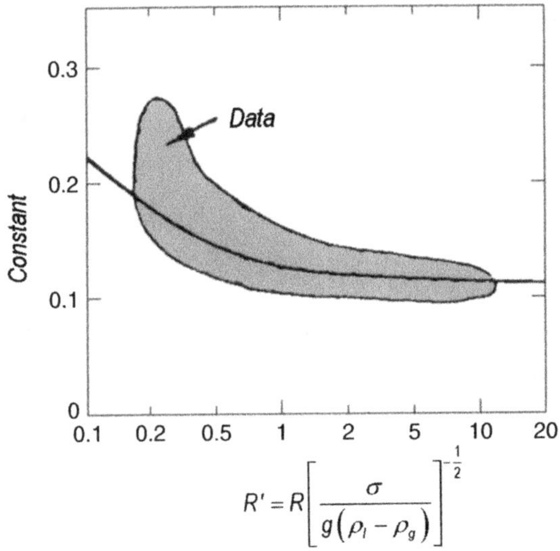

Fig. 6.13 Constant for CHF on a horizontal cylinder.

The constant C depends on radii

$$C = 0.116 + 0.3e^{-3.44\sqrt{R'}} \quad \text{and} \quad R' = \frac{R}{[g(\rho_L - \rho_G)]^{1/2}}$$

For industrial heaters (>8 mm in diameter), $R' > 1$, giving $C = 0.116$. Figure 6.13 shows the constant.

6.1.2. Flow or Convective Boiling

In most process equipment that involves boiling, the phase change takes place while flowing through a tube or a channel, due to heat transfer from the heated tube wall. Boiler tubes in water tube boilers and evaporator coils of refrigerators are typical examples. This is known as *flow* or *convective boiling*. Flow boiling is different from pool boiling in a number of ways:

1. The fluid pressure changes during flow, and hence the saturation temperature varies in the flow direction.
2. The quality (relative proportion of vapor) varies continuously in the flow direction.
3. There is no uniform pattern of nucleation and evaporation; it changes from section to section, depending on the heat and mass flux.

Compared with pool boiling, which depends primarily on heat flux, the most distinctive feature of flow boiling is that it depends more on mass flux. At low quality (vapor fraction), flow boiling is similar to nucleate boiling and depends mostly on heat flux and very weakly on mass flux or quality. But as quality increases it goes to convective boiling mode, where it depends very little on heat flux but depends strongly on mass flux. At high quality the wall dries out and the heat transfer coefficient decreases.

Two-Phase Flow To understand convective boiling it is most important to understand the features of two-phase flow in general. It must be understood that two-phase flow may occur even without boiling (adiabatic), as in the case of an air–water flow. However, boiling makes the process more complex, due to continuous change from one phase to other.

Flow pattern maps Hundreds of flow patterns have been observed by investigators, of which the following are important:

1. Vertical upward adiabatic flow
2. Vertical upward flow with heat transfer
3. Horizontal adiabatic flow
4. Horizontal flow with heat transfer
5. Flow in microchannels

Figure 6.14 shows typical flow patterns for these cases. Usually, these flow patterns contain bubbly flow, plug or slug flow, churn or semiannular flow, annular flow, and mist or spray flow where the wall dries up. The same characteristics can be better understood from Fig. 6.15, suggested by Collier and Thome (1994).

The analysis and correlations for pressure drop and heat transfer in flow boiling are quite complex. Since no work on the flow boiling of nanofluids has appeared in the literature, we do not explore these areas here.

6.2. POOL BOILING OF NANOFLUIDS

In Chapters 3 and 5 we noted that with a very small volume fraction of nanoparticles, thermal conductivity and convective heat transfer capability are enhanced significantly without the problems encountered in common slurries, such as clogging, erosion, sedimentation, and large increases in pressure drop. This naturally prompts us to ask whether such fluids can be used for two-phase applications; in other words, whether phase change in such suspensions will be helpful or detrimental to the process of heat transfer. In addition to this, when using nanofluids for convective cooling, one must also be aware of its boiling characteristics. This is because even if nanofluids are unattractive with respect to two (or rather, three)-phase applications, during convective heat transfer with high heat flux locally, the boiling limit may be reached. It is important that the behavior of nanofluids under such conditions be known accurately, to avoid unwanted effects

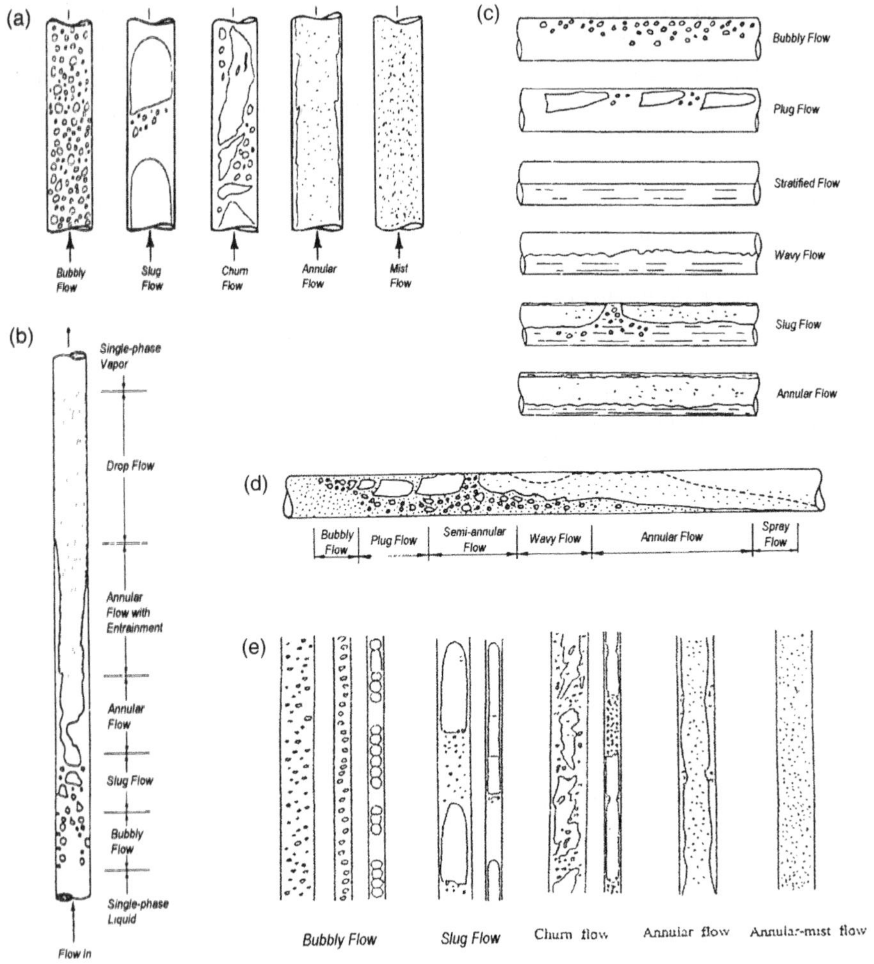

Fig. 6.14 Typical flow patterns: (a) vertical upward adiabatic; (b) vertical upward boiling; (c) horizontal adiabatic; (d) horizontal boiling; (e) microchannels.

such as local hot spots, which can cause significant deterioration in the reliability of components to be cooled.

The task of finding the boiling characteristics of nanofluids was first taken up by Das et al. (2003a). The paper presented an experimental study of the pool boiling characteristics of water–Al_2O_3 nanofluids under atmospheric conditions. The thrust of the experiment was to compare their pool boiling parameters with those of pure water, thus clarifying the applications and limitations of nanofluids under the conditions of phase change. The experimental setup was designed to give pool boiling on a horizontal tube. Special watch was kept so that the experiments for different nanofluids and water are performed under identical conditions. The test section (Fig. 6.16) consists of 120 mm × 100 mm × 200 mm

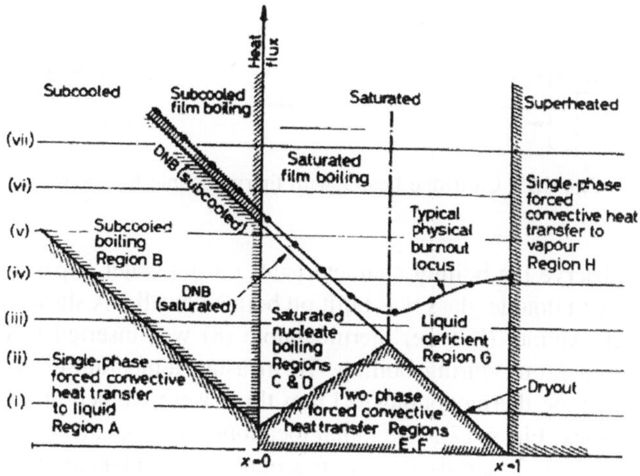

Fig. 6.15 Regions of flow boiling at different heat flux values.

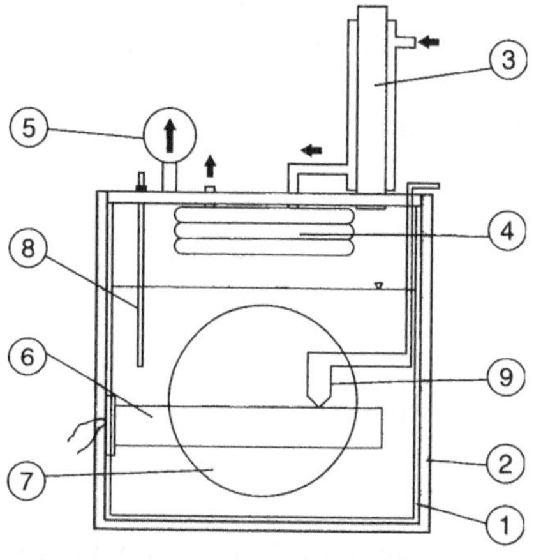

Fig. 6.16 Boiling vessel of Das et al. (2003a).

rectangular stainless steel vessels (1) with thick insulation (2) outside. The vessel has two cooling arrangements cascaded together. The first (3) is a countercurrent copper condenser that connects the vessel directly to the atmosphere, maintaining atmospheric pressure in it, and also serves the purpose of after-cooling of any vapor thet may try to escape as well as acting as a vent to noncondensable gases. The cooling water from this vertical condenser is then circulated through an oval copper coil (4) which condenses the bulk of the vapor produced. This coil hangs from the roof of the vessel. A pressure gauge (5) mounted at the top of the vessel checks the pressure at which boiling takes place. A cylindrical cartridge heater (6) 20 mm in diameter is used as the boiling surface. It is inserted from the

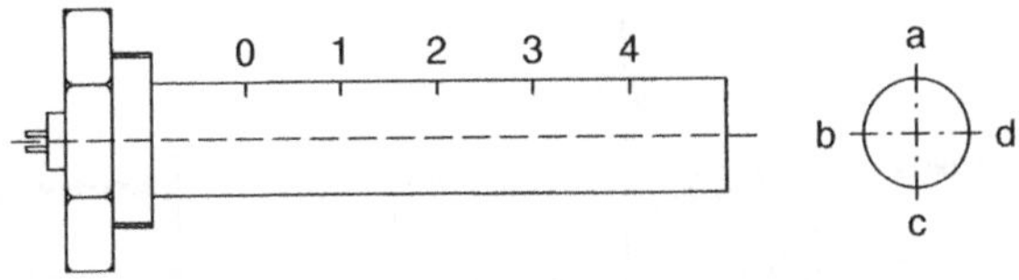

Fig. 6.17 Cartridge heater with thermocouple locations.

sidewall. To observe the boiling characteristics during water experiments, round windows (7), with double glass was built on both sidewalls. A sheathed 0.5- mm-thick Chromel–Alumel (K-type) thermocouple (8) was inserted to observe the bulk liquid temperature during boiling. To measure temperature on the heating cartridge, 10 K-type thermocouples 0.1 mm thick were welded at various radial and axial locations (Fig. 6.17). The radial locations are a, b, c, and d and the axial locations are 0, 1, 2, 3, and 4. The thermocouples were planted at locations 0 d, 1a, 1b, 2a, 2b, 2c, 2 d, 3c, 3 d, and 4a. The power supply to the heater was varied by a transformer and the power was recorded with a wattmeter. For characterizing pool boiling phenomena it is important to know the heater geometry and surface accurately. In this experiment, stainless steel heaters 20 mm in diameter with a rating of 420 V and 2.5 kW were used. The heater surface is machine drawn. The major parameter for the characterization of surface roughness are Ra and Rq (DIN 4762), which are defined as

$$\mathrm{Ra} = \frac{1}{L}\int_{0}^{L} |Z(x)|\, dx \tag{6.38}$$

$$\mathrm{Rq} = \sqrt{\frac{1}{L}\int_{0}^{L} Z(x)^2\, dx} \tag{6.39}$$

Experiments were carried out to evaluate pool boiling with nanofluids of 1%, 2%, and 4% Al_2O_3 nanoparticle concentration in water. The boiling characteristics of nanofluids were taken for which the q–ΔT results are shown in Fig. 6.18. This indicates clearly that the boiling performance of the base fluid (water) deteriorated with the addition of nanoparticles since the boiling curves are shifted to the right. This means that without changing the boiling temperature the nanofluid can cause harm to a cooled surface if the boiling limit is reached because it will result in a higher wall superheat, meaning a higher surface temperature compared to water at the same heat flux. It has been observed that a shift of the curve to the right is not proportional to the particle concentration. For example, in Fig. 6.18, for the smoother heater (surface, Fig. 6.19), a considerable shift of the curve

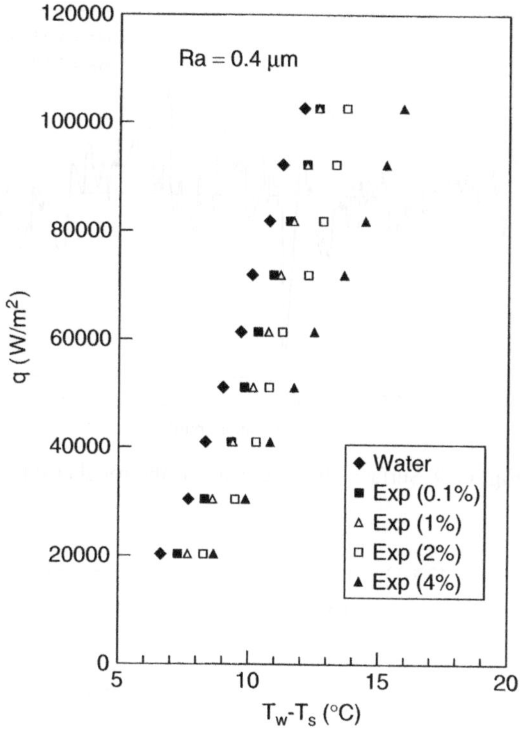

Fig. 6.18 Boiling curve for the smoother tube.

was observed with only 0.1% particle concentration, and thereafter from 1 to 4% concentration a regular shift of the curve was observed at lower heat fluxes. However, at the upper part of the curves the difference between wall superheats for various particle concentrations was found to increase with increasing heat flux. This depicts a regular but nonlinear tendency of deterioration in boiling character for nanofluids with increased particle concentration. To examine this deterioration under different heater surface conditions, the same experiments were repeated for a rougher heater, and the q–ΔT characteristics for this are shown in Fig. 6.20. Here also we see a shift of the boiling curve to the right, indicating deterioration of the boiling performance with particle concentration. However, the shift was found to be different than that for the smoother heater. In this case a more drastic increase in wall superheat was observed for nanofluids up to 1% particle concentration, after which it seemed to slow down up to 4% (measured range). This can be better understood from Fig. 6.21, where the boiling curves for two heaters are compared for pure water and 1% and 4% particle concentration. Here all the curves are shifted toward the left for the rougher heater, due to the

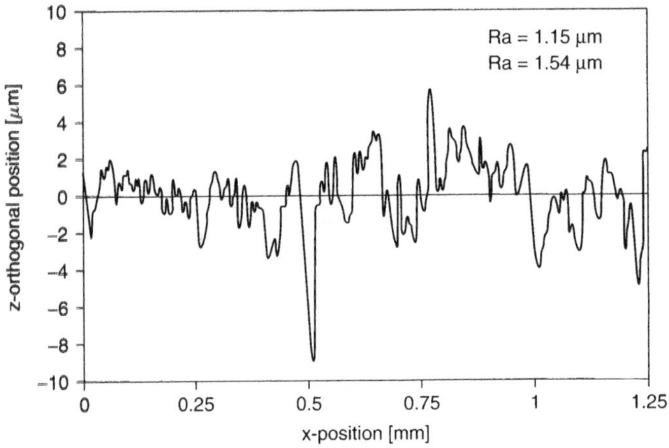

Fig. 6.19 Surface characteristics of the rougher tube.

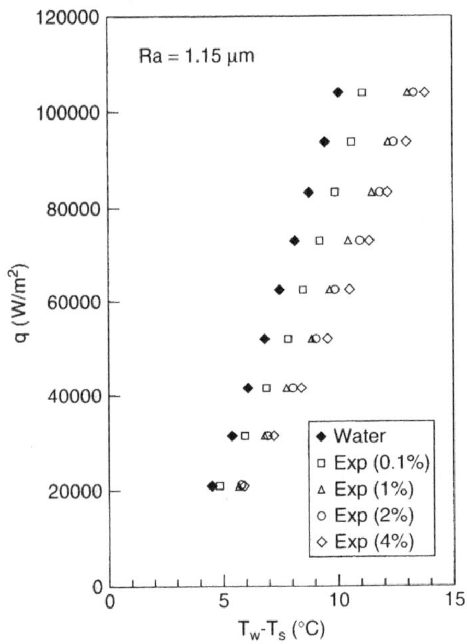

Fig. 6.20 Boiling curve for the rougher tube.

increase in surface roughness, but the extent of the shift for different particle concentration is different and depends on the heat flux.

The present results are somewhat contrary to expectations. Nanofluids show a substantial increase in the thermal conductivity of fluids with nanoparticles. The surface tension and latent heat remains unaffected and the only unfavorable

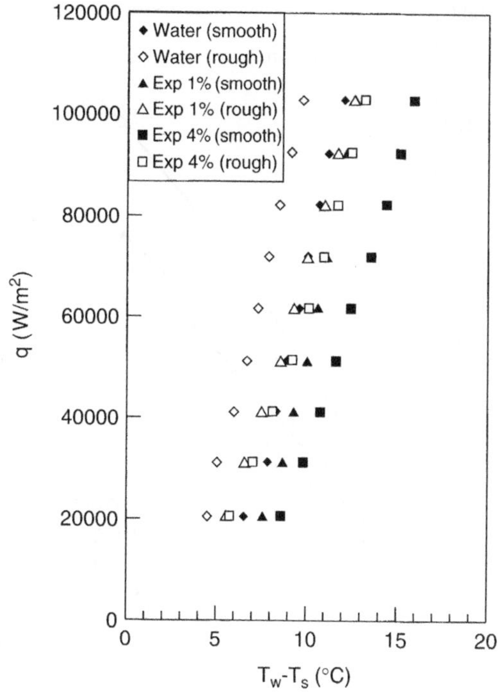

Fig. 6.21 Comparison of boiling characteristics of the rougher and smoother tube.

change in fluid property due to the presence of particles is the increase in viscosity. Since fluid conduction in microlayer evaporation under a bubble as well as in reformation of thermal boundary layer at the nucleation site (Benjamin and Balakrishnan, 1996) plays a major role in heat transfer during pool boiling, with such a substantial increase in thermal conductivity, nanofluids are expected to enhance heat transfer characteristics during pool boiling. For pool boiling on horizontal tubes at moderate heat flux, the series of works from Cornwell et al. (1989, 1990, and 2000) conclusively proved the importance of sliding bubbles, where conduction again plays an important role. Thus, both for stationary bubble development and the sliding bubble mechanism, the increase in thermal conductivity is expected to enhance heat transfer during boiling, which is contrary to what has been observed in the present set of experiments. The fact that the present increase of wall superheat in boiling, and as a consequence, decrease in boiling heat transfer coefficient, is an additional effect can be understood from Figs. 6.22 and 6.23. Here, in keeping with a Cornwell–Houston (1994) type of correlation, Nu–Re_b plots for both heaters are shown.

It is evident that for each particle concentration the Nu–Re_b characteristics are different and shifted downward. This conclusively brings out the fact that the change in boiling characteristics of nanofluids cannot be explained in terms of property change alone because the Nu–Re_b correlations are altered. The following

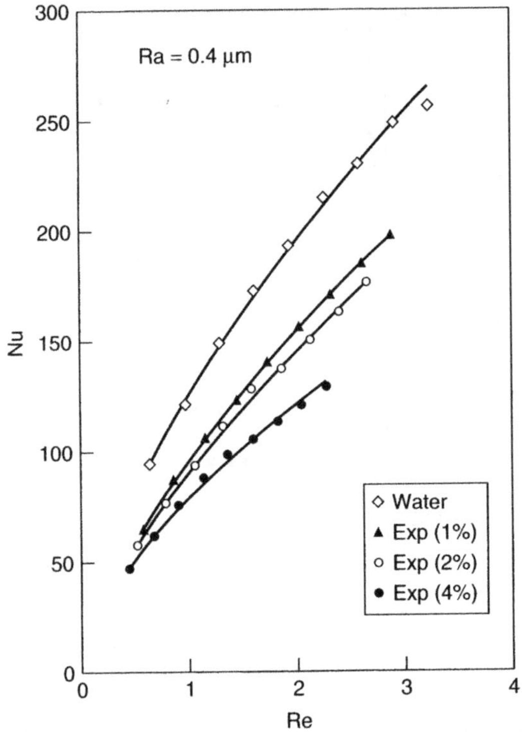

Fig. 6.22 Dimensionless boiling curve for the smoother tube.

correlations were obtained. For heater 1 (smooth),

$$\text{pure water Nu} = 97.9\ \text{Re}_b^{0.638}\text{Pr}^{0.4}$$

$$1\%\ \text{Al}_2\text{O}_3\ \text{Nu} = 78.84\ \text{Re}_b^{0.687}\text{Pr}^{0.4}$$

$$2\%\ \text{Al}_2\text{O}_3\ \text{Nu} = 72.39\ \text{Re}_b^{0.69}\text{Pr}^{0.4}$$

$$4\%\ \text{Al}_2\text{O}_3\ \text{Nu} = 67.56\ \text{Re}_b^{0.619}\text{Pr}^{0.4}$$

and for heater 2 (rough),

$$\text{pure water Nu} = 137\ \text{Re}_b^{0.526}\text{Pr}^{0.4}$$

$$1\%\ \text{Al}_2\text{O}_3\ \text{Nu} = 99.48\ \text{Re}_b^{0.503}\text{Pr}^{0.4}$$

$$2\%\ \text{Al}_2\text{O}_3\ \text{Nu} = 94.63\ \text{Re}_b^{0.495}\text{Pr}^{0.4}$$

$$4\%\ \text{Al}_2\text{O}_3\ \text{Nu} = 89.12\ \text{Re}_b^{0.490}\text{Pr}^{0.4}$$

The lines indicating these correlations are shown in Figures 6.22 and 6.23. For the smoother heating surface the shift in boiling character was found to be more

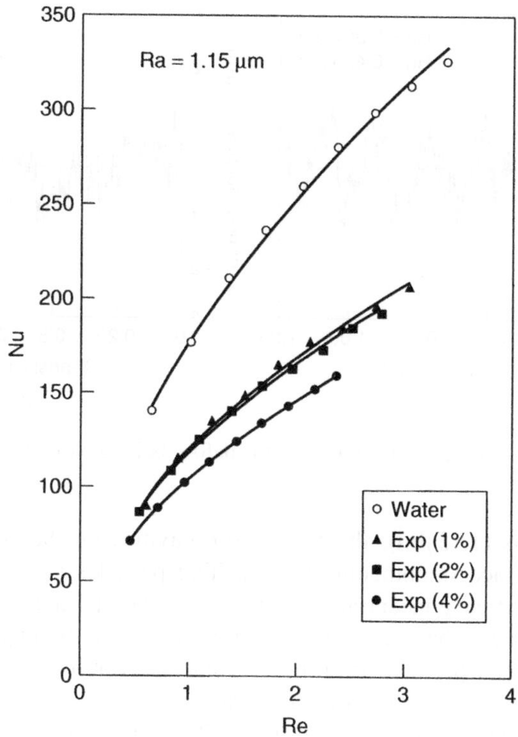

Fig. 6.23 Dimensionless boiling curve for the rougher tube.

or less uniform with concentration, whereas for the rougher surface it was found to be rapid at lower concentration and slower thereafter.

To determine which effect skirts property variation, the surface characteristics of the heaters were reexamined after the runs with nanofluids and before jet cleaning of the surfaces. It was found that a considerable reduction in surface roughness takes place, which returns the surface to close to its original condition after cleaning. As an example, the surface characteristics of the smooth heater [Fig. 6.24(a)] were changed to an even lower value [Fig. 6.24(b)] after boiling nanofluid, which indicates a probable cause for the deterioration in boiling characteristics. Because the nanoparticles are one to two orders of magnitude smaller (20 to 50 nm) than the roughness (0.2 to 1.2μm) of the heating surface, the particles sit on the relatively uneven surface during boiling. These trapped particles change characteristics of the surface, making it smoother. This causes degradation of the boiling characteristics. For higher particle concentration, the particle forms a virtualy layer on the heating surface, hindering the heat transfer. Thus, the small particle size causes surface skirting, which overshadows the thermal conductivity enhancement of nanofluids.

The large change in the boiling character of the rougher heater with smaller particle concentration can be explained similarly. In this case, due to higher

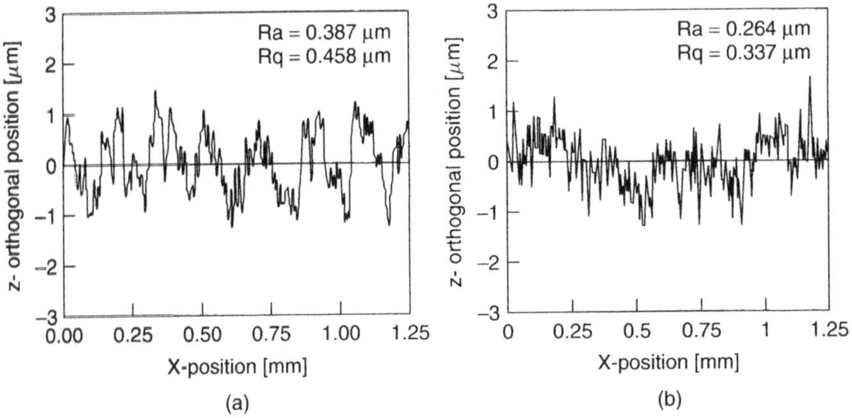

Fig. 6.24 Surface characteristics: (a) before boiling; (b) after boiling.

surface roughness (1.15 μm), there are more cavities on the surface, and as a consequence, surface smoothening by attaching particles is more abrupt. Thus, with a smaller particle concentration ($<1\%$), sufficient particles are deposited on the uneven surface to have a considerable effect on the boiling character. Any additional deposit of particles from higher concentration in the fluid brings only marginal deterioration of pool boiling characteristics.

Later, the same authors (Das et al., 2003b) showed that pool boiling of nanofluids on narrow horizontal tubes (4 and 6.5 mm in diameter) is qualitatively different from that of large-diameter tubes, due to the difference in the bubble sliding mechanism. It was found that in this range of narrow tubes, boiling performance of nanofluid deteriorates less than it does in large industrial tubes, which make them less susceptible to local overheating in convective applications. For boiling on tubes 4 and 6.5 mm in diameter the sliding seems to be less important mechanism for larger bubbles which are comparable to the size of bubbles of boiling on a 20-mm tube. This is because of the relatively small size of the tube, which produces a large curvature of the surface that prohibits the sliding of larger bubbles but induces direct departure. However, a large number of smaller bubbles are produced in a sustainable way and they slide, but a relatively shorter distance. Nu–Re_b data are plotted for these experimental runs with water. The data shows that for larger-diameter tubes the data fit quite well the convective-type correlation suggested by Cornwell and Houston (1994):

$$\mathrm{Nu} = C \cdot \mathrm{Re}_b^{0.67}\mathrm{Pr}^{0.4} \tag{6.40}$$

as well as the correlation suggested by Gorenflo (1991). Here for both Nu and Re_b the tube diameter was used as the characteristic length and the viscosity of the nanofluid measured at saturation temperature was used in the calculation of Re. The result shows that for tube diameters between 4 and 6.5 mm the correlation is not strong, which is expected because the correlations above were not developed

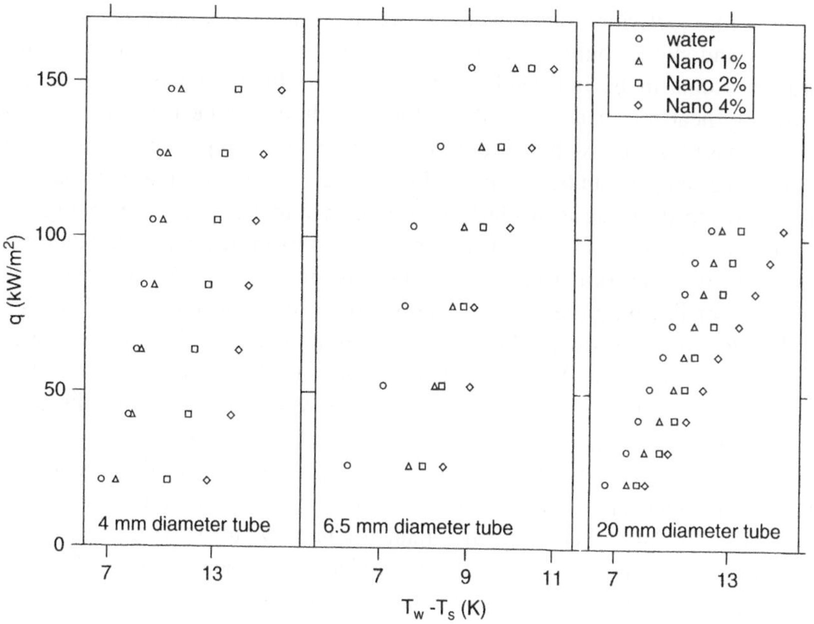

Fig. 6.25 q−ΔT curves for tubes of various diameters.

for narrow tubes. The results above clearly indicate the need to develop correlations for boiling on narrow tubes. Subsequently, experiments were carried out to evaluate pool boiling with nanofluids of 1, 2, and 4% Al_2O_3 nanoparticle concentration in water. Repeatable boiling characteristics of these nanofluids in q–ΔT form are presented in Fig. 6.25 for various tube diameters. These plots clearly indicate that, in general, the boiling performance of the base fluid deteriorates with the addition of nanoparticles pushing the boiling curves to the right, which means that the nanofluid can cause harm to cooled surfaces if boiling occurs, because it will produce a higher surface temperature than will water at the same heat flux, as observed earlier by Das et al. (2003a).

It has been observed that shifting the curve to the right is not proportional to the particle concentration and is strongly dependent on the tube diameter, even for similar values of surface roughness. For narrower heaters (4 and 6.5 mm) the curve shift is considerable and is of almost the same order over the entire range of heat flux. For a 20-mm tube at 1 to 4% concentration, a regular curve shift was observed at lower heat fluxes, but at the upper part of the curves the difference between wall superheat values for various particle concentrations was found to increase with increasing heat flux. This depicts the general tendency toward deterioration of the boiling character of nanofluids, generally increasing with particle concentration. However the nature of deterioration is different in the narrow tube regime from the large-diameter-tubes domain. This study indicates that in the region of narrow tubes, the tube diameter plays a crucial role in determining the nature of this deterioration, presumably centered around change in the bubble diameter and sliding bubble mechanism. In the narrow diameter range, the deterioration seems to be independent of heat flux, which for the larger diameter is strongly heat-flux dependent. Also, for pure fluids the heat transfer increases with increased viscosity, but in the present case two competing phenomena are taking place: increase in heat transfer due to viscosity, and decrease in heat transfer due to the decrease in nucleation site density that results from plugging of (micro) surface cavities by nanoparticles. The results indicate that the latter effect dominates over the former because the increase in viscosity is very marginal.

To further illustrate the effect of nanoparticles on heat transfer, a dimensionless Nu–Re_b plot is shown in Fig. 6.26. This figure indicates that for each particle concentration value the Nu–Re_b characteristics are different and shifted downward. This is a general observation for all tubes, which indicates that the change in boiling characteristics of nanofluids can be explained neither in terms of property change nor of changes in Nu and Re_b due to changes in the characteristic length (diameter). The change in Nu–Re_b correlations is more drastic at higher Re_b values for large-diameter tubes than for narrow tubes. This indicates that for high-heat-flux applications, the danger of local overheating when the boiling point is reached for a nanofluid is less for narrow tubes than for larger tubes.

These observations regarding deterioration of pool boiling in nanofluids were substantiated further by Bang and Chang (2005a) using an elaborate apparatus with visualization windows (Fig. 6.27). The distinctive feature of this experiment was that it used a plane horizontal heating surface devoid of such effects

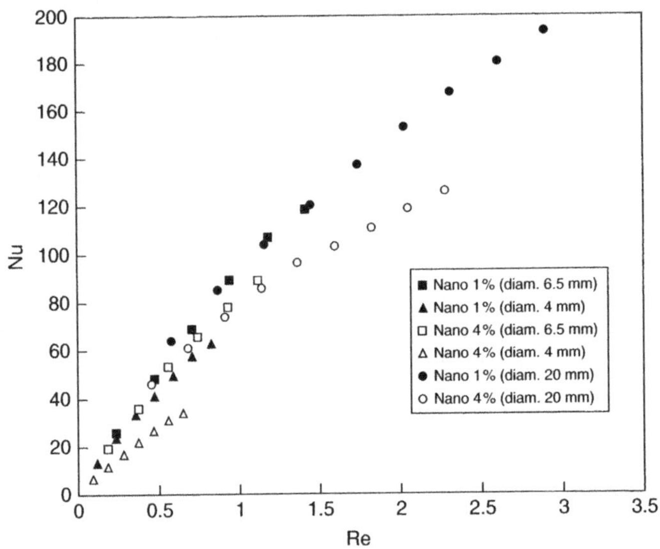

Fig. 6.26 Dimensionless boiling characteristics of nanofluid boiling on narrow tubes.

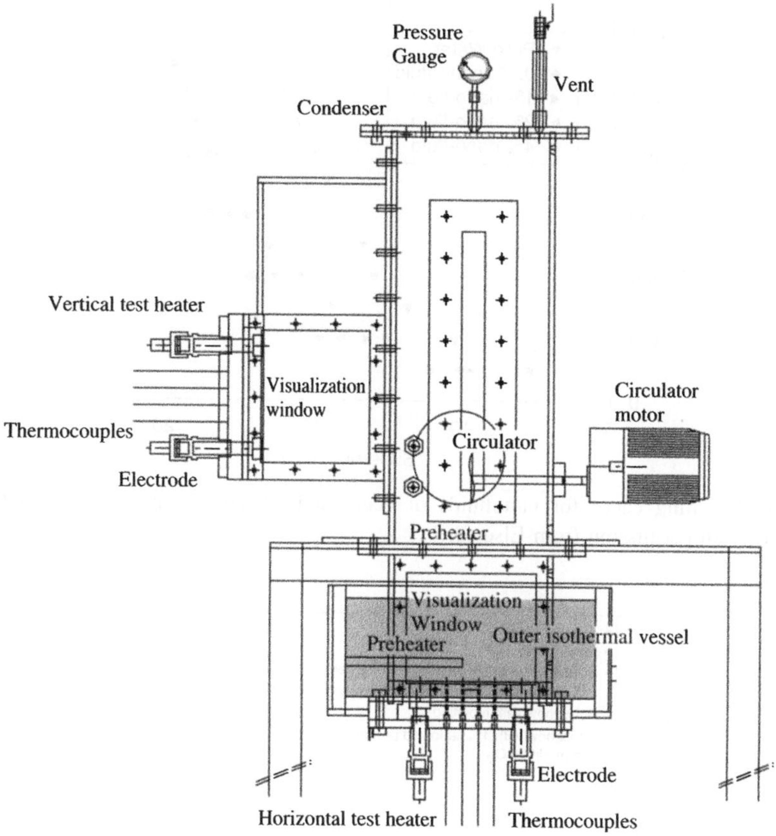

Fig. 6.27 Nanofluid boiling experimental facility of Bang and Chang. [From Bang and Chang (2005a), with permission from Elsevier.]

as bubble sliding. They also carried out a visualization for water and a dilute (0.5%) Al_2O_3–water nanofluid, but it did not reveal any substantial physics. They used a much smoother heater than that used by Das et al. (2003a): one having a surface roughness of ≈ 37 nm. They made a number of important observations regarding the boiling characteristics of nanofluids. Like Das et al. (2003a) , they observed deterioration of boiling with nanofluid concentration, but the rate of heat transfer was somewhat different, which they attributed to the difference in geometrical features of the heaters in the two studies. They could also identify a clear natural convection regime, followed by nucleate boiling (Fig. 6.28). They observed further that the experimental data do not conform to the Rohsenow correlation [(equation 6.31)] by simply using effective nanofluid properties to change the fluid properties. They tried different variations of the same correlation, such as using the Rohsenow correlation, changing only the effective conductivity or the constant C_{sf}. It was found that rather than changing the fluid properties, modification of the surface fluid combination factor C_{sf} provides a closer approximation

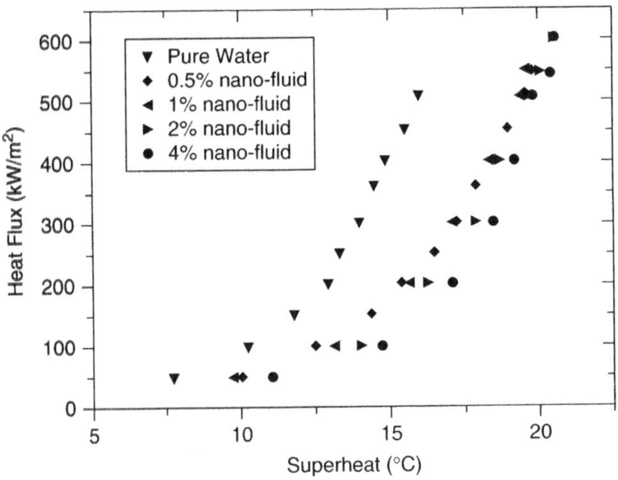

Fig. 6.28 Boiling carve for nanofluids of Bang and Chang. [From Bang and Chang (2005a), with permission from Elsevier.]

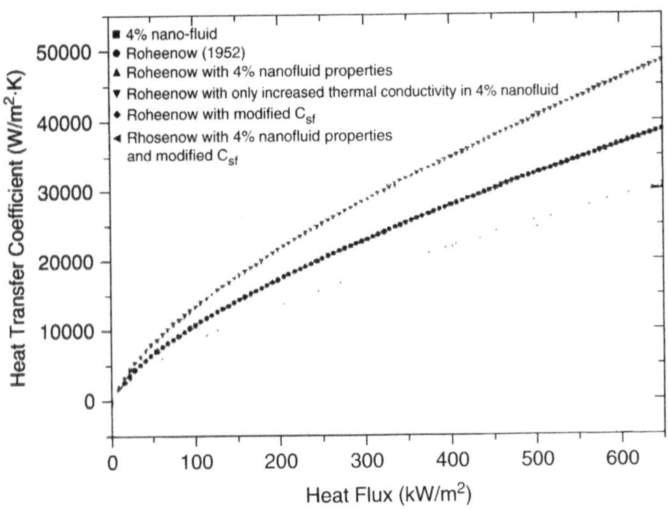

Fig. 6.29 Boiling 4% nanofluid and application of the Rohsenow correlation. [From Bang and Chang (2005a), with permission from Elsevier.]

to the experimental boiling data of nanofluids (Fig. 6.29). This indicates strongly that the modification of surface characteristics during nanofluids boiling might hold the key to explaining the deterioration in nanofluid boiling.

However, Bang and Chang's (2005a) explanation regarding surface modification was quite different. Contrary to the observations of Das et al. (2003a), they found that the roughness of the clear heater increased after boiling of 0.5%

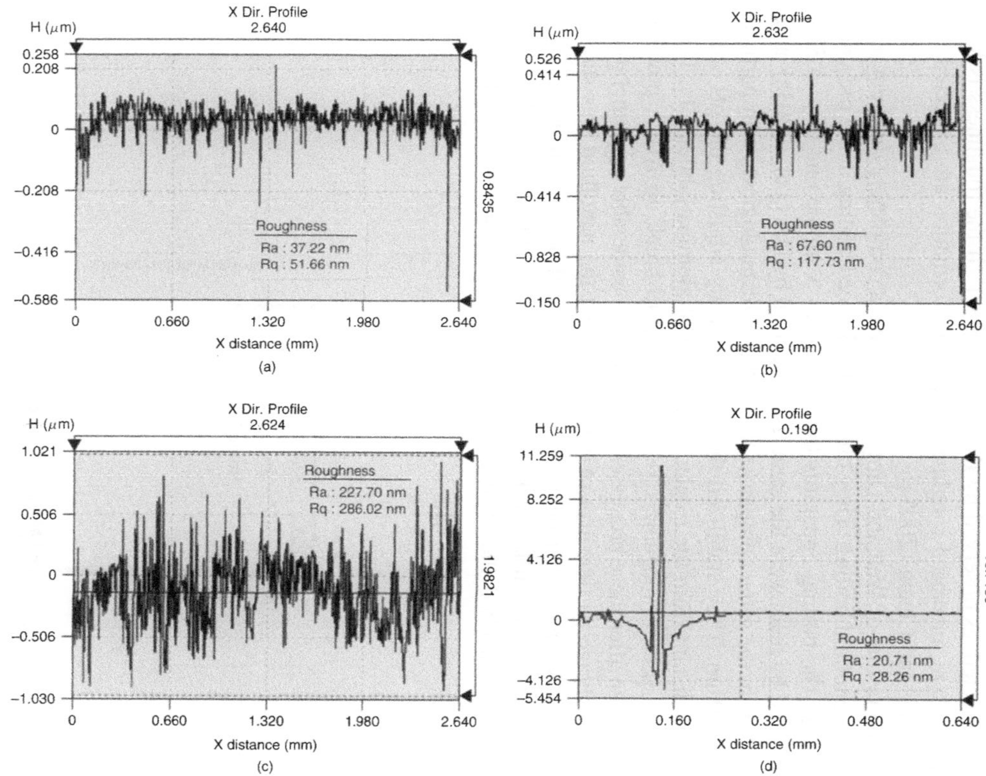

Fig. 6.30 Representative surface roughness:(a) clear heater:(b) heater submerged in 0.5% alumina nanofluid;(c) heater submerged in 4% alumina nanofluid;(d) locally smoothened heater in 0.5% alumina nanofluid. [From Bang and Chang (2005a), with permission from Elsevier.]

alumina nanofluid and increased even more in 4% alumina nanofluid. They also found a locally smoothened heater. The roughness profiles are shown in Fig. 6.30. They explained the result as due to the closeness of roughness and particle size. The fouling effect (particles forming a layer on the surface), here reduces heat transfer due to the poor conductivity of Al_2O_3. This reduces the heat transfer even though the surface roughness is increased. They commented that if the surface roughness is much higher than the particle size, the plugging effect of nucleation sites is to be expected, as observed by Das et al.(2003a) . If the roughness is less than the particle size, increased roughness and layer formation are expected, as observed by Bang and Chang (2005a).

However, the results of Wen and Ding (2005) gave a completely different picture regarding manofluid boiling. They observed enhanced boiling in the presence of nanoparticles. The boiling apparatus they used was a simple one (Fig. 6.31). The particles used by them were the same as those used by Das et al. (2003a) and were acquired from the same company (Nano Phase Technologies), with particle sizes of 10 to 50 nm. They stabilized the suspension by adjusting the pH value near 7, which is distant from the isoelectrical point of alumina (about 9.1). They also used a high-speed homogenizer (~24,000 rpm) to break up Al_2O_3 powder agglomerates. Even after these processes, they found considerable agglomeration, resulting in an average particle size of 167.54 nm, but the nanofluid was stable. They used a 2.4-kW ring heater below the stainless steel boiling surface.

Their results were quite different from those of earlier studies. Whereas their pure water results matched the traditional Rohsenow correlation, heat transfer

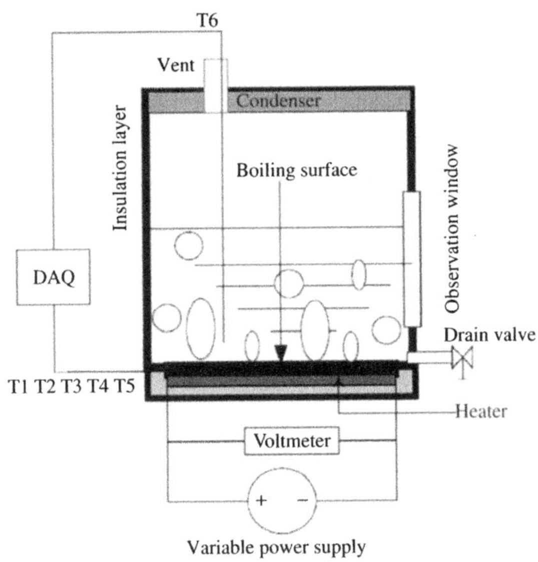

Fig. 6.31 Pool boiling setup of Wen and Ding. [From Wen and Ding (2005), with permission from Springer.]

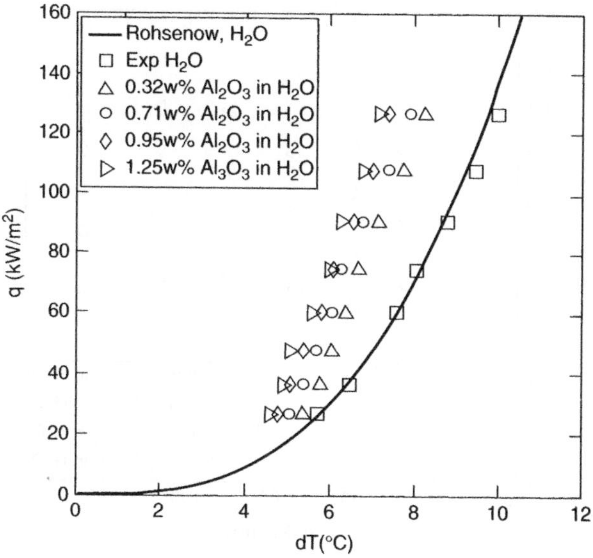

Fig. 6.32 Pool boiling data of Wen and Ding and comparison with the Rohsenow correlation. [From Wen and Ding (2005), with permission from Springer.]

with nanofluids showed enhanced heat flux at the same wall superheat, which increased with particle-volume fraction (Fig. 6.32).

They observed an increase as high as 40% in the heat transfer coefficient, as shown in a plot of the heat transfer coefficient ratio between nanofluids and pure water (Fig. 6.33). Also, the enhancement was with just 1.25 wt % of particles, which is about 0.3 vol %. They found that this increase is much more than the measured value of thermal conductivity enhancement, and hence boiling enhancement cannot be explained by conductivity enhancement alone. They also observed that apart from being strongly dependent on the particle concentration, enhancement, is also dependent on heat flux, giving higher values with increasing heat flux.

Explanations of the foregoing results were not conclusive because Wen and co-workers themselves only indicated several possibilities. They first indicated the possibility of agglomeration remaining in the fluids used by Das et al. (2003a) and Bang and Chang (2005a). This is certainly not confirmed because a TE micrograph of Das et al. (2003a) claims good dispersion. However, it must be indicated that the particle concentration of Wen and Chang (2005) was much less (an order of magnitude lower) than those of Das et al. (2003a) and Bang and Chang (2005a). Whereas Das et al. (2003a) used 1 to 4% particles by volume, Wen and Ding (2005) used 0.32 to 1.25 ut% (which is about 0.08 to 0.3 vol %). Hence, the fact that Wen and Ding (2005) did not observe any significant change in surface characteristics is understandable. In fact, looking at the results it may be said that although the results are different with respect to boiling enhancement

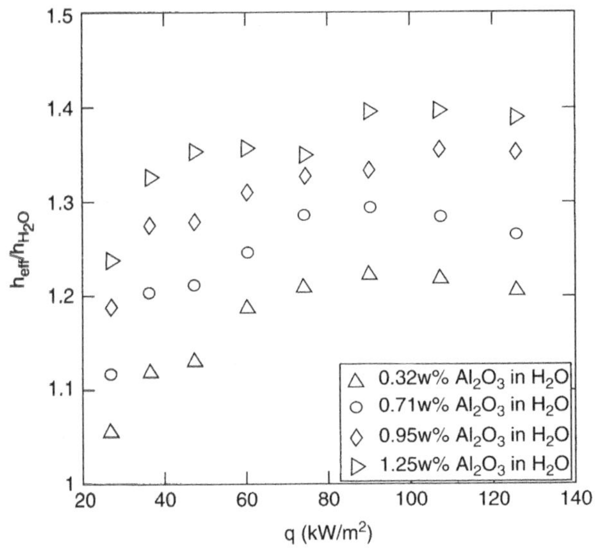

Fig. 6.33 Heat transfer enhancement in pool boiling. [From Wen and Ding (2005), with permission from Springer.]

(one positive, one negative), they need not negate each other; both results may be true in the respective particle concentration ranges.

These two ranges may be dominated by different phenomena, giving different heat transfer characteristics. Wen and Ding (2005) also indicated other reasons, such as surface characteristics other than roughness: surface wettability, effect of dispersants and surfactants, measurement techniques, and the characteristic size of the system. However, all these explanations are open to question until data are available from more systematic studies. Thus, it is clear that although the investigation of nanofluid boiling has begun, the early results are still sketchy because they are based on speculative reasoning rather than scientific evidence. Systematic experimentation over a wider range of parameters (e.g., heating surface geometry, surface characteristics, particle properties, suspension thermophysical properties, surfactants) is required.

6.3. CRITICAL HEAT FLUX IN POOL BOILING OF NANOFLUIDS

While differing trends in the nucleate pool boiling regimes are observed for nanofluids, the value of the critical heat flux (CHF) is enhanced in nanofluids. All investigators agree to it and this is also encouraging with regard to both the extent of CHF enhancement and its probable implications in a large variety of industrial processes.

You et al. (2003) were the first to investigate the CHF phenomenon of pool boiling of nanofluids. Their boiling apparatus was the usual apparatus with a

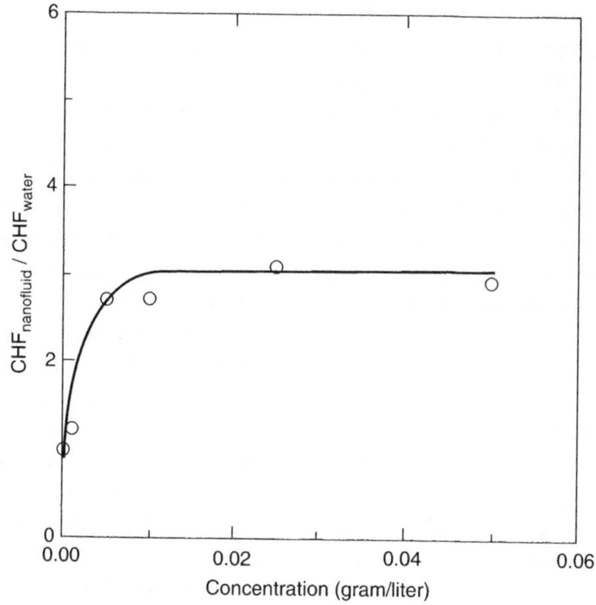

Fig. 6.34 Enhancement of CHF with particle concentration. [From You et al. (2003), with permission from the American Institute of Physics.]

horizontal heater. The results were astonishing, showing a 50 to 200% rise in CHF over pool boiling of pure water (Fig. 6.34). They used concentrations upto 0.05 g/l, which is 0.5 wt% and about 0.013 vol%. This dramatic increase in CHF increased sharply with concentration, and beyond about 0.2 wt% it remained constant at a value of about 300% of the value of pure water (Fig. 6.34).

They used high-speed photography to investigate the probable cause of this CHF enhancement by visual observation of bubble departure characteristics. To explain the observations they used the well-known *Zuber correlation*,

$$Q_{\mathrm{CHF}} = 0.131 h_{\mathrm{lv}} \rho_v^{1/2} [g\sigma(\rho_l - \rho_v)]^{1/2} \tag{6.41}$$

where h_{lv} is the latent heat of vaporization, σ the surface tension, and the subscripts v and l stand for vapor and liquid, respectively. They argue that since the nanofluids do not react chemically and particle concentration is very low, the density and latent heat in equation (6.46) are not changed by the presence of particles. However, effective surface tension gets changed since visual observation shows an increase in bubble departure size, and the Fritz formula [equation (6.15)] shows the dependence of surface tension on the bubble departure diameter. However, this dependence is only

$$\sigma \alpha d^2.$$

Thus, for the 30% increase in bubble departure diameter observed, yields only a 15% enhancement in the CHF according to Zuber's correlation, and hence the 200% increase in CHF remains largely unexplained.

The next work on CHF in the pool boiling of nanofluids was presented by Vassallo et al. (2004). In this work, a horizontal 18-gauge NiCr wire was used instead of a heating surface. The wire was long (~75 mm), and the wire temperature was evaluated from the resistance of the wire, which is a known function of temperature. They validated their experiment by boiling of pure water and comparing their results with the Nukiyama curve. They also compared the CHF of water with the predictions of the Zuber correlation [equation (6.41)] and found it to agree well. Subsequently, they carried out experiments with varying amounts of silica particles (between 2 and 9 vol %). Although they did not observe any definite effect (enhancement or deterioration) in the nucleate boiling characteristics of the nanofluids (rather, they observed a scatter), there was a definite and dramatic increase in the CHF of water for both 15- and 50-nm particles. For both particle sizes the enhancement was about 200% (Fig. 6.35), which is similar to the observations of You et al. (2003). To further investigate the effect of particle size, they also carried out experiments with large (~3 μm) particles. Here also enhancement was observed (about 100% enhancement), although little less than that with nanosuspensions. For large particles the wire failed before attaining film boiling, whereas for nanofluids failure was observed in the transition or the pure film boiling regime. They also observed a thin coating of silica particles on the

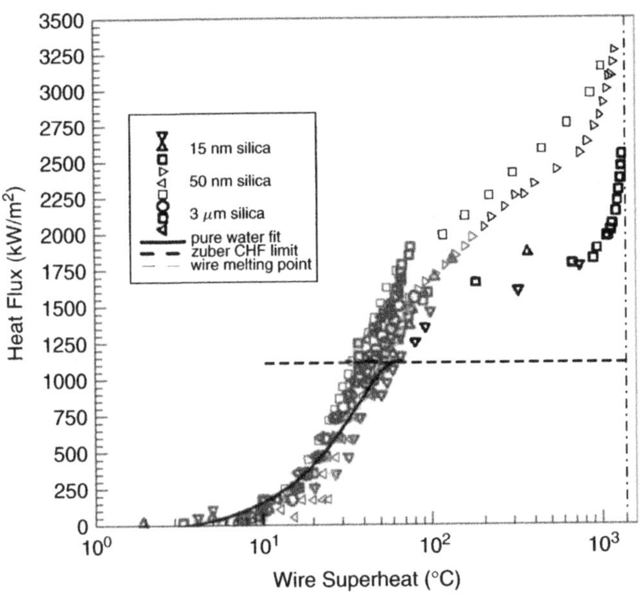

Fig. 6.35 Boiling curve of NiCr wire ($D = 0.4$ nm) in silica–water suspension. [From Vassallo et al. (2004), with permission from Elsevier.]

Table 6.2 CHF Enhancement of Nanofluids (MW/m^2)

	Prediction for Water	Pure Water	0.5% NF	1% NF	2% NF	4% NF
Horizontal test section ($\theta = 90$)	1.22	1.74	2.30	2.64	2.57	2.4
Vertical test section ($\theta = 0$)	0.88	1.2	1.36	1.36	1.36	1.36

Source: Bang and Chang (2005a), with permission from Elsevier.

wire after boiling but concluded that increased roughness alone cannot explain such an unusual rise in CHF.

In their study of nucleate boiling of nanofluids, Bang and Chang (2005a) also investigated CHF in the pool boiling of water–Al_2O_3 nanofluids. They investigated the CHF values on both horizontal and vertical surfaces and observed enhancement over pure water results. However, the enhancement they reported was much less then that reported by You et al. (2003) and Vassallo et al. (2004). To compare with theoretical predictions, they used a CHF correlation given by

$$Q_m = C_{\mathrm{CHF}}, f(\theta)\rho_g i_{fg}\left[\frac{\sigma(\rho_f - \rho_g)g}{\rho_g^2}\right]^{1/4} \tag{6.42}$$

where θ is the orientation angle, i_{fg} is the latent heat of vaporization, and the subscripts f and g indicate liquid and vapor phases. The constant $C_{\mathrm{CHF},f}$ is given by

$$C_{\mathrm{CHF}}, f = 0.034 + 0.0037(180 - \theta)^{0.656} \tag{6.43}$$

The enhancement reported by them was 32% for horizontal surfaces and 13% for vertical surfaces. Table 6.2 sums up their results. They attributed the difference between their results and those of Vassallo et al. (2004) to differentces in the size and properties of the particles as well as in heater geometry and surface conditions.

Thus, it is obvious that there are dramatically enhanced CHF values in pool boiling of nanofluids, ranging from 30 to 200% under different particle and surface conditions. This can be encouraging for such practical applications as nuclear reactors, thermal power generation, and heat pipe applications.

6.4. OTHER INVESTIGATIONS RELATED TO BOILING OF NANOFLUIDS

Apart from direct pool boiling characteristics and applications related to heat pipes (discussed in Chapter 7), there are a few other interesting studies. Liu and Qiu (2006) investigated the boiling of an impinging jet of nanofluid on a flat surface. They used CuO-water nanofluids of 0.1 to 2% particle concentration by weight with a jet impact velocity of 0.5 to 6.5 m/s. The heat transfer surface was 20 mm in diameter and the jet nozzle was 4 mm in diameter. They concluded that the boiling characteristics of nanofluids in jet impingement are considerably poorer then those of pure water, which is in line with the conclusions of Das et al. (2003a) and Bang and Chang (2005a). They showed that the boiling characteristics of nanofluids cannot be predicted by the traditional jet impingement boiling empirical equation,

$$q_w = \Delta T_{\text{sat}}^{2.9} \qquad \text{W/m}^2 \cdot \text{K} \tag{6.44}$$

Instead, from their boiling data, they predicted the correlation,

$$q_w = 1.4 \times 10^{-8} \Delta T_{\text{sat}}^{8} \qquad \text{W/m}^2 \cdot \text{K} \tag{6.45}$$

They also observed CHF enhancement by about 25% for both saturated and subcooled nanofluids and formation of a sorption layer on the surface during jet boiling that leads to surface roughness as well as a decreased contact angle. Thus, the key observations on nucleate boiling and in jet impingement are in line qualitatively with the observations of the pool boiling.

Another interesting study by Bang and Chang (2005b) was on the use of nanofluids to identify the existence of a liquid film separating a vapor bubble from a heated surface. This is not really a work on nanofluid but is an innovative way of characterizing two-phase phenomena using nanofluids.

Thus, the themes discussed in this chapter clearly indicate the interesting and often contradictory nature of the effects of nanoparticles in boiling. However, CHF enhancement in nanofluids seems to be a unanimous observation, the reason for which is still unclear.

REFERENCES

Bang, I. C., and S. H. Chang (2005a). Boiling heat transfer performance and phenomena of Al_2O_3–water nano-fluids from a plain surface in a pool, *Int. J. Heat Mass Transfer*, 48: 2407–2419.

Bang, I. C., and S. H. Chang (2005b). Direct observation of a liquid film under a vapor environment in a pool boiling using a nanofluid, *Appl. Phys. Lett.*, 86(13):134107.

Benjamin, R. J., and A. R. Balakrishnan (1996). Nucleate pool boiling heat transfer of pure liquids at low to moderate heat fluxes, *Int. J. Heat Mass Transfer*, 39: 2495–2504.

Collier, J. G., and J. R. Thome (1994). *Convective Boiling and Condensation*, 3rd ed., McGraw-Hill/Oxford University Press, London and New York, pp. 170, 175.

Cooper, M. G. (1984). Saturation nucleate pool boiling: a simple correlation, *Proe.First UK National Conference on Heat Transfer*, vol. 2, pp. 785–793.

Cornwell, K., and S. D. Houston (1994). Nucleate pool boiling on horizontal tubes: a convection based correlation, *Int. J. Heat Mass Transfer*, 37 (Suppl. 1): 303–309.

Das, S. K., and W. Roetzel (2002). Heat transfer model for pool boiling on a horizontal tube, presented at the International Heat Transfer Conference, Grenoble, France 2002.

Das, S. K., N. Putra, and W. Roetzel (2003a). Pool boiling characterization of nano-fluids, *Int. J Heat Mass Transfer*, 46: 851–862.

Das, S. K., N. Putra, and W. Roetzel (2003b). Pool boiling nano-fluids on horizontal narrow tubes, *Int. J Multiphase Flow*, 29: 1237–1247.

Dhir, V. K. (2000). On the use of numerical simulations to augment our understanding of boiling heat transfer, *Proc. National Heat Transfer Conference*, Pittsburgh, PA, pp. 1–21.

Fritz, W. (1935). Berechnung des Maximavolumens von Dampfblasen, *Phys. Z.*, 36: 379–384.

Gorenflo, D. (1991). Behältersieden, *VDI-Wärmeatlas*, Band 6, erweiterte Auflage, Ha 1 bis–Ha 26, VDI Verlag, Dusseldorf, germany.

Hsu, Y. Y., and R. W. Graham (1976). *Transport Processes in Boiling and Two-Phase Systems*, Hemisphere, New York, Chaps. 5 and 6.

Jakob, M. (1949). *Heat Transfer*, Vol. 1, Wiley, New York, Chap. 29.

Kutateladze, S. S. (1959). Critical heat flux during sub-cooled liquid flow [in Russian], *Energetica*, 7: 229–239, and *Izv. Akad. Nauk Otd. Tekh. Nauk*, 4: 529.

Liu, Z., and Y. Qiu (2006). Boiling heat transfer characteristics of nanofluids jet impingement on a plate surface, *Int. j. Heat Mass Transfer*, in press.

Mikic, B. B., and W. M. Rohsenow (1969). A new correlation of pool boiling data including the effect of heating surface characteristics, *J. Heat Transfer*, 91: 245–250.

Moore, F. D., and R. B. Mesler (1961). The measurement of rapid surface temperature fluctuations during nucleate boiling of water, *AIChE J*. 7: 620–624.

Mostinski, J. L. (1963). Application of the rule of corresponding states for the calculation of heat transfer and critical heat flux, *Teploenergetika*, 4: 66.

Nukiyama, S. (1934). The maximum and minimum values of heat transmitted from metal to boiling water under atmospheric pressure, *J. Jpn. Soc. Mech. Eng*., 37: 367 (translation: *Int. J. Heat Mass Transfer*, 9: 1419, 1966).

Rayleigh, J. W. S. (1917). *Philos. Mag*., XXXIV: 94, cited in H. Lamb, *Hydrodynamics*, Dover, New York, 1945, p. 122.

Rose, J., H. Uehara, S. Koyama, and T. Fujji (1999). Film condensation, in *Handbook of Phase Change: Boiling and Condensation*, S. G. Kandlikar, M. Shoji, and V. K. Dhir, Eds., Taylor & Francis, London Chap. 19.

Stephan, K. (1992). *Heat Transfer in Condensation and Boiling*, Springer-Verlag.

Stephan, K., and P. Preusser (1979). Wärmeübergang und maximale Wärmestromdichte beim Behältersieden binärer und ternärer Flussigkeitsgemische, *Chem. Ing. Tech*., 51, 37 (Synopse MS 649/79).

Vassallo, P., R. Kumar, and S. D'Amico (2004). Pool boiling heat transfer experiments in silica–water nano-fluids, *Int. J. Heat Mass Transfer*, 47(2): 407–411.

Wen, D., and Y. Ding (2005). Experimental investigation into the pool boiling heat transfer of aqueous based alumina nanofluids, *J. Nanopart. Res*., 7: 265–274.

Whalley, P. B. (1996). *Two-Phase Flow and Heat Transfer*, Oxford Science Publications, Oxford.

You, S. M., J. H. Kim, and K. M. Kim (2003). Effect of nanoparticles on critical heat flux of water in pool boiling of heat transfer, *Appl. Phys. Lett*. 83(16): 3374–3376.

Zuber, N. (1958). On the stability of boiling heat transfer, *J. Heat Transfer*, 80: 711

7 Applications and Future Directions

Today more than ever, many industries facing thermal challenges have a pressing need for ultrahigh-performance cooling. However, conventional coolants are inherently poor heat transfer fluids. Therefore, a strong need exists for new and innovative concepts to achieve ultrahigh-performance cooling in thermal management systems. Although particle-in-liquid suspensions or slurries are frequently used in industry, they are not suitable for heat transfer applications, due to severe problems caused by large particles in those suspensions or slurries. The major problem with traditional suspensions containing millimeter- or micrometer-sized particles is the rapid settling of these particles. If the fluid were kept circulating to prevent much settling, the microparticles would damage the walls of the pipe, wearing them thin. Other problems include large increases in pressure drop and clogging, particularly in small thermal control systems.

Nanofluids are a new type of heat transfer fluid engineered by uniform and stable suspension of nanometer-sized particles into liquids. Most nanofluids are very dilute suspensions of nanoparticles in liquids and contain a very small quantity, preferably less than 1% by volume, of nanoparticles. The average size of nanoparticles used in nanofluids may vary from 1 to 100 nm (preferably < 10 nm). Because nanoparticles are so small, they remain in suspension almost indefinitely and dramatically reduce erosion and clogging compared with the suspension of larger particles. Also, their larger surface area may improve heat transfer.

A number of experiments show that stable suspensions of a small amount of nanoparticles in traditional fluids produce dramatic changes in the thermal properties of base fluids. It has been shown that stable nanofluids have distinctive features such as high thermal conductivities at very low nanoparticle concentrations (Eastman et al., 2001; Patel et al., 2003), a nonlinear relationship between thermal conductivity and particle concentration (Choi et al., 2001; Hong et al., 2005; Murshed et al., 2005; Chopkar et al., 2006), strong temperature- and size-dependent conductivity (Das et al., 2003bb; Chon et al., 2005) and a threefold increase in critical heat flux in pool boiling compared to base fluids (You et al., 2003; Vassallo et al., 2004). Furthermore, recent experiments have shown that some nanofluids enhance the convective heat transfer coefficient by up to 100% compared to that of water (Xuan and Li, 2003;

Nanofluids: Science and Technology, By Sarit K. Das, Stephen U. S. Choi, Wenhua Yu, and T. Pradeep

Faulkner et al., 2004; Ding et al., 2006). These key features of nanofluids present a great opportunity for thermal scientists to explore new frontiers in wet nanotechnology and allow a variety of nanofluids, such as nanotechnology-based coolants, lubricants, hydraulic fluids, and metal-cutting fluids, to be used for a wide range of industrial applications. Therefore, nanofluids are not only of academic interest but also of industrial interest. Nanofluids can be used to improve heat transfer and energy efficiency in many thermal control systems. Nanofluids offer several benefits: for example, higher cooling rates, smaller and lighter cooling systems, reduced inventory of heat transfer fluids, decreased pumping-power needs, reduced friction coefficient, and improved wear resistance. Furthermore, nanofluids are being investigated for medical applications such as cancer therapy as well as for numerous engineering applications. For these applications it is highly desirable to achieve the highest possible thermal properties at the smallest possible concentrations of nanoparticles.

Much of the work in the field of nanofluids is being done in national laboratories and academia and is at a stage beyond discovery– research. However, a great number of companies in the United States and other countries have been showing great interest and suggesting a number of possible applications of nanofluids. This great industrial interest shows that nanofluids can be used for a wide variety of industries, ranging from electronics, transportation, HVAC, and process heating and cooling to energy conversion and supply and magnet cooling. Furthermore, the number of companies that see the practical potential of nanofluid technology and are in active development work in the area of nanofluids for specific industrial applications are increasing. For example, in the transportation industry, some of the leading motor companies have in-house applied research and development of nanofluids.

Liquid cooling is described in Section 7.1. Original and significant papers concerned with applications of nanofluids, and original contributions to the development of nanofluids, have been brought together in Section 7.2. Some challenges in applied nanofluid research and directions for applied research in the development of commercial nanofluids are presented in Section 7.3.

7.1. LIQUID COOLING

Cooling is a top technical challenge facing high-tech industries such as microelectronics, transportation, manufacturing, and metrology. For example, the heat generation rates and device temperatures of many electronic devices are increasing continuously due to trends toward higher levels of integration, faster clock speeds, and smaller feature size. Cooling is required to maintain various electronic products at desired operating temperatures for proper functioning and long life. Powerful chips are at the heart of electronic products. As chip performance continues to progress according to Moore's law, power densities and device temperatures reach levels that prevent their reliable operation. This trend presents cooling challenges to thermal scientists and engineers. Although thermal management covers the entire heat path from the chip to the ambient, removal of high heat fluxes from the heat source and cooling of local hot spots at the chip and package levels have become the most vexing problems in thermal management of electronic devices and systems.

Air cooling, consisting of a heat sink and a fan, is the most common method for cooling electronics. Some new techniques have emerged to extend the useful range of air cooling, such as piezofans (Acikalin et al., 2004) and synthetic jet cooling (Glezer and Mahalingam, 2003). For heat fluxes below 100 W/cm^2, air cooling may remain the cooling method of choice, but with unacceptable noise levels. The International Technology Roadmap for Semiconductors (ITRS) predicts that by 2018, high-performance integrated circuits will contain more than 9.8 billion transistors on a chip area of 280 mm^2: more than 40 times as many as on today's chips of 90-nm node size. Therefore, future processors for high-performance computers and servers are expected to dissipate higher power, in the range 100 to 300 W/cm^2. Since conventional forced-convection air-cooling techniques are reaching their limits and will no longer be enough for high-heat-flux devices, liquid cooling techniques that are capable of removing heat fluxes over 100 W/cm^2 are expected to take center stage in electronics cooling (Schmidt, 2005). Furthermore, it is possible that the next generation of computer chips will produce hot spots with a heat flux over 1000 W/cm^2. Hot spots increase the failure rate of microelectronics devices. Since air cooling makes it harder to avoid hot spots, liquid cooling is expected to become more important.

Liquid cooling technologies have been developed and applied. Some high-end desktop computers use small water cooling systems to beat the heat. Today, nearly all notebook computers use phase-change liquid cooling (heat pipes) to move heat from the CPU to the case. High-power-density devices such as power electronics have used liquid cooling.

A number of single- and two-phase liquid cooling techniques, such as liquid-cooled microchannel heat sinks, advanced heat pipes, immersion cooling involving pool boiling of a dielectric working fluid, and liquid jet impingement cooling, have been developed to handle high levels of heat dissipation in electronics at the chip or package level. Air-cooled aluminum heat pipes developed at Intel can remove a total power of 200 W with a heat flux up to 200 W/cm^2. Loop heat pipes or capillary pumped loops can operate at power levels in excess of 600 W/cm^2 without depriming the wick. A number of devices have been developed that are capable of removing heat fluxes in excess of 1000 W/cm^2, such as the microjet impingement cooling array (Wang et al., 2004). For some consumer electronics applications, such as high-power-density semiconductor devices (solid-state light sources) for projection television sets, heat fluxes exceed 2000 W/cm^2. Therefore, liquid cooling appears to be inevitable.

With nanofluids, the cooling performance could be improved (Lee and Choi, 1996). In the realm of electronics cooling, some companies are conducting research to use nanofluids instead of water. In addition, nanofluids could effectively remove hot spots and maintain components at uniform temperatures. Considering the range of efforts under way to extend liquid cooling technologies and the superior thermal properties of nanofluids, the future seems bright for using nanofluids for high-heat-flux and hot-spot cooling systems for computer, telecom, power and defense electronics uses, among others.

7.2. APPLIED RESEARCH IN NANOFLUIDS

Research on a variety of nanofluids applications is under way. Nanofluids find most of their applications in thermal management of industrial and consumer products. Efficient cooling is vital to realizing the functions and long-term reliability of a variety of industrial and consumer products and there are tribological and biomedical applications. Other potential applications are described briefly because there are no published papers. Recent studies have demonstrated the ability of nanofluids to improve the performance of real-world devices and systems such as automatic transmissions.

7.2.1. Cooling Applications

Crystal Silicon Mirror Cooling One of the first applications of research in the field of nanofluids is for developing an advanced cooling technology to cool crystal silicon mirrors used in high-intensity x-ray sources (Lee and Choi, 1996). Because an x-ray beam creates tremendous heat as it bounces off a mirror, cooling rates of 2000 to 3000 W/cm^2 should be achievable with the advanced cooling technology. Lee and Choi carried out analysis to estimate the performance of microchannel heat exchangers with water, liquid nitrogen, and nanofluids as the working fluid. For an optimized channel width that minimizes the thermal resistance of a microchannel heat exchanger, performance of a nanofluid-cooled microchannel heat exchanger has been compared with that of water-cooled and liquid-nitrogen-cooled microchannel heat exchangers. The results show that nanofluids can remarkably reduce the thermal resistances and increase the power densities, so they demonstrated the superiority of a nanofluid-cooled silicon microchannel heat exchanger. The benefits of using nanofluids as a room-temperature coolant are clear, including dramatic enhancement of cooling rates while operating the advanced cooling system at room temperature. Furthermore, the possibility of thermal distortion and flow-induced vibration will be eliminated by passing the nanofluids through microchannels within the silicon mirror itself.

The advanced cooling technology developed by Lee and Choi (1996) employs microchannels filled with nanofluids. The advanced cooling technology could provide more efficient cooling than that of other cooling technologies because the microchannels increase the effective heat transfer area, and the metallic nanoparticles increase the effective thermal conductivity of coolants. The advanced cooling technology may be used in cooling engines, superconducting magnets, and densely packed computer chips. Lee and Choi (1996) estimated that for high-aspect-ratio microchannels, power densities of $\approx$ 3000 W/cm^2 should be achievable using nanofluids. Therefore, future experimental work on nanofluid-cooled microchannel heat exchangers will advance the art of cooling high-heat-load devices.

Electronics Cooling Chien et al. (2003) were probably the first to show experimentally that the thermal performance of heat pipes can be enhanced by nearly a factor of 2 when nanofluids are used. They used water-based nanofluids containing 17-nm gold nanoparticles as the working fluid in a disk-shaped miniature heat pipe (DMHP). They measured the thermal resistance of the DMHP with both nanofluids and deionized (DI) water. The results show that the thermal resistance of a DMHP is reduced significantly (40%) when nanofluids are used instead of DI water.

Tsai et al. (2004) used gold nanofluids as the working fluid for a conventional meshed circular heat pipe. Monodispersed gold nanoparticles of various sizes (2 to 35 and 15 to 75 nm) were synthesized by the reduction of aqueous hydrogen tetrachloroaurate ($HAuCl_4$) with trisodium citrate and tannic acid. The heat pipe was designed as a heat spreader for a CPU in a notebook or desktop PC. A 200-mesh wire screen was used in the heat pipe being tested. They measured the thermal resistance of the meshed heat pipe with nanofluids and DI water. The thermal resistance of the meshed heat pipe with nanofluids is in the range 0.17 to 0.215°C/W, lower than that with DI water. The results show that at the same charge volume, there is a significant reduction (by as much as 37%) in the thermal resistance of heat pipe with nanofluid compared with DI water. The results also show that the thermal resistance of a vertical meshed heat pipe varies with the size of gold nanoparticles and that monodispersed nanoparticles are better than aggregated nanoparticles. The work clearly shows the advantages of a *conventional* circular heat pipe with nanofluids over that with DI water.

Kang et al. (2006) measured the temperature distribution and thermal resistance of a conventional grooved circular heat pipe with water-based nanofluids containing a tiny amount (1 to 50 ppm) of 35-nm silver nanoparticles. The measured wall temperature of the heat pipe is lower with nanofluids than with DI water and decreases with increasing concentration of silver nanoparticles, up to 50 ppm. The results also show that at the same charge volume, the thermal resistance of a heat pipe with nanofluids is reduced by 10 to 80% compared with that of DI water at an input power of 30 to 60 W. They compared their work with 35-nm silver nanofluids to that by Wei et al. (2005) with 10-nm silver nanofluids to show that the maximum reduction in the thermal resistance of the heat pipe is 50% for 10-nm silver nanoparticles and 80% for 35-nm silver nanoparticles. This nanoparticle-size-dependent performance of the heat pipe is very interesting and needs further experimental study.

Ma et al. (2006a; b) were first to develop an ultrahigh-performance chip cooling device called the nanofluid oscillating heat pipe (OHP). Conventional heat pipes with oscillating motions generated by a variable-frequency shaker dramatically increased the heat removal rate in capillary tubes. However, the use of mechanically driven shakers limits their application to chip cooling. Compared to regular heat pipes, OHPs, pioneered by Akachi (1990), have very attractive features: (1) there are no moving parts, because heat from heat sources such as computer chips drives oscillating flow inside the capillary tube; (2) the thermally driven oscillating flow enhances both forced-convection and phase-change heat transfer; and (3) the liquid flow does not interfere with vapor flow because the liquid and vapor phases both flow in the same direction.

However, the low conductivity of the working fluid and the thermocapillary flow in the thin-film region may result in local flow instability that limits its ability to remove ultrahigh heat fluxes over 1000 W/cm^2. No existing cooling technologies can effectively remove such heat fluxes. Ma et al. (2006a,b) estimated that an OHP with water-based nanofluids containing Al_2O_3 nanoparticles has the ability to remove heat in excess of 1000 W/cm^2, so they proposed the novel concept of combined nanofluids and OHPs for breakthrough chip cooling. Proof of concept work was done on water-based nanofluids containing 1 vol% of 20 to 50-nm diamond nanoparticles in a vertical OHP. The results clearly demonstrate that when the OHP is charged with nanofluids, its heat transport capability is enhanced significantly. For example, at an input power of 80.0 W, diamond nanofluids can reduce the temperature difference between an evaporator and condenser from 40.9 in an OHP with water to 24.3°C in an OHP with nanofluids. It is interesting to note that although diamond nanoparticles can settle in stationary water, the thermally excited oscillating motion in the OHP can keep them suspended. Nanofluid OHPs appear to be the most likely candidate cooling device for removing heat fluxes over 1000 W/cm^2. This innovative and interesting work will advance the state of the art in nanofluid applications and accelerate development of a highly efficient cooling device for ultrahigh-heat-flux electronic systems. More work will be necessary to demonstrate that a combination of OHP and nanofluid technologies can produce a heat removal rate over 1000 W/cm^2. Furthermore, it is vital to perform basic theoretical studies to understand the fundamentals of nanofluids in thin-film evaporation and oscillating motion.

Chien and Huang (2005) numerically investigated silion microchannel heat sink (MCHS) performance using nanofluids as the coolant. The nanofluids are a mixture of pure water and copper nanoparticles with various volume fractions in the range 0.3 to 2%. The nanofluids are treated as a single-phase fluid in the theoretical model of pressure drop for nanofluid flow in microchannels. Thermal dispersion due to random particle motion is included in the experimental heat transfer correlation for laminar flow of nanofluids in microchannels. For laminar fully developed flow in the two MCHS geometries studied, the analytical results show that compared with pure water, nanofluids can enhance MCHS performance. The performance enhancement is due to the increased thermal conductivity of nanofluids and nanoparticle thermal dispersion. Another finding is that nanoparticles do not produce an extra pressure drop because the nanoparticle is small and the particle volume fraction is low. This work demonstrates that nanofluids have the potential to enhance MCHS performance.

Koo and Kleinstreuer (2005) simulated and analyzed steady laminar flow of nanofluids in microchannels. New models for the effective thermal conductivity and dynamic viscosity of nanofluids are employed in the analysis. The results show that the addition of 20-nm copper nanoparticles at low volume fractions (1 to 4%) to high-Prandtl-number liquids significantly increases the heat transfer performance of a microchannel heat sink. As a result of the analysis, the following recommendations are made for microheat-sink performance improvements: use of high-Prandtl-number base fluids, nanoparticles of high thermal conductivity with a dielectric onstant very close to that of the base fluid, microchannels with a high aspect ratio, and channel walls treated to avoid nanoparticle accumulation. These recommendations could be significant for practical applications of nanofluids.

Chein and Chuang (2007) investigated MCHS performance analytically and experimentally using nanofluids as the coolant. They carried out a simple theoretical analysis and performed experiments to verify their theoretical predictions. They made the CuO–water nanofluids using an arc-submerged nanoparticle synthesis system. No dispersant was added in the nanofluids. The needle-shaped CuO nanoparticle sizes are quite uniform with average length and width 80 and 20 nm, respectively. The results show that when the flow rate is low, the amount of heat absorbed by water-based nanofluids containing CuO nanoparticles is greater than that absorbed by water and that the MCHS wall temperature is lower with nanofluds than with water. The results also show that although nanofluids have a higher viscosity than water, the pressure drop across the nanofluid-cooled MCHS increases only slightly compared with the water-cooled MCHS.

Palm et al. (2006) investigated the heat transfer enhancement capabilities of nanofluids inside typical radial flow impingement jet cooling systems. The laminar forced-convection flow of water-based nanofluids containing Al_2O_3 nanoparticles with volume fractions of 1 and 4% in a radial flow cooling system was considered using the temperature-dependent properties of nanofluids. The results show that nanofluids can increase the average wall heat transfer coefficient significantly and decrease the wall shear stress. Furthermore, the use of a temperature-dependent property model predicts much better thermal and hydraulic performance than that in previous predictions using constant properties (Roy et al., 2004). This is encouraging for the use of nanofluids in impinging jet cooling systems.

Zhou (2004) investigated the heat transfer characteristics of copper nanofluids with acoustic cavitation bubbles. Acetone-based nanofluids containing copper nanoparticles with average particle sizes in the range 80 to 100 nm were used in this study. Copper nanoparticles dispersed in acetone by acoustic cavitation bubble clusters are extremely stable. The two important findings of this study are that (1) with no acoustic field, copper nanoparticles enhance single-phase convection heat transfer and reduce boiling heat transfer, and (2) with an acoustic field, copper nanoparticles enhance both single-phase convection and pool boiling heat transfer. The second finding is substantially different from previous experimental studies such as Das et al. (2003aa) and Zhou and Liu (2004). Thus, Tzou showed, probably for the first time, that when copper nanoparticles and acoustic cavitation work together, both single-phase natural convection and pool boiling heat transfer were enhanced markedly and the boiling hysteresis disappeared. This interesting and significant work will accelerate practical applications of nanofluids.

Vehicle Cooling Nanoparticles can be dispersed not only in coolants and engine oils, but in transmission fluids, gear oils, and other fluids and lubricants. These nanofluids may provide better overall thermal management and better lubrication. Tzeng et al. (2005) were probably the first to apply nanofluid research in cooling a real-world automatic power transmission system. They dispersed CuO and Al_2O_3 nanoparticles into automatic transmission oil to investigate the optimum possible compositions of a nanofluid for higher heat transfer performance. The experimental platform is the real rotary blade coupling (RBC) of a power transmission system of a real-time four-wheel-drive vehicle. It adopts advanced RBC, where a high local temperature occurs easily at high rotating speed. RBC design is so precise that if the local temperature is higher than 266°F, excessive thermal stress may damage its rotating components. As a result, the power cannot be transmitted to the rear wheels, affecting vehicle performance severely. Moreover, the damaged RBC is not repairable and should be replaced. Therefore, it is imperative to improve the heat transfer efficiency to contain excessive thermal stress on the components of the power transmission system. They measured the temperature distribution of the RBC exterior at four different rotating speeds (400, 800, 1200, and 1600 rpm), simulating the conditions of a real car at various rotating speeds. The results show that CuO nanofluids have the lowest temperature distribution at both high and low rotating speed and accordingly, the best heat transfer effect. This work is significant because it shows a real-world application of nanofluids and so represents a giant step forward for industrial applications of nanofluids.

Transformer Cooling The power generation industry is interested in transformer cooling application of nanofluids for reducing transformer size and weight. The ever-growing demand for greater electricity production will require upgrades of most transformers at some point in the near future at a potential cost of millions of dollars in hardware retrofits. If the heat transfer capability of existing transformers can be increased, many of the upgrades may not be necessary. Xuan and Li (2000) and Yu et al. (2007) have demonstrated that the heat transfer properties of transformer oils can be improved by using nanoparticle additives.

The increased thermal transport of transformer oils translates into either a reduction in the size of new transformers at the same level of power transmitted or an increase in the performance of existing transformers. Keeping at the cutting edge of nanotechnology remains a top task for many companies and laboratories. Specifically, nanofluid-based transformer oil is likely to be the next-generation cooling fluid in transformers. The first key element in nanofluid technology is uniform dispersion of nonagglomerated nanoparticles. Homogeneity of the dispersion may be overcome by special mechanical dispersing techniques and the creative use of chemical dispersants. However, this goal is still challenging for new combinations of nanoparticle-based fluids. We need to focus more on dynamic interactions between nanoparticles and liquid molecules and interface structure and chemistry.

Space and Nuclear Systems Cooling You et al. (2003) and Vassallo et al. (2004) have discovered the unprecedented phenomenon that nanofluids can double or triple the CHF in pool boiling. Kim et al. (2006) found that the high surface wettability caused by nanoparticle deposition can explain this remarkable thermal properties of nanofluids. The work is important in developing realistic predictive models of the CHF in nanofluids. The ability to greatly increase the CHF, the upper heat flux limit in nucleate boiling systems, is of paramount practical importance to ultrahigh-heat-flux devices that use nucleate boiling, such as high-power lasers and nuclear reactor components. Therefore, nanofluids have opened up exciting possibilities for raising chip power in electronic devices or simplifying cooling requirements for space applications. Most of all, leading nuclear researchers are very much interested in the use of nanofluids with dramatically increased CHF values because it could enable very safe operation of commercial or military nuclear reactors. The Massachusetts Institute of Technology has established an interdisciplinary center for nanofluid technology for the nuclear energy industry. Currently, they are evaluating the potential impact of the use of nanofluids on the safety, neutronic, and economic performance of nuclear systems.

Defense Applications A number of military devices and systems, such as high-powered military electronics, military vehicle components, radars, and lasers, require high-heat-flux cooling, to the level of thousands of W/cm^2. At this level, cooling with conventional heat transfer fluids is difficult. Some specific examples of potential military applications include power electronics and directed-energy weapons cooling. Since directed-energy weapons involve heat sources with high heat fluxes (> 500 to $1000\,W/cm^2$), cooling of the direct-energy weapon and associated power electronics is critical and is further complicated by the limited capability of current heat transfer fluids. Nanofluids also provide advanced cooling technology for military vehicles, submarines, and high-power laser diodes. It appears that nanofluid research for defense applications considers multifunctional nanofluids with added thermal energy storage or energy harvesting through chemical reactions.

7.2.2. Tribological Applications

Nanofluid technology can help develop better oils and lubricants. Recent nanofluid activity involves the use of nanoparticles in lubricants to enhance tribological properties of lubricants, such as load-carrying capacity and antiwear and friction-reducing properties between moving mechanical components. In lubrication application it has been reported that surface-modified nanoparticles stably dispersed in mineral oils are very effective in reducing wear and enhancing load-carrying capacity (Que et al., 1997). Li et al. (2004) performed experiments on lubricant nanofluids containing IrO_2 and ZrO_2 nanopartcles. The results showed that nanoparticles decrease friction remarkably on the surface of 100 C6 steel.

7.2.3. Biomedical Applications

Nanofluids was originally developed primarily for thermal management applications such as engine, microelectronics, and photonics. However, nanofluids can be formulated for a variety of other uses for faster cooling. Nanofluids are now being developed for medical applications, including cancer therapy. Traditional cancer treatment methods have significant side effects. Iron-based nanoparticles can be used as delivery vehicles for drugs or radiation without damaging nearby healthy tissue by guiding the particles up the bloodstream to a tumor with magnets. Nanofluids could also be used for safer surgery by cooling around the surgical region, thereby enhancing a patient's chance of survival and reducing the risk of organ damage. In contrast to cooling, nanofluids could be used to produce higher temperatures around tumors, to kill cancerous cells without affecting nearby healthy cells (Jordan et al., 1999).

7.2.4. Other Potential Applications

Earlier we looked at a variety of nanofluid applications, of which electronics and engine cooling are two salient areas. Nanofluids would have a particularly high impact in these two areas. Therefore, more research is expected in vehicle cooling systems, including radiators, automatic transmission, and exhaust-gas recirculation heat exchangers. There will also be nanofluid applications in cooling fuel cells and power electronics for hybrid vehicles. Since there are reports that nanofluids reduce friction and wear, there would also be applications for oil and gas drilling.

Other possible areas for the application of nanofluids technology include cooling a new class of superpowerful and small computers and other electronic devices for use in military systems, airplanes, or spacecraft as well as for large-scale cooling. In the future, nanofluids could be used to maintain a high temperature gradient in thermoelectrics that would convert waste heat to useful electrical energy. In buildings, nanofluids could increase energy efficiency without the need to use a more powerful pump, so saving energy in a HVAC system and providing major environmental benefits. In the renewable energy industry, nanofluids could be utilized to enhance heat transfer from solar collectors to storage tanks and to increase the energy density. To this must be added cooling for major process industries, including materials, chemical, food and drink, oil and gas, paper and printing, and textiles.

Novel projected applications of nanofluids include sensors and diagnostics that instantly detect chemical warfare agents in water or water- or foodborne contamination; biomedical applications such as cooling medical devices, detecting unhealthy substances in the blood, cancer treatment, or drug delivery; and development of advanced technologies such as advanced vapor compression refrigeration systems. These are just a few of the almost endless variety of nanofluids applications. Therefore, nanofluids will be increasingly important for high-value-added niche applications as well as for high-volume applications.

7.3. FUTURE RESEARCH

Many industries have a strong need for improved fluids that can transfer heat more efficiently. Nanofluids transfer heat more efficiently than do conventional fluids. Therefore, when used to improve the design and performance of thermal management systems, nanofluids offer several benefits, including improved reliability, reduction in cooling system size, decreased pumping-power needs, increased energy and fuel efficiencies, and lower pollution. Thus, nanofluids can have a significant impact in cooling a number of high-heat-flux devices and systems used in consumer, industrial, and defense industries. Although nanofluids offer very promising opportunities, there are still a number of technical issues on the road to commercialization. Some technical barriers facing the development of commercially available nanofluid technology were identified in Chapter 1, where we suggested some research needed to overcome these barriers and to achieve cost-effective high-volume production of nanofluids. Nanofluids offer challenges related to production, properties, heat transfer, and applications. In this section we highlight some future directions in each of these challenging areas.

7.3.1. Production of Nanofluids

Nanofluids have been produced successfully in the laboratory. The challenge is now to develop techniques for cost-effective industrial-scale production of nanofluids because nanofluid production is a rate-limiting area in the introduction of nanofluids to commercial applications. Our focus should be on identifying promising methods that do not require a vacuum and that provide continuous fluid feed and extraction capabilities in a production system. New technologies for making stable nanofluids that do not require a vacuum and utilize a semicontinuous or continuous process will probably replace current methods of producing nanofluids. In the future, these new methods could lead to the ability to make nanofluids much faster and cheaper than with current methods. The critical technical breakthroughs in industrial-scale production of nanofluids necessary to bring nanofluids to commercialization are expected to be achieved through continued support of nanofluid R&D and collaboration with industrial partners.

7.3.2. Thermal Properties and Heat Transfer Performance of Nanofluids

In almost all cases, the thermal conductivity of conventional heat transfer fluids is improved by the addition of small amounts of nanoparticles. Nanofluids containing nonagglomerated nanoparticles produced by one-step methods are much more effective in their ability to conduct heat. Nanofluids containing carbon nanotubes in oil show the largest enhancement, as much as 2.5 times higher than that of plain base fluid at very low (< 1 vol%) concentrations.

Experimental investigations have demonstrated remarkable heat transfer enhancement when using nanofluids in forced convection: a 40% increase in turbulent convection heat transfer with the addition of 2.0 vol% Cu nanoparticles in water and roughly a twofold increase in laminar convection heat transfer by the addition of 1.1 vol% CNTs in water.

In the future, nanofluid properties and heat transfer performance should be tested under potential service conditions and all of the experimental data on nanofluids should be collated for designers of industrial thermal management systems. For example, designers of advanced engine cooling systems can use the database and assess the effects of the nanofluid's superior heat transfer characteristics on the size of the engine compartment and fuel economy of the vehicle.

7.3.3. Applications of Nanofluids

Applied research in nanofluids has demonstrated in the laboratory that nanoparticles can be used to enhance the thermal conductivity and heat transfer performance of conventional heat transfer fluids. Some researchers took the concept one step further into practical applications and demonstrated the ability of nanofluids to improve the performance of real-world devices and systems such as automatic transmissions. Thus, nanofluid research has made the initial transition from our laboratory to industrial research laboratories. This extremely important work has provided guidance as to the right direction. However, it is only the first step in the development of commercial nanofluid technology. The development and demonstration of nanofluid technology for commercialization are currently limited, but are expected to grow rapidly with strong collaboration between nanofluid researchers in academia and industry. Long-term suspension stability and homogeneity of dispersed nanoparticles are vital to commercial nanofluids. Protective coating of metallic nanoparticles may require methods to enhance oxidative stability for a long period of time. In the future, promising nanofluids should be studied not only under real-world conditions of use, but also over a longer period of time.

Nanofluid research could lead to a major breakthrough in developing next-generation coolants for numerous engineering and medical applications, as described earlier. Applied research in nanofluids would improve the competitive edge for a number of products only if applied research is based on the fundamental theories of nanofluids. Nanofluid theory is essential to applications because the formulation of nanofluids can be designed to optimize their use in specific applications. For example, a new theoretical model in which a key parameter is the particle size predicts strongly size-dependent conductivity (Jang and Choi, 2004). This size-as-a-parameter approach, instead of the conventional adding-more-particles approach, could lead to an important breakthrough in the manufacture of nanotechnology-based coolants. Probably one of the most fascinating feature of nanofluids is related to their strong temperature-dependent conductivity (Das et al., 2003b). This unique and unprecedented property could be utilized to develop smart nanofluids for removing hot spots in high-heat-flux microelectronics. Therefore, nanofluids may play a key role in the design of high-reliability high-performance electronic systems. Cooling hot spots requires the development of smart liquid coolants such as nanofluids or local cooling techniques such as microrefrigerators.

At present, a severe lack of modeling capabilities exist to predict the unprecedented thermal properties of nanofluids. Any one proposed theory or model of nanofluid is not able to explain all experimental data; additional research is needed. We need to make more accurate measurements not only at the macroscopic level, but also at the nanoscale level, particularly as a function of temperature, and generate more data to really determine the issues discussed in Chapter 1. Accurate and precise measurement methods and highly controlled experiments are critical for reliable thermophysical properties data for promising nanofluids, better theoretical understanding of nanoscale mechanisms for enhanced properties, and development of physics-based models. Combined experimental work and modeling is needed to understand the underlying physics of heat conduction in nanofluids. Critical experiments are also needed to allow development of theories of nanofluids that can provide an excellent description of how nanoparticles enhance thermal transport in nanofluids. Only realistic theories can guide the formulation of optimized nanofluids for practical applications. With the formulation of optimized nanofluids many people would fully appreciate the thermal benefits of nanofluids.

With continued collaboration between basic and applied nanofluids researchers in academia and industry on the thermal properties, performance, theory, mechanisms, modeling, and development and eventual commercialization of nanofluids, nanofluid research is expected to bring breakthroughs in nanotechnology-based cooling technology and have a strong impact on a wide range of engineering and biomedical applications of nanofluids.

REFERENCES

Acikalin, T., S. M. Wait, S. V. Garimella, and A. Raman (2004). Experimental investigation of the thermal performance of piezoelectric fans, *Heat Transfer Eng*., 25(1): 4–14.

Akachi, H. (1990). Structure of a heat pipe, U.S. patent 4,921,041.

Chein, R., and J. Chuang (2007). Experimental microchannel heat sink performance studies using nanofluids, *Int. J. Therm. Sci*., 46(1): 57–66.

Chein, R., and G. Huang (2005). Analysis of microchannel heat sink performance using nanofluids, *Appl. Therm. Eng*., 25: 3104–3114.

Chien, H. T., C. I. Tsai, P. H. Chen, and P. Y. Chen (2003). Improvement on thermal performance of a disk-shaped miniature heat pipe with nanofluid, *Proc. International*

Conference on Electronics Packaging Technology 2003, IEEE, Piscataway, NJ, pp. 389–391.

Choi, S. U. S., Z. G., Zhang, W., Yu, F. E., Lockwood, and E. A. Grulke (2001). Anomalous thermal conductivity enhancement in nano-tube suspensions, *Appl. Phys. Lett.*, 79: 2252–2254.

Chon C. H., K. D., Kihm, S. P. Lee, and S. U. S. Choi (2005). Empirical correlation finding the role of temperature and particle size for nanofluid (Al_2O_3) thermal conductivity enhancement, *Appl. Phys. Lett.*, 87: 153107.

Chopkar, M, P. K., Das, and I. Manna (2006). Synthesis and characterization of nanofluid for advanced heat transfer applications, *Scr. Mater.*, 55: 549–552.

Das, S. K., N. Putra, and W. Roetzel (2003a). Pool boiling characteristics of nano-fluids, *Int. J. Heat Mass Transfer*, 46(5): 851–862.

Das, S. K., N., Putra, P., Thiesen, and W. Roetzel (2003b). Temperature dependence of thermal conductivity enhancement for nanofluids, *J. Heat Transfer*, 125: 567–574.

Ding, Y., H., Alias, D., Wen, and Williams, R. A. (2006). Heat transfer of aqueous suspensions of carbon nanotubes (CNT nanofluids), *Int. J. Heat Mass Transfer*, 49: 240–250.

Eastman, J. A., S. U. S. Choi, S. Li, W. Yu, and L. J. Thompson (2001). Anomalously increased effective thermal conductivities of ethylene glycol– based nanofluids containing copper nanoparticles, *Appl. Phys. Lett.*, 78(6): 718–720.

Faulkner, D. J., D. R. Rector, J. J. Davidson, and R. Shekarriz (2004). Enhanced heat transfer through the use of nanofluids in forced convection, Paper. IMECE2004-62147, presented at the 2004 ASME International Mechanical Engineering Congress and RD&D Expo, Anaheim, CA, Nov. 13–19.

Glezer, A., and R. Mahalingam (2003). System and method for thermal management by synthetic jet ejector channel cooling techniques, U.S. patent 6,588,497.

Hong, T. K., H.S. Yang and C. J. Choi (2005). Study of the enhanced thermal conductivity of Fe nanofluids, *J. Appl. Phys.*, 97: 064311.

Jang, S. P., and S. U. S. Choi (2004). Role of Brownian motion in the enhanced thermal conductivity of nanofluids, *Appl. Phys. Lett.*, 84: 4316–4318.

Jordan, A., R. Scholz, P. Wust, H. Fähling, and R. Felix (1999). Magnetic fluid hyperthermia (MFH): cancer treatment with ac magnetic field induced excitation of biocompatible superparamagnetic nanoparticles, *J. Magn. Magn. Mater.*, 201(1–3): 413–419.

Kang, S. W., W. C. Wei, S. H. Tsai, and S. Y. Yang (2006). Experimental investigation of silver nano-fluid on heat pipe thermal performance, *Appl. Therm. Eng.*, 26: 2377–2382.

Kim, S. J., I. C. Bang, J. Buongiorno, and L. W. Hu (2006). Effects of nanoparticle deposition on surface wettability influencing boiling heat transfer in nanofluids, *Appl. Phys. Lett.*, 89: 153107.

Koo, J., and C. Kleinstreuer (2005). Laminar nanofluid flow in microheat-sinks, *Int. J. Heat Mass Transfer*, 48: 2652–2661.

Lee, S., and S. U. S. Choi (1996). Application of metallic nanoparticle suspensions in advanced cooling systems, in *Recent Advances in Solids/Structures and Application of Metallic Materials*, Y. Kwon, D. Davis, H. Chung, Eds., PVP-342/MD-72, American Society of Mechanical Engineers, New York, pp. 227–234.

Li, J. F., H. Liao, X. Y. Wang, B. Normand, V. Ji, C. X. Ding, and C. Coddet (2004). Improvement in wear resistance of plasma sprayed yttria stabilized zirconia coating using nanostructured powder, *Tribol. Int.*, 37: 77–84.

Ma, H. B., C. Wilson, B. Borgmeyer, K. Park, Q. Yu, S. U. S. Choi, and M. Tirumala (2006a). Effect of nanofluid on the heat transport capability in an oscillating heat pipe, *Appl. Phys. Lett.*, 88: 143116.

Ma, H. B., C. Wilson, Q. Yu, K. Park, S. U. S. Choi, and M. Tirumala (2006b). An experimental investigation of heat transport capability in a nanofluid oscillating heat pipe, *J. Heat Transfer*, 128(11): 1213–1216.

Murshed, S. M. S., K. C., Leong and C. Yang (2005). Enhanced thermal conductivity of TiO_2– water based nanofluids, *Int. J. Therm. Sci*, 44: 367–373.

Palm, S. J., G., Roy, and C. T., Nguyen (2006). Heat transfer enhancement with the use of nanofluids in radial flow cooling systems considering temperature-dependent properties, *Appl. Therm. Eng.*, 26(17– 18): 2209–2218.

Patel, H. E., S. K. Das, T. Sundararajan, N. A. Sreekumaran, B. George, and T. Pradeep (2003). Thermal conductivities of naked and monolayer protected metal nanoparticle based nanofluids: manifestation of anomalous enhancement and chemical effects, *Appl. Phys. Lett.*, 83(14): 2931–2933.

Que, Q., J. Zhang, and Z. Zhang (1997). Synthesis, structure and lubricating properties of dialkyldithiophosphate-modified Mo– S compound nanoclusters, *Wear*, 209(1– 2): 8–12.

Roy, G., C. T., Nguyen, and P., Lajoie (2004). Numerical investigation of laminar flow and heat transfer in a radial flow cooling system with the use of nanofluids, *Superlattices Microstruct.*, 35(3– 6): 497–511.

Schmidt, R. (2005). Liquid cooling is back, *Electron. Cool.*, 11(3): 34–38.

Tsai, C. Y., H. T. Chien, P. P. Ding, B. Chan, T. Y. Luh, and P. H. Chen (2004). Effect of structural character of gold nanoparticles in nanofluid on heat pipe thermal performance, *Mater. Lett.*, 58: 1461–1465.

Tzeng, S.-C., C.-W. Lin, and K. D. Huang (2005). Heat transfer enhancement of nanofluids in rotary blade coupling of four-wheel-drive vehicles, *Acta Mech.*, 179: 11–23.

Vassallo, P., R., Kumar, and S. D'Amico (2004). Pool boiling heat transfer experiments in silica– water nano-fluids, *Int. J. Heat Mass Transfer*, 47: 407–411.

Wang, E. N., L. Zhang, L. Jiang, J.-M. Koo, J. G. Maveety, E. A. Sanchez, K. E. Goodson, and T. W. Kenny (2004). Micromachined jets for liquid impingement cooling of VLSI chips, *J. MicroElectroMech. Syst.*, 13(5): 833–842.

Wei, W. C., S. H. Tsai, S. Y. Yang, S. W. Kang (2005). Effect of nano-fluid concentration on heat pipe thermal performance, *IASME Trans.*, 2: 1432–1439.

Xuan, Y., and Q. Li (2000). Heat transfer enhancement of nanofluids, *Int. J. Heat Fluid Flow*, 21(1): 58–64.

Xuan, Y., and Q. Li (2003). Investigation on convective heat transfer and flow features of nanofluids, *J. Heat Transfer*, 125: 151–155.

You, S. M., J. H., Kim, and K. M. Kim (2003). Effect of nanoparticles on critical heat flux of water in pool boiling of heat transfer, *Appl. Phys. Lett.*, 83: 3374–3376.

Yu, W., S. U. S. Choi, and J. Drobnik (2007). Temperature and concentration dependence of effective thermal conductivities of alumina-oil based nanofluids, presented at the

ECI Conference on Nanofluids: Fundamental and Applications, Copper Mountain, CO, Sept. 16–20.

Zhou, D. W. (2004). Heat transfer enhancement of copper nanofluid with acoustic cavitation. *Int. J. Heat Mass Transfer*, 47: 3109–3117.

Zhou, D. W., and D. Y. Liu (2004). Heat transfer characteristics of nanofluids in an acoustic cavitation field, *Heat Transfer Eng*., 25(6): 90–100.

APPENDIX: Nanoparticles Prepared by Various Routes

NOTES

acac: acetylacetonate
AOT: sodium bis(2-ethylhexyl)-sulfosuccinate
BSPP: bis(*p*-sulfonatophenyl) phenylphosphine dihydrate dipotassium salt solution
Cit: citrate
COD: 1,5-cyclooctadiene
COT: 1,3,5-cyclooctatriene
CTAB: cetyltrimethylammonium-bromide
CTAC: cetyltrimethylammonium-chloride
Cup: cupferron, $C_6H_5N(NO)O-$
DBA: dibenzylideneacetone
DBS: dodecylbenzenesulfonic acid
DEG: diethylene glycol
DMF: dimethylformamide
DPE: 1,1-diphenylethylene
DTAB: dodecyltrimethylammonium bromide
EG: ethylene glycol
FOD: 2,2-dimethyl-6,6,7,7,8,8,8-heptafluoro-3,5-octanedionate
Glyme: ethylene glycol dimethyl ether
GSH: glutathione
HDA: hexadecylamine
HDD: 1,2-hexadecanediol
HMT: hexamethylenetetramine
HOPG: highly oriented pyrolytic graphite
Igepal®: a surfactant
IPA: isopropyl alcohol
MTAB: myristyltrimethylammonium bromide
NaDDBS: sodium dodecyl-benzenesulfonate
NP-10: tergitol®, an ether
NTA: nitrilotriacetate
OA: oleyl amine
OAc: oleic acid
OTAB: octyltrimethylammonium bromide
PEGDE: poly(ethyleneglycol) dimethyl ether
PNIPAAm: poly(*N*-isopropylacrylamide)
PMMA: poly(methyl methacrylate)
PVP: poly(vinyl pyrrolidone)
RT: room temperature
SDBS: sodium dodecylbenzenesulfonate
SDS: sodium dodecylsulfonate
$TC_{12}AB$: tetradodecylammonium bromide
THF: Tetrahydrofuran
THF: tetrahydrofuran
TMAH: tetramethylammonium hydroxide
TMPD: tetramethyl *p*-phenylenediamine
TOA: triocyl amine
TOAB: tetraoctylammonium bromide
TOP: trioctyl phosphine
TOPO: tri-*n*-octylphosphine oxide ($C_{24}H_{51}OP$)
TPP: tripolyphosphate
(TOP)-Se: trioctylphosphine selenide
Triton® X-114: a surfactant

Nanofluids: Science and Technology, By Sarit K. Das, Stephen U. S. Choi, Wenhua Yu, and T. Pradeep

Metal Particle Prepared	Starting Material	Reducing Agent	Medium	Stabilizer	Phase Transfer	Condition	Redispersible	Mean Particle Diameter (nm)	Typical Application	Refs.
Ti	Titanium tetrachloride	$Ar-H_2$ plasma	Gas	No	No	Plasma	Not specified	2–10	Antimicrobial, antibiotic and antifungal agents	1
V	Bulk metal	Annealing	Not specified	No	No	300 K	Not specified	2–9	Multifunctional catalysis, photocatalysis	2
Cr	Chromium, Fischer carbene complex	TOP/ temperature	Biphase water/$CHCl_3$	TOP	No	300°C	Yes	2.5	Catalysis	3
Mn	Mn powder	Arc discharge	Ar gas	Not specified	No	Not specified	Yes	Up to 80	Material science, catalysis	4
	—	Bioreduction with hyacinth	—	—	—	—	—	1–4	—	5
Fe	$Fe(NO_3)_3 \cdot 9H_2O$	Hydrogen	Gas (fluidized bed reactor)	No	No	160–300°C	Not specified	300–500	Electronics, catalysis	6
	$FeCl_2$	$NaBH_4$	Triphase (xylene/ water/ pyridine)	TOPO	No	RT	Not specified	3	Electrocatalysis	7
	$Fe(CO)_5$	High temperature	Oleic acid	Oleic acid	No	300°C.	Yes	11–20	Electronics, drug delivery	8
Co	$Co(\eta^3\text{-}C_8H_{13})(\eta^4\text{-}C_8H_{12})$	Hydrogen	THF	PVP	No	0°C, 20°C, 60°C	Not specified	1–1.5	Catalysis	9
	$Co(AOT)_2$ (cobalt bis(2-ethylhexyl)-sulfosuccinate)	$NaBH_4$	Water/oil	TOP	No	RT	Yes	5.8	Data storage devices, sensors, catalysis	10

	$Co_2(CO)_8$	Thermolysis in the presence of aluminum alkyls	Toluene	Korantin SH, oleic acid, LP-4, AOT	No	110°C	Yes	10 ± 1.1	Technical and biomedical applications	11
	$CoCl_2$ vapor	H_2	Ar gas	Nil	No	800–950°C	Not specified	50–78	Electronic, magnetic, optical, and chemical properties	12
	Co^+ ions	Ion implantation	SiO_2 matrix	Nil	No	RT	No	1–10	Magnetic storage devices	13
	$CoCl_2$	Lithium hydrotriethylborate ($LiBH(C_2H_5)_3$)	THF	3-(*N*,*N*-Dimethyldodecylammonia)-propanesulfonate (SB12)	No	RT	Yes	< 5	Data storage devices, biomedical engineering	14
Ni	$Ni(COD)_2$	H_2	CH_2Cl_2	PVP	No	RT	Yes	20–30	Catalysts, engineering	15
	$NiCl_2$	Hydrazine	Water/CTAB/*n*-hexanol	Water/CTAB/*n*-hexanol	No	73°C	Yes	4.6	Catalysts, engineering materials, drug delivery	16
	$Ni(COD)_2$	Hydrazine	CH_2Cl_2	PVP	No	RT	Yes	30	Magnetic and electronic applications	17
Ni (rods)	$NiCl_2$	HDA/TOPO	THF	HDA	No	RT	Yes	—	Catalysis, drug delivery, electronics	18
	$NiCl_2$	Hydrazine	Aqueous	CTAB/$TC_{12}AB$	No	60°C	Yes	10–36	Catalysis, drug delivery, electronics	19

(*Continued*)

Metal Particle Prepared	Starting Material	Reducing Agent	Medium	Stabilizer	Phase Transfer	Condition	Redispersible	Mean Particle Diameter (nm)	Typical Application	Refs.
	$NiCl_2$	Hydrazine	Ethylene glycol	Nil	No	60°C	Yes	9.2	Catalysis, drug delivery, electronics	20
	$NiCl_2$	H_2	Ar gas	Oxide layer formed on the surface	No	800–950°C	Not mentioned	31–106	drug delivery, electronics	21
	$Ni(NO_3)_2$	CNT	N_2 atmosphere	CNT	No	600°C	Yes	10–50	Catalyst and conducting or magnetic materials	22
Cu	Cu^{2+}	$NaBH_4$/ hydrazine	Water/AOT	AOT reverse micelles	No	RT	Not specified	2–10	Catalysis	23
	Copper(II) acetate	Hydrazine	Water and 2-ethoxy-ethanol	PVP	No	Refluxing condition	Yes	6.6–30.2	Catalysis	24
	$CuCl_2$	$NaBH_4$	Water-in-oil microemulsions	Without micelles	No	RT	Not specified	5–15	Catalysis	25
	Copper(II) hydrazine carboxylate $Cu(N_2H_3COO)_2{\cdot}2H_2O$	Thermal/ sono-chemical	Water	Nil	No	~ 80°C	Not specified	200–250	Catalysis	26
	Cu salt	H_2	Water	Carbon nanotube template	No	Below 773 K	Yes	100 nm to several micro meters	Catalysis	27
	$Cu(NO_3)_2$	$NaBH_3CN$/ TMPD	Water in supercritical fluid micro-emulsion	Perfluoro-polyether phosphate	No	38°C	Not mentioned	5–15	Catalysis	28

	Cupric nitrate $[Cu(NO_3)_2 \cdot 2.5H_2O]$	IPA	Water–IPA mixture	CTAB	No	Ambient conditions	Not mentioned	5–20	Catalysis	29
	$CuSO_4$	Hydrazine	Ethylene glycol	Ethylene glycol	No	Microwave irradiation	Yes	~15	Catalysis	30
	$CuSO_4 \cdot 5H_2O$	$NaH_2PO_2 \cdot H_2O$	Ethylene glycol	Ethylene glycol	Polyvinyl-pyrrolidone	Microwave irradiation	Yes	~20	Thermal conductivity	31
Zn	$[Zn(C_6H_{11})_2]$	Thermal reduction	Anisole/water	PVP	No	130°C	Not specified	6–17	Sensors, transducers, photocells	32
Ga	Pure Ga metal	Evaporation–condensation	Sapphire	SiO_x	No	High temperature	No	10–18	Electronics, photonics	33
	Bulk gallium	Evaporation–condensation	Al_2O_3/SiO_x	Al_2O_3/SiO_x	No	High temperature	No	10–60	Electronics, photonics	34
Ga	Bulk gallium	Light-assisted self-assembly technique	High vacuum	Not specified	No	100 K	Not specified	50 ± 14	Electronics and photonics	35
Ga	$GaCl_3$	In situ reduction	Benzene	SBA-H (mesoporous silica)	No	Not specifed	No	Not mentioned	Photonics	36
Ge (nanowires)	$GeCl_4$/phenyl-$GeCl_3$	Na metal	Pentane/hexane	Alkyl group	No	275°C	Yes	7–30	Optoelectronics	37
Ge	$GeCl_4$	NaK alloy	Heptane	R-$GeCl_3$	No	270°C during crystalization	Yes	6–20	Optoelectronics	38
Ge	$Me_3GeS(CH_2)_3$ $Si(OMe)_3$	H_2	Air/H_2	SiO_2 xero gel	No	900°C	Not specified	2.5–14.5	Optoelectronics	39
Ge	$GeCl_4$	H_2	H_2 in argon	Zeolite Y	No	470°C	Not specified	3	Optoelectronics, photonics	40
Ge	Sodium germanide (NaGe)/$GeCl_4$	Metathesis reaction	Diglyme/glyme/triglyme	Methyllithium butyllithium, or octylmagnesium bromide	No	Refluxing conditions	Yes	6.2–6.5	Optoelectronics and photonics	41

(*Continued*)

Metal Particle Prepared	Starting Material	Reducing Agent	Medium	Stabilizer	Phase Transfer	Condition	Redispersible	Mean Particle Diameter (nm)	Typical Application	Refs.
Ge	GeI_4	$LiAiH_4$	CTAB/toluene	*n*-Alkene	No	Refluxing conditions	Yes	2–7	Optoelectronics, photonics	42
Ge (nano-cubes)	$GeCl_4$/phenyl-$GeCl_3$	Na	Hexane	Heptaethylene glycol monododecyl ether ($C_{12}E_7$)	No	280°C	No	100 ± 20	Optoelectronics, photonics	43
Ge	$Ge[N(SiMe_3)_2]_2$	Thermal reduction	Octadecene	Oleylamine	No	285°C	Yes	7 ± 4	Optoelectronics	44
Ge	$GeCl_4$	Sodium naphthalide	Glyme	Butyl ($-C_4H_9$)	No	0°C	Yes	6.1 ± 2	Photonics	45
Se	Selenious acid (H_2SeO_3)	Hydrazine	Heptane/water	AOT	No	RT	Not specified	4–300	Rectifiers, solar cells, photographic exposure meters, xerography	46
Se (nano-wires)	Sodium selenite (Na_2SeO_3)	Glutathione (GSH)	Water	Nil	No	RT	No	60 ± 5	Photographic exposure meters	47
Y	Bulk Y	Pulsed laser deposition	Not mentioned	Nil	No	Not mentioned	No	32	Television and laser systems	48
Mo	$Mo(CO)_6$	Thermal decomposition	Octyl ether solution/N_2 atmosphere	Octanoic acid/bis-2-ethyl-hexyl-amine	No	Not mentioned	Yes	3–14	Catalysis, magnetism, electronics	49
Ru	$RuCl_3$	Ethylene glycol	Ethylene glycol	PVP	No	150°C	Yes	2	Catalysis	50

	Ru(COD)(COT)	H_2	THF	PVP and cellulose acetate	No	RT	Yes	2–3	Catalysis	51
	$RuCl_3$	Ethylene glycol	Ethylene glycol	Ethylene glycol	No	453 K	No	5	Catalysis	52
	$RuCl_3$	$NaBH_4$	Water	Ethylene diamine	No	RT	Yes	~2.1	Catalysis	53
	Ru(COD)(COT)	Dihydrogen	THF–methanol solution	Chiral N-donor ligands	No	RT	Yes	1.6–2.5	Catalysis	54
Rh	$RhCl_3 \cdot 3H_2O$	High temperature, high pressure	PVP	Water–ethanol mixture	No	473 K	No	2.7–4.6	Environmental, chemical, and sensing	55
	$RhCl_3$	Lithium triethylborohydride [$LiB(C_2H_5)_3H$]	1-Dodecanethiol [$CH_3(CH_2)_{11}$-SH]	THF	No	60°C	Yes	1–3	Catalysis, chemical sensing, nanoscale capacitors, semiconductor devices	56
Pd	$[PdCl_4]^{2-}$	Photosensitized reduction	PVP	Alcohols	No	RT	Yes	1.7–3	Catalysis	57
	$PdCl_2$	$NaBH_4$	TOAB	Toluene-aqueous	Yes	RT	Yes	1–5	Catalysis	58
	$Pd(NH_3)_4Cl_2$	Hydrazine	Water-in-oil microemulsions	CTAB	No	RT	Yes	4 ± 0.9	Catalysis	59
	Pd surfactant complex	Thermal decomposition	—	Surfactant	—	—	Yes	3–7	—	60
	K_2PdCl_4	$NaBH_4$	Water	Dendrimers	No	Not specified	Yes	1.4, 1.7 ± 0.4	Catalysis	61

(*Continued*)

Metal Particle Prepared	Starting Material	Reducing Agent	Medium	Stabilizer	Phase Transfer	Condition	Redispersible	Mean Particle Diameter (nm)	Typical Application	Refs.
	—	—	—	TPP and TOP	No	—	Yes	—	Catalysis	62
	$Pd(FOD)_2$	Thermally induced reduction	*o*-Xylene/DMF/1-octanol	CTAB/DTAB/MTAB/OTAB	No	Refluxing condition	Yes	6.2–18.5	Catalysis	63
	$Pd(NO_3)_2$	Photoexcited Keggin ions	Water	Keggin ions	Yes	RT	Yes	4 ± 2	Catalysis	64
	$Pd(NO_3)_2$	Sonochemical reduction	Water	PVP	No	Not specified	Yes	3–6	Catalysis	65
Ag	Silver bis(2-ethyl-hexyl)sulfo-succinate, Ag(AOT)	Hydrazine	Water/AOT	Dodecane thiol	No	RT	Yes	3.4	Catalysis, optical and electronic devices	66
	$AgNO_3$	Trisodium citrate	Water	Trisodium citrate	No	Boiling condition	No	40–60	Electronics	67
	$AgNO_3$	$NaBH_4$	Biphase (toluene/water)	1-Dodecane thiol	Yes	RT	Yes	5–8	Catalysis, optical and electronic devices	68
	$AgNO_3$	Ultrasound	Water	NTA	No	20°C	Yes	20	Catalysis, optical and electronic devices	69
	$AgNO_3$	Amine	Ethanol	Amine	No	60°C	Yes	1–2	Catalysis	70
Ag (nano-wires)	$AgNO_3$	$NaBH_4$/ascorbic acid	Water	CTAB	No	RT	Yes	42 ± 3	Catalysis, optical and electronic devices	71
In	$[In(\eta 5\text{-}C_5H_5)]$	—	—	—	—	—	—	—	—	72
	$[In(\eta 5\text{-}C_5H_5)]$	—	Toluene	—	—	—	—	15 ± 2	—	73

	Bulk indium	Ultrasound irradiation	Paraffin oil	Nil	No	473 K	Not specified	50–2000	Photonic devices	74
	Anhydrous indium trichloride	Sodium metal	DMF or TOP	TOP	No	120–150°C	Yes	15–50	Bionanotechnology, nanoxerography	75
Sn	$SnCl_2 \cdot 2H_2O$	Mg	Water		No	RT	No	Not specified	Rechargeable batteries, gas sensors	76
	$SnCl_4/Mg_2Sn$	Metathesis reaction	Ethylene glycol dimethyl ether (glyme)	*n*-Butyl group	No	Refluxing conditions	No	6.5 ± 1.7	Rechargeable batteries, gas sensors	77
	Bulk tin	Dispersion method	Paraffin oil	Oxide formed on the surface	No	240°C	Not specified	30–40	Gas sensors	78
	$SnCl_4$	KBH_4	Water	Cellulose fibers	No	RT	Yes	20–30	Lithium–ion batteries	79
	$SnCl_4$	$NaBH_4$	1,2-Dimethoxyethane/deoxygenated water (argon atmosphere)	Hydrobenzamide, citrate, and poly(vinyl pyrrolidone)	No	RT	No	~50/~100–300	Batteries	80
Sb	Bulk metal	Thermal evaporation	Ultra high vacuum	HOPG surface	No	830 K	Not specified	120	Flame retardants	81
SB (nanowire)	$SbCl_3$	$NaBH_4$	DMF	PVP	No	RT	Not specified	~20	Catalysis	82
Te	Orthotelluric acid (H_6TeO_6) or tellurium dioxide (TeO_2)	Hydrazine	Water, EG, and water–EG	Nil	No	20–200°C	No	50–100	Optoelectronic devices	83

(*Continued*)

Metal Particle Prepared	Starting Material	Reducing Agent	Medium	Stabilizer	Phase Transfer	Condition	Redispersible	Mean Particle Diameter (nm)	Typical Application	Refs.
Te (rods)	$(NH_4)_2TeS_4$	Na_2SO_3	Water	NaDDBS	No	RT	No	14	Optoelectronic devices	84
Te (rods)	$(NH_4)_2TeS_4$	Sodium sulfite (Na_2SO_3)	Water	SDBS/SDS/ PVP	No	RT	No	10–40	Optoelectronic devices	85
Te (nano-tubes)	Sodium tellurate ($Na_2TeO_4 \cdot 2H_2O$)	Formamide ($HCONH_2$)	—	—	—	—	—	200–600	Photoconducting devices	86
Sm	$Sm(NO_3)_3 \cdot 5H_2O$	Bioreduction alfalfa (*Medicago sativa*)	Water	Nil	No	25°C	Not specified	10	Drug delivery, medicines	87
Eu	$EuCl_3 \cdot 6H_2O$	—	Water in oil	(3-Amino-propyl)tri-ethoxy-silane	No	RT	No	36 ± 4	Biological detection, biotechnology	88
Gd	$GdCl_3$	Alkalide reduction	THF	Nil	No	RT	Yes	12	Drug delivery	89
	Gadolinium hexane-dione (GdH)	Polyoxyl 20–stearyl ether	Oil-in-water micro-emulsions	Emulsifying wax micro-emulsion	No	55°C,	No	85 ± 9	Drug delivery (cancer therapy)	90
Tb	Bulk metal	Sputtering	Cr or W matrix	Cr or W matrix	No	Not specified	Not specified	8	Biological detection	91
	Tb^{3+} chelate	$NH_3/H_2O/$ Trixon-X-100	Oil-in-water micro-emulsions	Water-in-oil micro-emulsion	No	RT	No	45 ± 3	Biological detections	92
Dy	Bulk metal	Vapor deposition/ sputtering	Gas	Nil	No	Not mentioned	No	4–12	Magnetic applications	93
	$DyCl_3$	Alkalide reduction	THF	Crown ether (15-crown-5)	No	Not mentioned	No	8–16	Magnetic applications	94

Yb	Yb salt	Bioreduction	Water	Nil	No	RT	No	2–10	Fiber amplier and fiber optic technologies	95
Ta	Bulk tantalum	Hydrogen arc plasma method	—	Nil	No	10,000°C	No	Less than 10	Superconductor, as a dopant in photoelectrode materials	96
W	WO_3 powder	Electron beam irradiation	Nil	Nil	No	RT	No	2–6	Semiconductor devices	97
	Tungsten hexacarbonyl	Thermal decomposition	—	—	—	—	—	15–60	Semiconductor devices	98
Re	Re_2CO_{10}	Thermal decomposition	—	—	—	—	—	6–12.8	—	99
Ir	Hexachloroidic acid	Ethylene glycol	Ethylene glycol	Nil	No	100°C	Yes	3	Catalysts, electrocatalysts, chemical synthesis	100
Pt	—	—	—	—	—	RT	—	—	—	101
	$Pt(dba)_2$	CO	Toluene	CO/phosphine ligands/ solvent	No	RT	Yes	1–2	Catalysis	102
	$H_2PtCl_6{\cdot}6H_2O$	Ethanol reduction	Ethanol/water	PNIPAAm/ PVP	No	Refluxing conditions	Yes	0.5–4.5	Catalysis	103
	$H_2PtCl_6{\cdot}6H_2O$/ K_2PtCl_4	Lithium triethylborohydride	THF	Octadecane thiol	No	RT/35°C	Yes	~3	Hydrogen storage, electronics	104
	$H_2PtCl_6{\cdot}6H_2O$	H_2	Acetone	Nanostructured carbon	No	300°C	Yes	~3	Fuel cell systems	105

(*Continued*)

Metal Particle Prepared	Starting Material	Reducing Agent	Medium	Stabilizer	Phase Transfer	Condition	Redispersible	Mean Particle Diameter (nm)	Typical Application	Refs.
	Disodium hexahydroxyplatinate [$Na_2Pt(OH)_6$]	$NaBH_4$	Water	Thiol-functionalized ionic liquids (TFILs)	No	RT	Yes	2–4.2	Electronics, catalysis	106
Au	$HAuCl_4$	Na_3Cit	Water	Cit^{3-}	No	RT	No	12–60	Sols	107–109
	$HAuCl_4$	Ascorbic acid	Water	Cit^{3-}	No	RT	No	12	Biology	110
Au_{11}	Au-arylphosphine complexes	$NaBH_4$	Water	Aryl phosphine	No	Inert	Yes	0.82	Biology, TEM labeling	111
Au	$HAuCl_4$	Ethyl alcohol	Water		No	Ultrasonication	No	10	Biology	112
	$HAuCl_4$	P (white) in ether	Water		No	Boiling	No	3–5	Biology	113, 114
	$HAuCl_4$	NaSCN	Water	SCN	No	RT	No	2.6	Biology	115
	$HAuCl_4$	Na_3Cit/tannic acid	Water	Cit^{3-}, tannic acid	No	Heating	No	3–17	Biology	116, 117
	$HAuCl_4$	$NaBH_4$	Water	Cit^{3-}	No	4°C	No	4	Sols	118
	$HAuCl_4$	$NaBH_4$	Biphase (toluene/water)	Thiol	Yes	RT	Yes	3	Powder, solution, chemistry	119
	$HAuCl_4$	$Na_3Cit/NaBH_4$	Water	Cit^{3-}	No	0°C	No	4	Sols	120
Au_{55}	$HAuCl_4$	$(Ph_3P)AuCl$	Benzene	B_2H_6	Yes	Not specified	Yes	1.4	Catalysis, sensors, molecular electronics	121

Hg	$Hg(ClO_4)_2$	γ-Irradiation	Water	γ-Radiolytic reduction	No	γ-Irradiation	Not mentioned	~ 100	Optical devices	122
Pb	—	Melt-spinning and ball-milling techniques	—	—	—	—	—	5–30	Biological and chemical sensors	123
	$[Pb\{N(SiMe_3)_2\}_2]$	$[H_2Al(OtBu)]_2$,	THF/acetonitrile	Not specified	No	– 100°C	Yes	10–200	Biological and chemical sensors	124
Bi	Bismuth (III) citrate	$NaBH_4$	Poly (oxyethylene)$_9$ nonyl phenol ether, poly (oxyethylene)$_5$ nonyl phenol ether/water emulsion	PVP	No	RT	Yes	18–105	Electronics	125
	Bismuth 2-ethyl-hexanoate	$LiBEt_3H$	Dioctyl ether	TOP/oleic acid	No	175°C	Yes	15 ± 2	Thermoelec-tronics	126
	Bulk metal	High-energy electron beam	Ar and He gas	Nil	No	High temper-ature	No	4.5–10	Thermoelec-tronics	127
	Bulk bismuth	Solution dispersion method	Paraffin oil	Nil	No	280°C	Yes	40–50	"Green" lubricant materials, electronics	128

*Arranged in the order of atomic numbers.

OXIDE NANOPARTICLES

Aqueous Media

Nanomaterial	Starting Material	Precipitating	Stabilizing Agent	Conditions	Size (nm)	Refs.
VO_2(B)	NH_4VO_3	$N_2H_4 \cdot H_2O$	None	Calcined 300°C	35	129
Cr_2O_3	$K_2Cr_2O_7$	$N_2H_4 \cdot H_2O$	None	Calcined 500°C	30	129
γ-Mn_2O_3	$KMnO_4$	$N_2H_4 \cdot H_2O$	none	—	8	129
$Ni_{0.5}Zn_{0.5}Fe_2O_4$	Ni, Zn, Fe nitrates	NaOH	—	Annealed 300° C	9–90	130
$MgFe_2O_4$ nitrates	NaOH	—	—	—	—	131
$Sm_{1-x}Sr_xFeO_{3-\delta}$	—	—	—	—	—	132
$Ce_{1.8}Y_{0.2}O_{1.9}$	Nitrates ·	Oxalic acid	—	Annealed 500, 1000°C	10–100	133
CeO_2	Nitrate	$(NH_4)_2CO_3$	—	Annealed 300°C	6	134
NiO	Ni^{2+} salts	$(NH_4)_2CO_3$	—	Annealed 400°C	10–15	135
$Bi_4Ti_3O_{12}$	Basic TiO_2, $Bi(NO_3)_3$	H^+	—	500–800°C	16–48	136
TiO_2	$TiCl_3$	NH_4OH	PMMA	RT	50–60	137
Fe_3O_4	Fe^{2+}, Fe^{3+}	NaOH	—	70° C	—	138
$MnFe_2O_4$	Mn^{2+}, Fe^{2+}	NaOH	—	< 100°C	5–25	139
Pr-doped CeO_2	$Ce(NO_3)_3$, $PrCl_3$	—	HMT	100°C	13	140
$CoFe_2O_4$	Fe^{3+}, Co^{2+}	NaOH	—	< 100°C	14–18	141
$CoFe_2O_4$	—	—	H^+	—	—	142
Fe_3O_4	—	—	—	—	—	143, 144
$MnFe_2O_4$	$MnCl_2$, $FeCl_3$	NaOH	None	100°C	5–25	145
Fe_3O_4	$FeCl_3$, $FeCl_2$	NH_4OH	H+	N_2 atm.	8–50	146
NiO	$NiCl_2$	NH_4OH	CTAB	Annealed 500°C	22–28	147
ZnO	$ZnCl_2$	NH_4OH	CTAB	Annealed 500°C	40–60	147
SnO_2	$SnCl_4$	NH_4OH	CTAB	Annealed 500°C	11–18	147
Sb_2O_3	$SbCl_3$	NaOH	PVA	Annealed 350°C	10–80	148

Nonaqueous Media

Nanomaterial Agent	Starting Material	Medium	Precipitating	Stabilizing Agent	Conditions	Size (nm)	Refs.
$LiCoO_2$	$LiNO_3$, $Co(NO_3)_2$	Ethanol	KOH	—	400–700°C annealed	12–41	149
RuO_2	$RuNO(NO_3)_2$	Ethanol	TMAH	—	90°C for precipitation, 500°C annealed	14	150
γ-Fe_2O_3	$Fe(NO_3)_3$	Steric acid	Steric acid	—	125°C ppt, 200°C cal.	5–20	151, 152
$BaTiO_3$	$BaTi(O_2C(CH_3)_6 \cdot CH_3)[OCH(CH_3)_2]_5$	DPE	H_2O_2	Oelic acid	100° C	6–12	153
MFe_2O_4 (M = Mn, Fe, Co, Ni, Zn)	Chlorides	DEG	—	DEG	Heat	3–7	154
Fe_3O_4	Fe(acac)	Diethyl ether	—	Ethanol	HDD, OA, OAc	4	155
γ-Fe_2O_3	$FeCup_3$	OA	TOA	OA	300°C	4–10	156
MnO	$MnCup_2$						
Cu	CuCup						

NANOPARTICLES USING MICROEMULSION-BASED METHODS

Metals

Metal	Starting Material	Surfactant	Reductant	Reaction Conditions	Product Size (nm)	Refs.
Co	$CoCl_2$	AOT	$NaBH_4$	—	< 1	157
Ni	$NiCl_2$	CTAB	$N_2H_4 \cdot H_2O$	pH ~ 13	4	16
Cu	$Cu(AOT)_2$	AOT	N_2H_4	—	2–10	158, 159
	$Cu(AOT)_2$	AOT	$NaBH_4$	—	20–28	158, 159
Se	H_2SeO_3	AOT	$N_2H_4 \cdot HCl$	—	4–300	45
Rh	$RhCl_3$	PEGDE	H_2	—	3	160
Pd	$PdCl_2$	PEGDE	$N_2H_4 \cdot H_2O$	pH ~ 7	4	160
Ag	$AgNO_3$	PEGDE	$NaBH_4$	—	3–9	161
Ir	$IrCl_3$	PEGDE	H_2	70°C	3	160
Pt	H_2PtCl_6	PEGDE	$N_2H_4 \cdot H_2O$	—	3	160
Bi	$BiOClO_4$	AOT	$NaBH_4$	Ar atm.	2–10	161
FePt Fe_2Pt $FePt_3$	Fe^{2+}, Pt^{2+}	—	$NaBH_4$	—	—	162, 163

Oxides

Oxide	Starting Material	Surfactant	Precipitating Agent	Reaction Conditions	Size(nm)	Refs.
$LiNi_{0.8}Co_{0.2}O_2$	$LiNO_3$ $Ni(NO_3)_2$ $Co(NO_3)_2$	NP-10	Kerosene	Calcined 400–800°C	19–100	164
Al_2O_3	$AlCl_3$	Triton X-114	NH_4OH	Calcined 600–900°C	50–60	165
TiO_2	$Ti(OiPr)_4$	AOT	H_2O	—	20–200	166
$Mn_{1-x}Zn_xFe_2O_4$	$Mn(NO_3)_2$ $Zn(NO_3)_2$ $Fe(NO_3)_3$	AOT	NH_4OH	Calcined 300–600°C	5–37	167
Fe_3O_4	$FeCl_2$ $FeCl_3$	AOT	NH_4OH	—	~2	168
Fe_3O_4	$FeSO_4$	AOT	NH_4OH	—	10	169
$CoCrFeO_4$	$CoCl_2$ $CrCl_3$ $Fe(NO_3)_3$	SDS	CH_3NH_2	Calcined 600°C	6–16	170
$CoFe_2O_4$	$CoCl_2$ $FeCl_3$	SDS	CH_3NH_2	Dried 100°C	6–9	171
$Ni_{1-x}Zn_xFe_2O_4$	$Ni(NO_3)_2$ $Zn(NO_3)_2$ $Fe(NO_3)_3$	AOT	NH_4OH	Calcined 300–600°C	5–30	169
CuM_2O_5 (M) Ho, Er	$Cu(NO_3)_2$ $Ho(NO_3)_3$ $Er(NO_3)_3$	CTAB	$(NH_4)_2CO_3$	Calcined 900°C	25–30	172
$Y_3Fe_5O_{12}$ $Fe(NO_3)_3$	$Y(NO_3)_3$	Igepal CA-520	NH_4OH + $(NH_4)_2CO_3$	Calcined 600–1000°C	3	173
$YBa_2Cu_3O_{7-\delta}$	$Y(OAc)_3$ $BaCO_3$ $Cu(OAc)_2$	Igepal CA-430	Oxalic acid	—	3–12	174
SnO_2	$SnCl_4$	AOT	NH_4OH	Calcined 600°C	30–70	175
$BaFe_{12}O_{19}$	$Ba(NO_3)_2$ $Fe(NO_3)_3$	CTAB	$(NH_4)_2CO_3$	Calcined 950°C	5–25	176, 177
CeO_2	$Ce(NO_3)_3$	CTAB	NH_4OH	Calcined 500–700°C	6–10	178

Chalcogenides

Chalcogenide	Starting Material	Precipitating Agent	Size (nm)	Ref.
PbS, PbSe	$Pb(NO_3)_2$	Na_2S	2–4	179
ZnS (Mn doped)	$ZnCl_2$,$MnCl_2$	Na_2S	~5	180

ANISOTROPIC NANOPARTICLES

One-Dimensional Nanoparticles: Rods

Material	Starting Material	Seed	Reducing Agent	Stabilizer	Medium, Condition	Redispersibility	Particle Dimensions (nm)	Typical Applications	Refs.
Au	$HAuCl_4$	Au, 4 nm	$NaBH_4$/ascorbic acid	CTAB	Aqueous, RT	Yes	*d*: 15 *l*: variable	Nanolaser optics, therapeutics, sensor devices	181
	Au	—	Au (anode) and Pt (cathode)	CTAB	Aqueous, 42°C	Yes	*d*: 10 *l*: variable	Biological labels, sensor devices	182, 183
	$HAuCl_4$	—	$NaBH_4/Na_3Cit$	CTAB	Aqueous, laser irradiation	Yes	*d*:15 *l*: variable		184
Ag	$AgNO_3$	Ag, 4 nm	$NaBH_4$/ascorbic acid	CTAB	Aqueous, RT	Yes	*d*: 30 *l*: variable	SERS	71, 185
CdS	$CdCl_2$, Na_2S	—	—	CTAB	Aqueous, 40°C	Yes	*d*:10 *l*: variable	Semiconductors	186
	Cd, S	—	Ethylenediamine	Ethylenediamine	Organic, 120–190°C	Yes	*d*: 10 *l*: variable	Light-emitting devices	187

(Continued)

Material	Starting Material	Seed	Reducing Agent	Stabilizer	Medium, Condition	Redispersibility	Particle Dimensions (nm)	Typical Applications	Refs.
CdSe	$CdCl_2$, Na_2Se	—	—	CTAB	Aqueous 40°C	No	d:10 l:variable	Semiconductors, light-emitting devices	186
	$CdCl_2$, Se, Na	—	Ethylene-diamine	Ethylene-diamine	Organic 80–100°C	No	d: 20 l: variable	Semiconductor	188
	$Cd(CH_3)_2$, Se	—	Hexylphos-phonic acid	TOPO	Organic, 290°C	No	d: 5 l: variable	Light-emitting diodes, photovoltaic devices	189
ZnO	$ZnAc_2$	—	Hydrazine monohy-drate	DBS	Organic Heated up to boiling point of xylene	Not mentioned	d:150 l: 2170	Potential applica-tions(solar cells), nanolasers	190
	$ZnAc_2$	—	Hydrazine hydrate	SDS	Organic 90°C	Not mentioned	d: 80 l:variable	Semiconductors	191
	$Zn(NO_3)_2$	—	NaOH	Ethylene-diamine	Aqueous 180°C	Not mentioned	d: 45 l: 1540	Photonic and Electronic material	192

Three-Dimensional Nanoparticles: Triangles

Material	Starting Material	Seed	Reducing Agent	Stabilizer	Medium, Condition	Redispersibility	Edge Length Dimensions (nm)	Typical Applications	Refs.
Au	$HAuCl_4$	Au, 4–6 nm	$NaBH_4$/ascorbic acid	CTAB	Aqueous, RT	No	35	Sensors	193
Au	Au	—	Lemon grass extract	—	Aqueous, RT	No	440 thickness: 8–14	Therapeutics	194
Ag	$AgNO_3$	Ag, 2–15 nm	$NaBH_4$/Na_3 citrate	Na_3 citrate	Aqueous, laser irradiaiton	No	40–110	SERS	195

Material	Starting Material	Seed	Reducing Agent	Stabilizer	Medium, Condition	Redispersibility	Particle Dimensions	Typical Applications	Refs.
Au (prisms)	$HAuCl_4$	Au 4–6 nm	$NaBH_4$/ascorbic acid	CTAB	Aqueous, RT	Not mentioned	144	Atomic probes, SERS	196
Ag (prisms)	$AgNO_3$	Ag 8 nm	$NaBH_4/Na_3Cit$	BSPP	Aqueous, laser Irradiation	Not mentioned	Variable edge length	SERS	197
Au (plates)	$HAuCl_4$	—	Na_3Cit	PVP	Aqueous, heat	Not mentioned	Width 310, thickness 28	SERS	198
Ag (plates)	$AgNO_3$	Ag 15 nm	$NaBH_4/Na_3Cit$	CTAB	Aqueous, RT	Not mentioned	Width 200, thickness 20	SERS	199
Pd (triangular and hexagonal plates)	Na_2PdCl_4	—	Ethylene glycol	PVP	Organic, 85°C	Not mentioned	Edge length 28, thickness 5	Colorimetric sensing	200
PbSe (cubes)	Pb-$(Ac)_2 \cdot 3H_2O$, (TOP)-Se	—	Oleic acid	TOP	Organic, 230°C	No	Smallest cube 3–5 and variable	Semiconductors	201
Au (star shape)	$HAuCl_4$	Au, 4–6 nm	$NaBH_4$/ascorbic acid	CTAB	Aqueous, RT	No	Edge length 66	SERS	193
	$HAuCl_4$	—	Ascorbic acid	PVP	Aqueous, RT	No	Edge length 83, thickness 25	SERS	202
Au (bipod, tripod, tetrapod)	$HAuCl_4$	—	Ascorbic acid	CTAB	Aqueous, RT	No	Variable	Interconnectors in nano-devices	203
Co(nanocubes)	$Co_2(CO)_8$	—	*O,O*-bis(2-aminopropyl)poly (propylene glycol)	Hexane	Organic, 187°C	Not mentioned	Edge length 50–60	Magnetic nanoparticles	204

REFERENCES

1. A. B. Murphy, Formation of titanium nanoparticles from a titanium tetrachloride plasma, *J. Phys. D Appl. Phys.*, 37 (2004), 2841–2847.
2. W. Y. Hu, S. G. Xiao, J. Y. Yang, and Z. Zhang, Melting evolution and diffusion behavior of vanadium nanoparticles, *Eur. Phys. J. B*, 45 (2005), 547–554.
3. S. U. Son, Y. J. Jang, K. Y. Yoon, C. H. An, Y. Hwang, J. G. Park, H. J. Noh, J. Y. Kim, J. H. Park, and T. Hyeon, Synthesis of monodisperse chromium nanoparticles from the thermolysis of a Fischer carbene complex, *Chem. Commun.* (2005), 86–88.
4. P. Z. Si, E. Bruck, Z. D. Zhang, O. Tegus, W. S. Zhang, K. H. J. Buschow, and J. C. P. Klaasse, Structural and magnetic properties of Mn nanoparticles prepared by arc-discharge, *Mater. Res. Bull.*, 40 (2005) 29–37.
5. G. R. Ortega, P. S. Retchkiman, C. Zorrilla, H. B. Liu, G. Canizal, P. A. Perez, and J. A. Ascencio, Synthesis and characterization of Mn quantum dots by bioreduction with water hyacinth, *J. Nanosci. Nanotechnol.*, 6 (2006), 151–156.
6. E. Bermejo, T. Becue, C. Lacour, and M. Quarton, Synthesis of nanoscaled iron particles from freeze-dried precursors, *Powder Technol.*, 94 (1997), 29–34.
7. L. Guo, Q. J. Huang, X. Y. Li, and S. H. Yang, Iron nanoparticles: synthesis and applications in surface enhanced Raman scattering and electrocatalysis, *Phys. Chem. Chem. Phys.*, 3 (2001), 1661–1665.
8. T. Hyeon, S. S. Lee, J. Park, Y. Chung, and H. B. Na, Synthesis of highly crystalline and monodisperse maghemite nanocrystallites without a size-selection process, *J. Am. Chem. Soc.*, 123 (2001), 12798–12801.
9. J. Osuna, D. DeCaro, C. Amiens, B. Chaudret, E. Snoeck, M. Respaud, J. M. Broto, and A Fert, Synthesis, characterization, and magnetic properties of cobalt nanoparticles from an organometallic precursor, *J. Phys. Chem.*, 100 (1996), 14571–14574.
10. C. Petit, A. Taleb, and M. P. Pileni, Cobalt nanosized particles organized in a 2D superlattice: synthesis, characterization, and magnetic properties, *J. Phys. Chem. B*, 103 (1999), 1805–1810.
11. H. Bonnemann, W. Brijoux, R. Brinkmann, N. Matoussevitch, N. Waldofner, N. Palina, and H. Modrow, A size-selective synthesis of air stable colloidal magnetic cobalt nanoparticles, *Inorg. Chim. Acta*, 350 (2003), 617–624.
12. H. D. Jang, D. W. Hwang, D. P. Kim, H. C. Kim, B. Y. Lee, and I. B. Jeong, Preparation of cobalt nanoparticles by hydrogen reduction of cobalt chloride in the gas phase, *Mater. Res. Bull.*, 39 (2004), 63–70.
13. L. G. Jacobsohn, M. E. Hawley, D. W. Cooke, M. F. Hundley, J. D. Thompson, R. K. Schulze, and M. Nastasi, Synthesis of cobalt nanoparticles by ion implantation and effects of post implantation annealing, *J. Appl. Phys.*, 96 (2004), 4444–4450.
14. Y. J. Song, H. Modrow, L. L. Henry, C. K. Saw, E. E. Doomes, V. Palshin, J. Hormes, and C. S. S. R. Kumar, Microfluidic synthesis of cobalt nanoparticles, *Chem. Mater.*, 18 (2006) 2817–2827.
15. T. O. Ely, C. Amiens, B. Chaudret, E. Snoeck, M. Verelst, M. Respaud, and J. M. Broto, Synthesis of nickel nanoparticles: influence of aggregation induced by modification of poly(vinylpyrrolidone) chain length on their magnetic properties, *Chem. Mater.*, 11 (1999), 526–529.

16. D. H. Chen, and S. H. Wu, Synthesis of nickel nanoparticles in water-in-oil microemulsions, *Chem. Mater.*, 12 (2000), 1354–1360.
17. M. P. Zach and R. M. Penner, Nanocrystalline nickel nanoparticles, *Adv. Mater.*, 12 (2000), 878–883.
18. N. Cordente, M. Respaud, F. Senocq, M. J. Casanove, C. Amiens, and B. Chaudret, Synthesis and magnetic properties of nickel nanorods, *Nano Lett.*, 1 (2001), 565–568.
19. D. H. Chen and C. H. Hsieh, Synthesis of nickel nanoparticles in aqueous cationic surfactant solutions, *J. Mater. Chem.*, 12 (2002), 2412–2415.
20. S. H. Wu and D. H. Chen, Synthesis and characterization of nickel nanoparticles by hydrazine reduction in ethylene glycol, *J. Colloid Interface Sci.*, 259 (2003) 282–286.
21. Y. J. Suh, H. D. Jang, H. K. Chang, D. W. Hwang, and H. C. Kim, Kinetics of gas phase reduction of nickel chloride in preparation for nickel nanoparticles, *Mater. Res. Bull.*, 40 (2005), 2100–2109.
22. J. P. Cheng, X. B. Zhang, and Y. Ye, Synthesis of nickel nanoparticles and carbon encapsulated nickel nanoparticles supported on carbon nanotubes, *J. Solid State Chem.*, 176 (2006), 91–95.
23. M. P. Pileni and I. Lisiecki, Nanometer metallic copper particles synthesis in reverse micelles, *Colloids Surf. A Physicochem. Eng. Aspects*, 80 (1993), 63–68.
24. H. H. Huang, F. Q. Yan, Y. M. Kek, C. H. Chew, G. Q. Xu, W. Ji, P. S. Oh, and S. H. Tang, Synthesis, characterization, and nonlinear optical properties of copper nanoparticles, *Langmuir*, 13 (1997), 172–175.
25. L. M. Qi, J. M. Ma, and J. L. Shen, Synthesis of copper nanoparticles in nonionic water-in-oil microemulsions, *J. Colloid Interface Sci.*, 186 (1997), 498–500.
26. N. A. Dhas, C. P. Raj, and A. Gedanken, Synthesis, characterization, and properties of metallic copper nanoparticles, *Chem. Mater.*, 10 (1998), 1446–1452.
27. P. Chen, X. Wu, J. Lin, and K. L. Tan, Synthesis of Cu nanoparticles and microsized fibers by using carbon nanotubes as a template, *J. Phys. Chem. B*, 103 (1999), 4559–4561.
28. H. Ohde, F. Hunt, and C. M. Wai, Synthesis of silver and copper nanoparticles in a water-in-supercritical-carbon dioxide microemulsion, *Chem. Mater.*, 13 (2001), 4130–4135.
29. A. A. Athawale, P. P. Katre, M. Kumar, and M. B. Majumdar, Synthesis of CTAB-IPA reduced copper nanoparticles, *Mater. Chem. Phys.*, 91 (2005), 507–512.
30. H. T. Zhu, C. Y. Zhang, and Y. S. Yin, Novel synthesis of copper nanoparticles: influence of the synthesis conditions on the particle size, *Nanotechnology*, 16 (2005), 3079–3083.
31. Hai-tao Zhu, Yu-sheng Lin, and Yan-sheng Yin, A novel one-step chemical method for preparation of copper nanofluids, *J. Colloid Interface Sci.*, 277 (2004), 100–103.
32. F. Rataboul, C. Nayral, M. J. Casanove, A. Maisonnat, and B. Chaudret, Synthesis and characterization of monodisperse zinc and zinc oxide nanoparticles from the organometallic precursor $[Zn(C_6H_{11})_2]$, *J. Organomet. Chem.*, 643 (2002), 307–312.
33. P. Tognini, A. Stella, P. Cheyssac, and R. Kofman, Surface plasma resonance in solid and liquid Ga nanoparticles, *J. Non-Cryst. Solids*, 249 (1999), 117–122.

34. M. Nisoli, S. Stagira, S. DeSilvestri, A. Stella, P. Tognini, P. Cheyssac, and R. Kofman, Ultrafast electronic dynamics in solid and liquid gallium nanoparticles, *Phys. Rev. Lett.*, 78 (1997), 3575–3578.

35. K. F. MacDonald, V. A. Fedotov, and N. I. Zheludev, Optical nonlinearity resulting from a light-induced structural transition in gallium nanoparticles, *Appl. Phys. Lett.*, 82 (2003), 1087–1089.

36. L. Li and J. L. Shi, Insitu reduction and nitrification method for the synthesis of Ga and GaN quantum dots in the channels of mesoporous silicon materials, *Nanotechnology*, 17 (2006), 344–348.

37. J. R. Heath and F. K. LeGoues, A liquid solution synthesis of single crystal germanium quantum wires, *Chem. Phys. Lett.*, 208 (1993), 263–268.

38. J. R. Heath, J. J. Shiang, and A. P. Alivisatos, Germanium quantum dots: optical properties and synthesis, *J. Chem. Phys.*, 101 (1994), 1607–1615.

39. J. P. Carpenter, C. M. Lukehart, D. O. Henderson, R. Mu, B. D. Jones, R. Glosser, S. R. Stock, J. E. Wittig, and J. G. Zhu, Formation of crystalline germanium nanoclusters in a silica xerogel matrix from an organogermanium precursor, *Chem. Mater.*, 8 (1996), 1268–1274.

40. H. Miguez, V. Fornes, F. Meseguer, F. Marquez, and C. Lopez, Low-temperature synthesis of Ge nanocrystals in zeolite Y, *Appl. Phys. Lett.*, 69 (1996), 2347–2349.

41. B. R. Taylor, S. M. Kauzlarich, G. R. Delgado, and H. W. H. Lee, Solution synthesis and characterization of quantum confined Ge nanoparticles, *Chem. Mater.*, 11 (1999), 2493–2500.

42. E. Fok, M. L. Shih, A. Meldrum, and J. G. C. Veinot, Preparation of alkyl-surface functionalized germanium quantum dots via thermally initiated hydrogermylation, *Chem. Commun*. (2004), 386–387.

43. W. Z. Wang, J. Y. Huang, and Z. F. Ren, Synthesis of germanium nanocubes by a low-temperature inverse micelle solvothermal technique, *Langmuir*, 21 (2005), 751–754.

44. H. Gerung, S. D. Bunge, T. J. Boyle, C. J. Brinker, and S. M. Han, Anhydrous solution synthesis of germanium nanocrystals from the germanium(II) precursor $Ge[N(SiMe_3)_2]_2$, *Chem. Commun*. (2005), 1914–1916.

45. H. W. Chiu and S. M. Kauzlarich, Investigation of reaction conditions for optimal germanium nanoparticle production by a simple reduction route, *Chem. Mater.*, 18 (2006), 1023–1028.

46. J. A. Johnson, M. L. Saboungi, P. Thiyagarajan, R. Csencsits, and D. Meisel, Selenium nanoparticles: a small-angle neutron scattering study, *J. Phys. Chem. B*, 103 (1999), 59–63.

47. X. Y. Gao, T. Gao, and L. D. Zhang, Solution-solid growth of alpha-monoclinic selenium nanowires at room temperature, *J. Mater. Chem.*, 13 (2003), 6–8.

48. G. Bour, A. Reinholdt, A. Stepanov, C. Keutgen, and U. Kreibig, Optical and electrical properties of hydrogenated yttrium nanoparticles, *Eur. Phys. J. D*, 16 (2001), 219–223.

49. Y. Li, J. Liu, Y. Q. Wang, and Z. L. Wang, Preparation of monodispersed Fe–Mo nanoparticles as the catalyst for CVD synthesis of carbon nanotubes, *Chem. Mater.*, 13 (2001), 1008–1014.

50. Y. Motoyama, M. Takasaki, K. Higashi, S. H. Yoon, I. Mochida, and H. Nagashima, Highly-dispersed and size-controlled ruthenium nanoparticles on carbon nanofibers: preparation, characterization, and catalysis, *Chem. Lett* 35 (2006), 876–877.
51. C. Pan, K. Pelzer, K. Philippot, B. Chaudret, F. Dassenoy, P. Lecante, and M. J. Casanove, Ligand-stabilized ruthenium nanoparticles: synthesis, organization, and dynamics, *J. Am. Chem. Soc.* 123 (2001), 7584–7593.
52. A. Miyazaki, I. Balint, K. Aika, and Y. Nakano, Preparation of Ru nanoparticles supported on γ-Al_2O_3 and its novel catalytic activity for ammonia synthesis, *J. Catal.*, 204 (2001), 364–371.
53. J. Y. Lee, J. Yang, T. C. Deivaraj, and H. P. Too, A novel synthesis route for ethylenediamine-protected ruthenium nanoparticles, *J. Colloid Interface Sci.*, 268 (2003), 77–80.
54. S. Jansat, D. Picurelli, K. Pelzer, K. Philippot, M. Gomez, G. Muller, P. Lecante, and B. Chaudret, Synthesis, characterization and catalytic reactivity of ruthenium nanoparticles stabilized by chiral N-donor ligands, *New J. Chem.*, 30 (2006), 115–122.
55. M. Harada, D. Abe, and Y. Kimura, Synthesis of colloidal dispersions of rhodium nanoparticles under high temperatures and high pressures, *J. Colloid Interface Sci.*, 292 (2005), 113–121.
56. M. Marin-Almazo, J. A. Ascencio, M. Perez-Alvarez, C. Gutierrez-Wing, and M. Jose-Yacaman, Synthesis and characterization of rhodium nanoparticles using HREM techniques, *Microchem. J.*, 81 (2005), 133–138.
57. T. Teranishi and M. Miyake, Size control of palladium nanoparticles and their crystal structures, *Chem. Mater.*, 10 (1998), 594–600.
58. S. W. Chen, K. Huang, and J. A. Stearns, Alkanethiolate-protected palladium nanoparticles, *Chem. Mater.*, 12 (2000), 540–547.
59. C. C. Wang, D. H. Chen, and T. C. Huang, Synthesis of palladium nanoparticles in water-in-oil microemulsions, *Colloids Surf. A Physicochem. Eng. Aspects*, 189 (2001), 145–154.
60. S. W. Kim, J. Park, Y. Jang, Y. Chung, S. Hwang, T. Hyeon, and Y. W. Kim, Synthesis of monodisperse palladium nanoparticles, *Nano Lett.*, 3 (2003), 1289–1291.
61. R. W. J. Scott, H. C. Ye, R. R. Henriquez, and R. M. Crooks, Synthesis, characterization, and stability of dendrimer-encapsulated palladium nanoparticles, *Chem. Mater.*, 15 (2003), 3873–3878.
62. S. U. Son, Y. Jang, K. Y. Yoon, E. Kang, and T. Hyeon, Facile synthesis of various phosphine-stabilized monodisperse palladium nanoparticles through the understanding of coordination chemistry of the nanoparticles, *Nano Lett.*, 4 (2006), 1147–1151.
63. P. F. Ho and K. M. Chi, Size-controlled synthesis of Pd nanoparticles from beta-diketonato complexes of palladium, *Nanotechnology*, 15 (2004) 1059–1064.
64. S. Mandal, A. Das, R. Srivastava, and M. Sastry, Keggin ion mediated synthesis of hydrophobized Pd nanoparticles for multifunctional catalysis, *Langmuir*, 21 (2005), 2408–2413.
65. A. Nemamcha, J. L. Rehspringer, and D. Khatmi, Synthesis of palladium nanoparticles by sonochemical reduction of palladium(II) nitrate in aqueous solution, *J. Phys. Chem. B*, 110 (2006), 383–387.

66. A. Taleb, C. Petit, and M. P. Pileni, Synthesis of highly monodisperse silver nanoparticles from AOT reverse micelles: a way to 2D and 3D self-organization, *Chem. Mater.*, 9 (1997), 950–959.

67. P. V. Kamat, M. Flumiani, and G. V. Hartland, Picosecond dynamics of silver nanoclusters: photoejection of electrons and fragmentation, *J. Phys. Chem. B*, 102 (1998), 3123–3128.

68. B. A. Korgel, S. Fullam, S. Connolly, and D. Fitzmaurice, Assembly and self-organization of silver nanocrystal superlattices: ordered soft spheres, *J. Phys. Chem. B*, 102 (1998), 8379–8388.

69. J. J. Zhu, S. W. Liu, O. Palchik, Y. Koltypin, and A. Gedanken, Shape-controlled synthesis of silver nanoparticles by pulse sonoelectrochemical methods, *Langmuir*, 16 (2000), 6396–6399.

70. A. Frattini, N. Pellegri, D. Nicastro, and O. de Sanctis, Effect of amine groups in the synthesis of Ag nanoparticles using aminosilanes, *Mater. Chem. Phys*; 94 (2005), 148–152.

71. N. R. Jana, L. Gearheart, and C. J. Murphy, Wet chemical synthesis of silver nanorods and nanowires of controllable aspect ratio, *Chem. Commun.* (2001), 617–618.

72. K. Soulantica, A. Maisonnat, M. C. Fromen, M. J. Casanove, P. Lecante, and B. Chaudret, Synthesis and self-assembly of monodisperse indium nanoparticles prepared from the organometallic precursor [In(eta(5)-C_5H_5)], *Angew. Chem. Int. Ed.*, 40 (2001), 448–451.

73. K. Soulantica, L. Erades, M. Sauvan, F. Senocq, A. Maisonnat, and B. Chaudret, Synthesis of indium and indium oxide nanoparticles from indium cyclopentadienyl precursor and their application for gas sensing, *Adv. Funct. Mater.*, 13 (2003), 553–557.

74. Z. W. Li, X. J. Tao, Y. M. Cheng, Z. S. Wu, Z. J. Zhang, and H. X. Dang, A simple and rapid method for preparing indium nanoparticles from bulk indium via ultrasound irradiation, *Mater. Sci. Eng. A*, 407 (2005), 7–10.

75. P. K. Khanna, K. W. Jun, K. B. Hong, J. O Baeg, R. C. Chikate, and B. K. Das, Colloidal synthesis of indium nanoparticles by sodium reduction method, *Mater. Lett.*, 59 (2005), 1032–1036.

76. N. Avramova, N. S. Neykov, and S. K. Peneva, A calorimetric study of tin grown by reduction of $SnCl_2$ with Mg, *J. Phys. D Appl. Phys.*, 29 (1996), 1300–1305.

77. C. S. Yang, Q. Liu, S. M. Kauzlarich, and B. Phillips, Synthesis and characterization of Sn/R, Sn/Si-R, and Sn/SiO_2 core/shell nanoparticles, *Chem. Mater.*, 12 (2000), 983–988.

78. Y. B. Zhao, Z. J. Zhang, and H. X. Dang, Preparation of tin nanoparticles by solution dispersion, *Mater. Sci. Eng. A*, 359 (2003), 405–407.

79. A. Caballero, J. Morales, and L. Sanchez, Tin nanoparticles formed in the presence of cellulose fibers exhibit excellent electrochemical performance as anode materials in lithium-ion batteries, *Electrochem. Solid-State Lett.*, 8 (2005), A464–A466.

80. Y. Kwon, M. G. Kim, Y. Kim, Y. Lee, and J. P. Cho, Effect of capping agents in tin nanoparticles on electrochemical cycling, *Electrochem. Solid-State Lett.*, 9 (2006), A34–A38.

81. B. Stegemann, C. Ritter, B. Kaiser, and K. Rademann, Crystallization of antimony nanoparticles: pattern formation and fractal growth, *J. Phys. Chem. B*, 108 (2004), 14292–14297.
82. Y. W. Wang, B. H. Hong, J. Y. Lee, J. S. Kim, G. H. Kim, and K. S. Kim, Antimony nanowires self-assembled from Sb nanoparticles, *J. Phys. Chem. B.*, 108 (2004), 16723–16726.
83. B. Mayers and Y. N. Xia, One-dimensional nanostructures of trigonal tellurium with various morphologies can be synthesized using a solution-phase approach, *J. Mater. Chem.*, 12, (2002), 1875–1881.
84. Z. P. Liu, Z. K. Hu, Q. Xie, B. J. Yang, J. Wu, and Y. T. Qian, Surfactant-assisted growth of uniform nanorods of crystalline tellurium, *J. Mater. Chem.*, 13 (2003), 159–162.
85. Z. P. Liu, Z. K. Hu, J. B. Liang, S. Li, Y. Yang, S. Peng, and Y. T. Qian, Size-controlled synthesis and growth mechanism of monodisperse tellurium nanorods by a surfactant-assisted method, Langmuir, 20 (2004), 214–218.
86. G. C. Xi, Y. Y. Peng, W. C. Yu, and Y. T. Qian, Synthesis, characterization, and growth mechanism of tellurium nanotubes, *Cryst. Growth Des.*, 5 (2005), 325–328.
87. J. A. Ascencio, A. C. Rincon, and G. Canizal, Synthesis and theoretical analysis of samarium nanoparticles: perspectives in nuclear medicine, J. Phys. Chem. B, 109 (2005), 8806–8812.
88. M. Q. Tan, G. L. Wang, X. D. Hai, Z. Q. Ye, and J. L. Yuan, Development of functionalized fluorescent europium nanoparticles for biolabeling and time-resolved fluorometric applications, *J. Mater. Chem.*, 14 (2004), 2896–2901.
89. J. A. Nelson, L. H. Bennett, and M. J. Wagner, Solution synthesis of gadolinium nanoparticles, *J. Am. Chem. Soc.*, 124 (2002), 2979–2983.
90. M. O. Oyewumi, R. A. Yokel, M. Jay, T. Coakley, and R. J. Mumper, Comparison of cell uptake, biodistribution and tumor retention of folate-coated and PEG-coated gadolinium nanoparticles in tumor-bearing mice, *J. Controlled Release*, 95 (2004), 613–626.
91. Z. C. Yan, Y. H. Huang, Y. Zhang, H. Okumura, J. Q. Xiao, S. Stoyanov, V. Skumryev, G. C. Hadjipanayis, and C. Nelson, Magnetic properties of gadolinium and terbium nanoparticles produced via multilayer precursors, *Phys. Lett. B*, 67 (2003), 054403.
92. Z. Q. Ye, M. Q. Tan, G. L. Wang, and J. L. Yuan, Development of functionalized terbium fluorescent nanoparticles for antibody labeling and time-resolved fluoroimmunoassay application, *Talanta*, 65 (2005), 206–210.
93. N. B. Shevchenko, J. A. Christodoulides, and G. C. Hadjipanayis, Preparation and characterization of Dy nanoparticles, *Appl. Phys. Lett.*, 74 (1999), 1478–1480.
94. J. A. Nelson, L. H. Bennett, and M. J. Wagner, Dysprosium nanoparticles synthesized by alkalide reduction, *J. Mater. Chem.*, 13 (2003), 857–860.
95. J. A. Ascencio, A. C. Rodriguez-Monroy, H. B. Liu, and G. Canizal, Synthesis and structure determination of ytterbium nanoparticles, *Chem. Lett.*, 33 (2004), 1056–1057.
96. Y. Wang, Z. L. Cui, and Z. K. Zhang, Synthesis and phase structure of tantalum nanoparticles, *Mater. Lett* 58 (2004), 3017–3020.
97. Y. Tamou and S. Tanaka, Formation and coalescence of tungsten nanoparticles under electron beam irradiation, *Nanostruct. Mater*, 12 (1999), 123–126.

98. M. H. Magnusson, K. Deppert, and J. O. Malm, Single-crystalline tungsten nanoparticles produced by thermal decomposition of tungsten hexacarbonyl, *J. Mater. Res.*, 15 (2000), 1564–1569.

99. G. H. Lee, S. H. Huh, S. H. Kim, B. J. Choi, B. S. Kim, and J. H. Park, Structure and size distribution of Os, Re, and Ru nanoparticles produced by thermal decomposition of Os-3(CO)(12), Re-2(CO)(10), and Ru-3(CO)(12), *J. Korean Chem. Soc.*, 42 (2003), 835–837.

100. F. Bonet, V. Delmas, S. Grugeon, R. H. Urbina, P. Y. Silvert, and K. T. Elhsissen, Synthesis of monodisperse Au, Pt, Pd, Ru and Ir nanoparticles in ethylene glycol, *Nanostruct. Mater* 11 (1999), 1277–1284.

101. T. S. Ahmadi, Z. L. Wang, T. C. Green, A. Henglein, and M. A. El Sayed, Shape-controlled synthesis of colloidal platinum nanoparticles, *Science*, 272 (1996), 1924–1926.

102. A. Rodriguez, C. Amiens, B. Chaudret, M. J. Casanove, P. Lecante, and J. S. Bradley, Synthesis and isolation of cuboctahedral and icosahedral platinum nanoparticles: ligand-dependent structures, *Chem. Mater.*, 8 (1996), 1978–1986.

103. C. W. Chen and M. Akashi, Synthesis, characterization, and catalytic properties of colloidal platinum nanoparticles protected by poly(*N*-isopropylacrylamide), *Langmuir*, 13 (1997), 6465–6472.

104. C. Yee, M. Scotti, A. Ulman, H. White, M. Rafailovich, and J. Sokolov, One-phase synthesis of thiol-functionalized platinum nanoparticles, *Langmuir*, 15 (1999), 4314–4316.

105. S. H. Joo, S. J. Choi, I. Oh, J. Kwak, Z. Liu, O. Terasaki, and R. Ryoo, Ordered nanoporous arrays of carbon supporting high dispersions of platinum nanoparticles, *Nature*, 412 (2001), 169–172.

106. K. S. Kim, D. Demberelnyamba, and H. Lee, Size-selective synthesis of gold and platinum nanoparticles using novel thiol-functionalized ionic liquids, *Langmuir*, 20 (2004), 556–560.

107. J. Turkevich, P. Stevenson, and J. Hillier, A study of the nucleation and growth processes in the synthesis of colloidal gold, *Discuss. Faraday Soc.*, 11 (1951), 55–75.

108. G. Frens, Controlled nucleation for the regulation of the particle size in monodisperse gold suspensions, *Nature*, 241 (1973), 20–22.

109. J. De Mey, The preparation and use of gold probes, in *Immunocytochemistry: Modern Methods and Applications*, 2nd ed., J. M. Polak and S. Van Norden, Eds., John Wright, Bristol, England 1986 pp. 115–145.

110. E. C. Stathis and A. Fabrikanos, Preparation of colloidal gold, *Chem. Ind. (London)*, 27 (1958), 860.

111. P. A. Bartlett, B. Bauer, and S. Singer, Synthesis of water-soluble undecagold cluster compounds of potential importance in electron microscopic and other studies in biological systems, *J. Am. Chem. Soc.*, 100 (1978), 5085–5089.

112. C. L. Baigent and G. Muller, A colloidal gold prepared with ultrasonics, *Experimentia*, 36 (1980), 472.

113. J. Roth, The preparation of protein A-gold complexes with 3 nm and 15 nm gold particles and their use in labeling multiple antigens on ultra-thin sections, *Histochem. J.*, 14 (1982), 791.

114. W. P. Faulk and G. M. Taylor, An immunocolloid method for the electron microscope, *Immunochemistry*, 8 (1971), 1081–1083.

115. W. Baschong, J. M. Lucocq, and J. Roth, Thiocyanate gold: small (2–3 nm) colloidal gold for affinity cytochemical labeling in electron microscopy, *Histochemistry*, 83 (1985), 409.

116. H. Mühlpfordt, The preparation of colloidal gold particles using tannic acid as an additional reducing agent, *Experimentia*, 38 (1982), 1127.

117. J. W. Slot and H. J. Geuze, A new method for preparing gold probes for multiple labeling cytochemistry, *Eur. J. Cell Biol.*, 38 (1985), 87.

118. G. B. Birrell, K. K. Hedberg, and O. H. Griffith, Pitfalls of immunogold labeling: analysis by light microscopy, transmission electron microscopy and photoelectron microscopy, *J. Histochem. Cytochem.*, 35 (1987), 843–853.

119. M. Brust, M. Walker, D. Bethell, D. J. Schiffrin, and R. Whyman, Synthesis of thiol derivatised gold nanoparticles in a two-phase liquid–liquid system, *Chem. Commun.* (1994), 801–802.

120. S. O. Obare, R. E. Hollowell, and C. J. Murphy, Sensing strategy for lithium ion based on gold nanoparticles, *Langmuir*, 18 (2002), 10407–10410.

121. G. Schmid, R. Boese, R. Pfeil, F. Bandermann, S. Meyer, G. H. M. Calis, and J. W. A. van der Velden, $Au_{55}[P(C_6H_5)_3]_{12}Cl_6$ - ein Goldcluster ungewöhnlicher Größe($Au_{55}[P(C_6H_5)_3]_{12}Cl_6$ [a gold cluster of an exceptional size]), *Chem. Ber.*, 114 (1981), 3634-3642.

122. A. Henglein and M. Giersig, Optical and chemical observations on gold–mercury nanoparticles in aqueous solution, *J. Phys. Chem. B*, 104 (2000), 5056–506.

123. H. W. Sheng, G. Ren, L. M. Peng, Z. Q. Hu, and K. Lu, Superheating and melting-point depression of Pb nanoparticles embedded in Al matrices, *Philos. Mag. Lett.*, 73 (1996), 179–186.

124. M. Veith, S. Mathur, P. Konig, C. Cavelius, J. Biegler, A. Rammo, V. Huch, S. A. Hao, and G. Schmid, Template-assisted ordering of Pb nanoparticles prepared from molecular-level colloidal processing, *Compt. Rend. Chim.*, 7 (2005), 509–519.

125. J. Y. Fang, K. L. Stokes, J. A. Wiemann, W. L. Zhou, J. B. Dai, F. Chen, and C. J. O'Connor, Microemulsion-processed bismuth nanoparticles, *Mater. Sci. Eng. B*, 83 (2001), 254–257.

126. J. Y. Fang, K. L. Stokes, W. L. L. Zhou, W. D. Wang, and J. Lin, Self-assembled bismuth nanocrystallites, *Chem. Commun.* (2001), 1872–1873.

127. A. Wurl, M. Hyslop, S. A. Brown, B. D. Hall, and R. Monot, Structure of unsupported bismuth nanoparticles, *Eur. Phys. J. D*, 16 (2001), 205–208.

128. Y. B. Zhao, Z. J. Zhang, and H. X. Dang, A simple way to prepare bismuth nanoparticles, *Mater. Lett.*, 58 (2004), 790–793.

129. Z. Gui, R. Fan, W. Mo, X. Chen, L. Yang, and Y. Hu, Synthesis and characterization of reduced transition metal oxides and nanophase metals with hydrazine in aqueous solution, *Mater. Res. Bull.*, 38 (2003), 169–176.

130. A. S. Albuquerque, J. D. Ardisson, and W. A. Macedo, Nanosized powders of NiZn ferrite: synthesis, structure, and magnetism, *J. Appl. Phys.*, 87 (2000), 4352–4357.

131. Q. Chen, A. J. Rondinone, B. C. Chakoumakos, and Z. J. Zhang, Synthesis of superparamagnetic $MgFe_2O_4$ nanoparticles by coprecipitation, *J. Magn. Magn. Mater.*, 194 (1999), 1–7.

132. J. F. Wang, C. B. Ponton, and I. R. Harris, Ultrafine SrM particles with high coercivity by chemical coprecipitation, *J. Magn. Magn. Mater.*, 242–245 (2002), 1464–1467.

133. Y. Gu, G. Z. Li, G. Meng, and D. Peng, Sintering and electrical properties of coprecipitation prepared $Ce_{0.8}Y_{0.2}O_{1.9}$ ceramics, *Mater. Res. Bull.*, 35 (2000), 297–304.

134. J.-G. Li, T. Ikegami, Y. Wang, and T. Mori, Reactive ceria nanopowders via carbonate precipitation, *J. Am. Ceram. Soc.*, 85 (2002), 2376–2379.

135. L. Xiang, X. Y. Deng, and Y. Jin, Experimental study on synthesis of NiO nanoparticles, *Scr. Mater.*, 47 (2002), 219–224.

136. Y. Du, J. Fang, M. Zhang, J. Hong, Z. Yin, and Q. Zhang, Phase character and structural anomaly of $Bi_4Ti_3O_{12}$ nanoparticles prepared by chemical coprecipitation, *Mater. Lett.*, 57 (2002), 802–806.

137. P. H. Borse, L. S. Kankate, F. Dassenoy, W. Vogel, J. Urban, and S. K. Kulkarni, Synthesis and investigations of rutile phase nanoparticles of TiO_2, *J. Mater. Sci. Mater. Electron.*, 13 (2002), 553–559.

138. P. C. Kuo and T. S. Tsai, New approaches to the synthesis of acicular alpha-FeOOH and cobalt modified iron-oxide nanoparticles, *J. Appl. Phys.*, 65 (1989), 4349–4356.

139. Z. X. Tang, C. M. Sorensen, K. J. Klabunde, and G. C. Hadjipanayis, Preparation of manganese ferrite fine particles from aqueous solution, *J. Colloid Interface Sci.*, 146 (1991), 38–54.

140. T. C. Rojas and M. Ocana, Uniform nanoparticles of Pr(III)/ceria solid solutions prepared by homogeneous precipitation, *Scr. Mater.*, 46 (2002), 655–660.

141. C. N. Chinnasamy, B. Jeyadevan, O. Perales-Perez, K. Shinoda, K. Tohji, and A. Kasuya, Growth dominant Co-precipitation process to achieve high coercivity at room temperature in $CoFe_2O_4$ nanoparticles, *IEEE Trans. Magn.*, 38 (2002), 2640–2642.

142. J. Li, D. Dai, B. Zhao, Y. Lin, and C. Liu, Properties of ferrofluid nanoparticles prepared by coprecipitation and acid treatment, *J. Nanopart.*, Res., 4 (2002), 261–264.

143. K. T. Wu, P. C. Kuo, Y. D. Yao, and E. H. Tsai, Magnetic and optical properties of Fe_3O_4 nanoparticle ferrofluids prepared by coprecipitation technique, *IEEE Trans. Magn.*, 37 (2001), 2651–2653.

144. D. K. Kim, Y. Zhang, W. Voit, K. V. Rao, and M. Muhammed, *J. Mag. and Mag. Mater.*, 225 (2001), 30–36.

145. Z. X. Tang, C. M. Sorensen, K. J. Klabunde, and G. C. Hadjipanayis, Size-dependent Curie temperature in nanoscale $MnFe_2O_4$ particles, *Phys. Rev. Lett.*, 67 (1991), 3602–3605.

146. Z. L. Liu, Y. J. Liu, K. L. Yao, Z. H. Ding, J. Tao, and X. Wang, Synthesis and magnetic properties of Fe_3O_4 nanoparticles, *J. Mater. Synth. Process.*, 10 (2002), 83–87.

147. Y. Wang, C. Ma, X. Sun, and H. Li, Preparation of nanocrystalline metal oxide powders with the surfactant-mediated method, *Inorg. Chem. Commun.*, 5 (2002), 751–755.

148. Z. Zhang, L. Guo, and W. Wang, Synthesis and characterization of antimony oxide nanoparticles, *J. Mater. Res.*, 16 (2001), 803–805.

149. H. Chen, X. Qiu, W. Zhu, and P. Hagenmuller, Synthesis and high rate properties of nanoparticled lithium cobalt oxides as the cathode material for lithium-ion battery, *Electrochem. Commun.*, 4 (2002), 488–491.

150. S. Music, S. Popovic, M. Maljkovic, K. Furic, and A. Gajovic, Influence of synthesis procedure on the formation of RuO_2, *Mater. Lett.*, 56 (2002), 806–811.

151. P. Deb, T. Biswas, D. Sen, A. Basumallick, and S. Mazumder, Characteristics of Fe_2O_3 nanoparticles prepared by heat treatment of a nonaqueous powder precipitate, *J. Nanopart. Res.*, 4 (2002), 91–97.

152. G. Ennas, G. Marongiu, A. Musinu, A. Falqui, P. Ballirano, and R. Caminiti, Characterization of nanocrystalline γ-Fe_2O_3 prepared by wet chemical method, *J. Mater. Res.*, 14 (1999), 1570–1575.

153. S. O'Brien, L. Brus, and C. B. Murray, Synthesis of monodisperse nanoparticles of barium titanate: toward a generalized strategy of oxide nanoparticle synthesis, *J. Am. Chem. Soc.*, 123 (2001), 12085–12086.

154. D. Caruntu, Y. Remond, N. H. Chou, M.-J. Jun, G. Caruntu, J. He, G. Goloverda, C. O'Connor, and V. Kolesnichenko, Reactivity of 3d transition metal cations in diethylene glycol solutions: synthesis of transition metal ferrites with the structure of discrete nanoparticles complexed with long-chain carboxylate anions, *Inorg. Chem.*, 41 (2002), 6137–6146.

155. S. Sun and H. Zeng, Size-controlled synthesis of magnetite nanoparticles, *J. Am. Chem. Soc.*, 124 (2002), 8204–8205.

156. J. Rockenberger, E. C. Scher, and A. P. Alivisatos, A new nonhydrolytic single-precursor approach to surfactant-capped nanocrystals of transition metal oxides, *J. Am. Chem. Soc.*, 121 (1999), 11595–11596.

157. J. P. Chen, K. M. Lee, C. M. Sorensen, K. J. Klabunde, and G. C. Hadjipanayis, Magnetic properties of microemulsion synthesized cobalt fine particles, *J. Appl. Phys.*, 75 (1994), 5876–5878.

158. I. Lisiecki and M. P. Pileni, Synthesis of copper metallic clusters using reverse micelles as microreactors, *J. Am. Chem. Soc.*, 115 (1993), 3887–3896.

159. M. Boutonnet, J. Kizling, P. Stenius, and G. Maire, The preparation of monodisperse colloidal metal particles from microemulsions, *Colloids Surf.*, 5 (1982), 209–225.

160. P. Barnickel, A. Wokaun, W. Sager, and H.-F. Eicke, Size tailoring of silver colloids by reduction in W/O microemulsions, *J. Colloid Interface Sci.*, 148 (1992), 80–90.

161. E. E. Foos, R. M. Stroud, A. D. Berry, A. W. Snow, and J. P. Armistead, Synthesis of nanocrystalline bismuth in reverse micelles, *J. Am. Chem. Soc.*, 122 (2000), 7114–7115.

162. E. E. Carpenter, J. A. Sims, J. A. Wienmann, W. L. Zhou, and C. J. O'Connor, Magnetic properties of iron and iron platinum alloys synthesized via microemulsion techniques, *J. Appl. Phys*, 87 (2000), 5615–5617.

163. E. E. Carpenter, A. Kumbhar, J. A. Wiemann, H. Srikanth, J. Wiggins, W. Zhou, and C. J. O'Connor, Synthesis and magnetic properties of gold–iron–gold nanocomposites, *Mater. Sci. Eng. A*, 286 (2000), 81–86.

164. C.-H. Lu and H.-C. Wang, Synthesis of nano-sized $LiNi_{0.8}Co_{0.2}O_2$ via a reverse-microemulsion route, *J. Mater. Chem.*, 13 (2003), 428–431.

165. Y.-X. Pang and X. Bao, Aluminium oxide nanoparticles prepared by water-in-oil microemulsions, *J. Mater. Chem.*, 12 (2002), 3699–3704.

166. P. D. Moran, J. R. Bartlett, G. A. Bowmaker, J. L. Woolfrey, and R. P. Cooney, Formation of TiO_2 sols, gels and nanopowders from hydrolysis of $Ti(OiPr)_4$ in AOT reverse micelles, *J. Sol-Gel Sci. Technol.*, 15 (1999), 251–262.

167. D. O. Yener and H. Giesche, Processing of pure and Mn, Ni, and Zn doped ferrite particles in microemulsions, *Ceram. Trans.*, 94 (1999), 407–418.

168. H. S. Lee, W. C. Lee, and T. Furubayashi, A comparison of coprecipitation with microemulsion methods in the preparation of magnetite, *J. Appl. Phys.*, 85 (1999), 5231–5233.

169. C. J. O'Connor, C. T. Seip, E. E. Carpenter, S. Li, and V. T. John, Synthesis and reactivity of nanophase ferrites in reverse micellar solutions, *Nanostruct. Mater.*, 12 (1999), 65–70.

170. C. R. Vestal and Z. J. Zhang, Synthesis of $CoCrFeO_4$ nanoparticles using microemulsion methods and size-dependent studies of their magnetic properties, *Chem. Mater.*, 14 (2002), 3817–3822.

171. C. Liu, A. J. Rondinone, and Z. J. Zhang, Synthesis of magnetic spinel ferrite $CoFe_2O_4$ nanoparticles from ferric salt and characterization of the size-dependent superparamagnetic properties, *Pure Appl. Chem.*, 72 (2000), 37–45.

172. F. Porta, C. Bifulco, P. Fermo, C. L. Bianchi, M. Fadoni, and L. Prati, Synthesis of spherical nanoparticles of $Cu_2L_2O_5$ (L = Ho, Er) from W/O microemulsions, *Colloids Surf. A*, 160 (1999), 281–290.

173. P. Vaqueiro, M. A. Lopez-Quintela, and J. Rivas, Synthesis of yttrium iron garnet nanoparticles via coprecipitation in microemulsion, *J. Mater. Chem.*, 7 (1997), 501–504.

174. P. Kumar, V. Pillai, S. R. Bates, and D. O. Shah, Preparation of $YBa_2Cu_3O_{7-x}$ superconductor by coprecipitation of nanosize oxalate precursor powder in microemulsions, *Mater. Lett.*, 16 (1993), 68–74.

175. K. C. Song and J. H. Kim, Synthesis of high surface area tin oxide powders via water-in-oil microemulsions, *Powder Technol.*, 107 (2000), 268–272.

176. V. Pillai, P. Kumar, M. S. Multani, and D. O. Shah, Structure and magnetic properties of nanoparticles of barium ferrite synthesized using microemulsion processing, *Colloids Surf. A*, 80 (1993), 69–75.

177. B. J. Palla, D. O. Shah, P. Garcia-Casillas, and J. Matutes-Aquino, Preparation of nanoparticles of barium ferrite from precipitation in microemulsions, *J. Nanopart. Res.*, 1 (1999), 215–221.

178. Z. Wu, J. Zhang, R. E. Benfield, Y. Ding, D. Grandjean, Z. Zhang, and J. Xin, Structure and chemical transformation in cerium oxide nanoparticles coated by surfactant cetyltrimethylammonium bromide (CTAB): an x-ray absorption spectroscopic study, *J. Phys. Chem. B*, 106 (2002), 4569–4577.

179. T. Hirai, Y. Tsubaki, H. Sato, and I. Komasawa, Mechanism of formation of lead sulfide ultrafine particles in reverse micellar systems, *J. Chem. Eng. Jpn.*, 28 (1995), 468–473.

180. L. M. Gan, B. Liu, C. H. Chew, S. J. Xu, S. J. Chua, G. L. Loy, and G. Q. Xu, Enhanced photoluminescence and characterization of Mn-doped ZnS nanocrystallites synthesized in microemulsion, *Langmuir*, 13 (1997), 6427–6431.

181. N. R. Jana, L. Gearheart, and C. J. Murphy, Wet chemical synthesis of high aspect ratio cylindrical gold nanorods, *J. Phys. Chem. B*, 105, (2001) 4065–4067.

182. Y. Y. Yu, S. S. Chang, C. L. Lee, and C. R. C. Wang, Gold nanorods: electrochemical synthesis and optical properties, *J. Phys. Chem. B*, 101 (1997), 6661–6664.

183. S. Link, M. B. Mohamed, and M. A. El-Sayed, Simulation of the optical absorption spectra of gold nanorods as a function of their aspect ratio and the effect of the medium dielectric constant, *J. Phys. Chem. B*, 103 (1999), 3073–3077.

184. F. Kim, J. H. Song, and P. Yang, Photochemical synthesis of gold nanorods, *J. Am. Chem. Soc.*, 124 (2002), 14316–14317.

185. K. Aslan, Z. Leonenko, J. R. Lakowicz, and C. D. Geddes, Fast and slow deposition of silver nanorods on planar surfaces: application to metal-enhanced fluorescence, *J. Phys. Chem. B*, 109 (2005), 3157–3162.

186. C. C. Chen, C. Y. Chao, and Z. H. Lang, Simple solution-phase synthesis of soluble CdS and CdSe nanorods, *Chem. Mater.*, 12 (2000), 1516–1518.

187. Y. D. Li, H. W. Liao, Y. Ding, Y. T. Qian, L. Yang, and G. E. Zhou, Nonaqueous synthesis of CdS nanorod semiconductor, *Chem. Mater.*, 10 (1998), 2301–2303.

188. W. Wang, Y. Geng, P. Yan, F. Liu, Y. Xie, and Y. Qian, Synthesis and characterization of MSe (M = Zn, Cd) nanorods by a new solvothermal method, *Inorg. Chem. Commun.*, 2 (1999), 83–85.

189. L. Manna, E. C. Scher, and A. P. Alivisatos, Synthesis of soluble and processable rod-, arrow-, teardrop-, and tetrapod-shaped CdSe nanocrystals, *J. Am. Chem. Soc.*, 122 (2000), 12700–12706.

190. L. Guo, Y. L. Ji, and H. Xu, Regularly shaped, single-crystalline ZnO nanorods with Wurtzite structure, *J. Am. Chem. Soc.*, 124 (2002), 14864–14865.

191. A. Dev, S. K. Panda, S. Kar, S. Chakrabarti, and S. Chaudhuri, Surfactant-assisted route to synthesize well-aligned ZnO nanorod arrays on sol-gel-derived ZnO thin films, *J. Phys. Chem. B*, 110 (2006), 14266–14272.

192. B. Liu and H. Chun Zeng, Hydrothermal synthesis of ZnO nanorods in the diameter regime of 50 nm, *J. Am. Chem. Soc.*, 125 (2003), 4430–4431.

193. T. K. Sau and C. J. Murphy, Room temperature, high-yield synthesis of multiple shapes of gold nanoparticles in aqueous solution, *J. Am. Chem. Soc.*, 126 (2004), 8648–8649.

194. S. S. Sankar, A. Rai, B. Ankamwar, A. Singh, A. Ahmad, and M. Sastry, Biological synthesis of triangular gold nanoprisms, *Nature Mater.*, 3 (2004), 482–488.

195. H. Jia, W. Xu, J. An, D. Li, and B. Zhao, A simple method to synthesize triangular silver nanoparticles by light irradiation, *Spectrochim. Acta A*, 64 (2006), 956–960.

196. J. E. Millstone, S. Park, K. L. Shuford, L. Qin, G. C. Schatz and C. A. Mirkin, Observation of a quadrupole plasmon mode for a colloidal solution of gold nanoprisms, *J. Am. Chem. Soc.*, 127 (2005), 5312–5313.

197. R. Jin, Y. W. Cao, C. A. Mirkin, K. L. Kelly, G. C. Schatz, and J. G. Zheng, Photoinduced conversion of silver nanospheres to nanoprisms, *science*, 294 (2001), 1901–1903.

198. C. S. Ah, Y. J. Yun, H. J. Park, W. J. Kim, D. H. Ha, and W. S. Yun, Size-controlled synthesis of machinable single crystalline gold nanoplates, *Chem. Mater.*, 17 (2005), 5558–5561.

199. S. Chen and D. L. Carroll, Silver nanoplates: size control in two dimensions and formation mechanisms, *J. Phys. Chem. B*, 108 (2004) 5500–5506.

200. Y. Xiong, J. M. McLellan, J. Chen, Y. Yin, Z. Y. Li, and Y. Xia, Kinetically controlled synthesis of triangular and hexagonal nanoplates of palladium and their SPR/SERS properties, *J. Am. Chem. Soc.*, 127 (2005) 17118–17127.

201. Y. Ding, Z. L. Wang, W. Lu, and J. Fang, Spontaneous fractal aggregation of gold nanoparticles and controlled generation of aggregate-based fractal networks at air/water interface, *J. Phys. Chem. B*, 109 (2005) 19213–19218.

202. M. Yamamoto, Y. Kashiwagi, T. Sakata, H. Mori, and M. Nakamoto, Synthesis and morphology of star-shaped gold nanoplates protected by poly(*N*-vinyl-2-pyrrolidone), *Chem. Mater.*, 17 (2005), 5391–5393.

203. S. Chen, Z. L. Wang, J. Ballato, S. H. Foulger, and D. L. Carroll, Monopod, bipod, tripod, and tetrapod gold nanocrystals, *J. Am. Chem. Soc.*, 125 (2003), 16186–16187.

204. C. P. Graf, R. Birringer, and A. Michels, Synthesis and magnetic properties of cobalt nanocubes, *Phys Rev. B*, 73 (2006), 212401–212404.

INDEX

Nanofluids: Science and Technology, By Sarit K. Das, Stephen U. S. Choi, Wenhua Yu, and T. Pradeep

I

J

K

L

T

Printed and bound by CPI Group (UK) Ltd, Croydon, CR0 4YY
18/09/2026
14964971-0001